# A Course in Mathematical Biology

# Mathematical Modeling and Computation

## About the Series

The SIAM series on Mathematical Modeling and Computation draws attention to the wide range of important problems in the physical and life sciences and engineering that are addressed by mathematical modeling and computation; promotes the interdisciplinary culture required to meet these large-scale challenges; and encourages the education of the next generation of applied and computational mathematicians, physical and life scientists, and engineers.

The books cover analytical and computational techniques, describe significant mathematical developments, and introduce modern scientific and engineering applications. The series will publish lecture notes and texts for advanced undergraduate- or graduate-level courses in physical applied mathematics, biomathematics, and mathematical modeling, and volumes of interest to a wide segment of the community of applied mathematicians, computational scientists, and engineers.

Appropriate subject areas for future books in the series include fluids, dynamical systems and chaos, mathematical biology, neuroscience, mathematical physiology, epidemiology, morphogenesis, biomedical engineering, reaction-diffusion in chemistry, nonlinear science, interfacial problems, solidification, combustion, transport theory, solid mechanics, nonlinear vibrations, electromagnetic theory, nonlinear optics, wave propagation, coherent structures, scattering theory, earth science, solid-state physics, and plasma physics.

# A Course in Mathematical Biology

## Quantitative Modeling with Mathematical and Computational Methods

**Gerda de Vries**
University of Alberta
Edmonton, Alberta, Canada

**Thomas Hillen**
University of Alberta
Edmonton, Alberta, Canada

**Mark Lewis**
University of Alberta
Edmonton, Alberta, Canada

**Johannes Müller**
Technical University Munich
Munich, Germany

**Birgitt Schönfisch**
University of Tübingen
Tübingen, Germany

**siam**

Society for Industrial and Applied Mathematics
Philadelphia

MAPLE is a registered trademark of Waterloo Maple, Inc.

Mathematica is a registered trademark of Wolfram Research, Inc.

MATLAB is a registered trademark of The MathWorks, Inc. and is used with permission. The MathWorks does not warrant the accuracy of the text or exercises in this book. This book's use or discussion of MATLAB software or related products does not constitute endorsement or sponsorship by The MathWorks of a particular pedagogical approach or particular use of the MATLAB software. For MATLAB information, contact The MathWorks, 3 Apple Hill Drive, Natick, MA, 01760-2098 USA, Tel: 508-647-7000, Fax: 508-647-7001 info@mathworks.com, www.mathworks.com

UNIX is a registered trademark of The Open Group in the United States and other countries.

Windows is a registered trademark of Microsoft Corporation in the United States and/or other countries.

Figure 3.10 is reprinted with permission from page 190, Figure 5.14 of Edelstein-Keshet, L., *Mathematical Models in Biology*, Classics in Applied Mathematics 46, SIAM, Philadelphia, 2005.

Figure 9.3 is reprinted with permission of Kluwer Academic from Figure 4 of Soll, D. R., Behavioral studies into the mechanism of eukaryotic chemotaxis, *J. Chemical Ecology*, 16: 133–150, 1990.

Figures 9.4 and 9.5 are reprinted with permission of the University of Chicago Press from Figure 1 and Table 1, respectively, of Lubina, J. A. and Levin, S. A., The spread of a reinvading species, *AM NAT*, 151(4): 526–543, 1988.

Figure 9.6 is reprinted with permission from Figure 11 on page 94 and Figure 12 on page 95 of Stark, L. W., *Neurological Control Systems: Studies in Bioengineering*, Kluwer Academic/Plenum Publishers, 1968.

Figure 9.8 is reprinted with permission of the Royal Society of London from Figure 24 of Hubel, D. H. and Wiesel, T. N., Functional architecture of Macaque monkey visual cortex. *Proc. Roy. Soc. Lond. B.*, 198: 1–59, 1977.

Cover art courtesy of Dr. R. Wolff, www.rwphoto.de (green iguana), and Dr. B. Schönfisch (mollusks).

**Library of Congress Cataloging-in-Publication Data**
A course in mathematical biology : quantitative modeling with mathematical and computational methods / Gerda de Vries... [et al.]
    p.cm. – (Mathematical modeling and computation)
    Includes bibliographical references (p.  ).
    ISBN 978-0-898716-12-2 (pbk.)
        1. Biology–Mathematical models. I. Vries, Gerda de. II. Series.

QH323.5.C69 2006
570.1'5118–dc22

                                                                            2006044305

 Partial royalties from the sale of this book are placed in a fund to help students attend SIAM meetings and other SIAM-related activities. This fund is administered by SIAM, and qualified individuals are encouraged to write directly to SIAM for guidelines.

**siam** is a registered trademark.

# Contents

# Preface

Mathematical biology is growing rapidly. Mathematics has long played a dominant role in our understanding of physics, chemistry, and other physical sciences. However, wholesale application of mathematical methods in the life sciences is relatively recent. Now questions about infectious diseases, heart attacks, cell signaling, cell movement, ecology, environmental changes, and genomics are being tackled and analyzed using mathematical and computational methods.

While the application of quantitative analysis in the life sciences has borne fruit in the research arena, only recently has it impacted undergraduate education. Until a few years ago, the number of undergraduate texts in mathematical biology could be counted on one hand. Now this has changed dramatically. Recent undergraduate texts range from simple introductions to biological numeracy (Burton [35, 36]), freshman calculus for students in the life sciences (Adler [1], Neuhauser [125]), modeling with differential equations (Taubes [155], Edelstein-Keshet [51], Britton [29]), computer algebra (Yeargers, Shonkwiler, and Herod [168]), and dynamical computer-based systems (Hannon and Ruth [78]), to name but a few.

Despite the plentitude of new books, mathematical biology is still rarely offered as an undergraduate course. This book is designed for undergraduate students. Our target audience are students in mathematics, biology, physics, or other quantitative sciences at the sophomore or junior level. Our aim is to introduce students to problem solving in the context of biology. The focus in our presentation is on integrating analytical and computational tools in the modeling of biological processes.

The book stems from pedagogic material developed by the authors for a 7–11 day workshop in mathematical biology, which has been taught since 1995 at the University of Tübingen (Germany) and since 2001 at the University of Alberta (Canada). Additional material has been added to make the book suitable for use in a full-term course in mathematical biology.

There are three parts to this book: (I) analytical modeling techniques, (II) computational modeling techniques, (III) problem solving.

Part I covers basic analytical modeling techniques. We discuss the formulation of models using difference equations, differential equations, probability theory, cellular automata, as well as model validation and parameter estimation. We emphasize the modeling process and qualitative analysis, rather than explicit solution techniques (which can be found in other textbooks). Classical models for disease, movement, and population dynamics are derived from first principles. Each section provides a number of biologically motivated exercises.

Part II introduces computational tools used in the modeling of biological problems. Students are guided through symbolic and numerical calculations with Maple (for readers who prefer an alternative software package, such as *Mathematica* or MATLAB, see "How to Use This Book" below). Many of the examples and exercises of this part relate directly to the models discussed in Part I. This part of our book has been designed such that students can work through the material independently and at their own pace. Readers without any programming background will pick up valuable computational skills. Readers who already have programming background will be able to skip some elementary exercises and focus attention on the biological applications.

Part III provides open-ended problems from epidemiology, ecology, and physiology. Each problem is formulated in a way that makes it accessible to students. In most cases, questions will guide the student through the modeling process. These problems can be used as the basis for extended investigation, for example, as a term project or as a team project. We conclude Part III with a detailed presentation of two projects (cell competition and the chemotactic paradox) based on solutions developed by teams of undergraduate students who participated in one of our workshops.

The field of mathematical biology is, admittedly, immense. This book does not attempt to achieve a comprehensive introduction to the field. Subjects are tempered by the test of being able to teach them effectively in a short period of time. Problems are biased towards the authors' interests, but are sufficiently wide-ranging to include something of interest for most students. Ultimately, we hope that this book offers the first step into a detailed modeling of problems in the life sciences.

## How to Use This Book

We envision that this book can be used in a number of ways. Here we list some ideas about how a course could be designed based on the material of this book.

**Full-Term Course:** During a full-term course, material from Part I can be covered. Students should have access to computers to complete Part II and for the project work of Part III. Although students can work through the computer tutorial on their own, we recommend a two-hour computer lab during which an instructor is available to help the students get started. Projects from the open-ended problems from Part III may be assigned early in the course, with students submitting a written report, or presenting the project in class (or both) towards the end of the course.

**10-Day Workshop:** During the first half of the workshop, the focus should be on learning modeling with analytical and computational tools, based on Parts I and II of this book. Ideally a mixture of discrete-time equations, differential equations, and stochastic models should be covered. Specific topics would depend on the background of the instructor(s). We feel that Sections 2.2, 3.1–3.4, 4.3, and 5.1–5.6 should be included in any course. Lectures on these topics may be supplemented by homework. In our experience, students need about 15 hours to work through the computer tutorial of Part II.

During the second half of the workshop, students should work in teams of two (maybe three) on one of the open-ended problems from Part III. Under the guidance of an instructor, students develop a model, analyze and/or simulate the model, and prepare a presentation. We have found it important to stress that problems are open-ended, and have no "right solution" *per se*. It is the process of model development that is most important, not necessarily the end product. In many cases, students will need to simplify their problem and build a hierarchy of models, each model incorporating additional realism from the original problem.

**Substituting Maple with Other Software:** Although we have based Part II of this book on Maple, we do not wish to give the impression that Maple is necessarily the ideal software to be used. In fact, we believe that it does not really matter which software package is used. Instructors or students proficient with other software, such as *Mathematica* or MATLAB, will readily be able to adapt the examples and exercises of Part II for the alternative software. A version of Part II in Mathematica is available at http://www.siam.org/books/mm12.

**Working on Open-Ended Projects:** Since the problems from Part III are open ended there is a danger of aiming too high. Some of the problems are currently being studied by experienced researchers, and it would be impossible to follow all the relevant literature. For a beginning modeler, we give the following guidelines.

From the project description, readers should be able to understand the biological problem at hand to a certain extent. Some reading of supplemental material might be useful. For most projects, a specific reference is given, and the Internet is always a good resource. It is not required to study the biological topic at length. Initial efforts in mathematical modeling require only the identification of basic mechanisms.

When the biological problem at hand is understood, students should determine first which of the model classes presented in Part I might be useful (discrete/continuous, deterministic/stochastic). With the help of an instructor, they then proceed to develop a mathematical model. After the students and the instructors agree upon a reasonable model, the students work on it, do the analysis, and write the software. Many projects are accompanied with data. In that case, data fitting will be an important element of the project. Last but not least, the model should be used to explain important aspects of the biological phenomenon and to make predictions for other experiments or observations.

**Internet Resources:** A webpage related to this book that contains solutions to most of the exercises and the computer tutorial of Part II in *Mathematica* can be found at http://www.siam.org/books/mm12.

## Acknowledgments

Although five authors worked on this text, it would not have been written without support from various sources. We express our thanks to Andrew Beltaos, an undergraduate student who helped with editing, making figures, and solving exercises. We are grateful to the Universities of Tübingen and Alberta for providing the environment to run our workshops. We also are grateful to the Pacific Institute for the Mathematical Sciences (PIMS), the Mathematics of Information Technology and Complex System (MITACS), and the Department of Mathematical and Statistical Sciences at the University of Alberta for significant financial support for the Alberta workshops. Moreover, we express our thanks to our colleagues who have taught with us in our workshops; in alphabetical order: Gary de Young, Leah Edelstein-Keshet, K. P. Hadeler, Christina Kuttler, Michael Li, Frithjof Lutscher, Michael Mackey, Annette Räbiger, Rebecca Tyson, and Pauline van den Driessche. Thanks to Michael Baake, Fred Brauer, three anonymous reviewers, and many of our students for valuable comments on earlier versions of this text. We would particularly like to thank all student participants of our workshops. Their enthusiasm and interest in mathematical modeling was our motivation to write this book. Patient help and guidance from the SIAM editorial staff was invaluable in the finishing stages of this work. We thank our families for their encouragement and support.

June 2006

**Gerda de Vries** is Associate Professor in the Department of Mathematical and Statistical Sciences at the University of Alberta, Canada.

**Thomas Hillen** is Professor in the Department of Mathematical and Statistical Sciences at the University of Alberta, Canada.

**Mark Lewis** is Professor and Senior Canada Research Chair in Mathematical Biology in the Department of Mathematical and Statistical Sciences and Department of Biological Sciences at the University of Alberta, Canada.

**Johannes Müller** is Professor of Mathematical Methods in Molecular Biology and Biochemistry in the Centre of Mathematical Sciences at the Technical University Munich, Germany.

**Birgitt Schönfisch** is a Scientific Employee in the Department of Medical Biometry at the University of Tübingen, Germany.

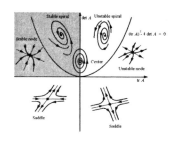

# Part I
# Theoretical Modeling Tools

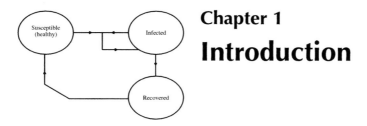

# Chapter 1

# Introduction

## 1.1 The Modeling Process

Mathematical biology dwells at the interface of two fields: applied/computational mathematics and biology. Individually, these fields are growing quickly due to rapidly changing technology and newly emerging subdisciplines. Coupled together, the fields provide the basis for the emerging scientific discipline of mathematical biology, whose focus is interdisciplinary scientific problems in quantitative life sciences.

What can biology offer mathematics and computation? Biological models offer a seemingly endless supply of challenging and interesting nonlinear problems to solve. These nonlinear problems can provide a testing ground for applied mathematical and computational methods, and generate the impetus to develop new mathematical and computational methods and approaches.

What can mathematics and computation offer biology? Mathematics and computation can help solve a growing problem in biological research. Data collection, varying from gene sequencing to remote sensing via satellites, is now inundating biologists with complex patterns of observations. The ability to collect new data outstrips our ability to heuristically reason mechanisms of cause and effect in complex systems. It is the analysis of mathematical models that allows us to formalize the cause and effect process and tie it to the biological observations.

The mathematical model describes interactions between biological components. Analysis of the model, via computational and applied mathematical methods, allows us to deduce the consequences of the interactions. For example, voltage-dependent data on movement of electrically charged ions across a nerve membrane are inputs for models of electrophysiology. The output is a prediction of the dynamics of electrical activity in nerves. The behavior and survival of newly infected individuals are inputs to disease models. The output is a prediction of when and where the disease will outbreak, and how it can be controlled.

To become a successful modeler, modeling tools are required. The first part of this book gives an introduction to some of the more powerful modeling tools, such as discrete models, ordinary differential equations, partial differential equations, stochastic models,

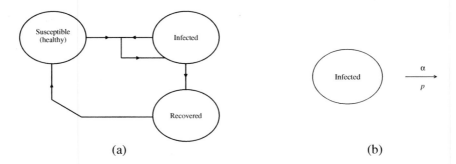

(a)                                                                              (b)

**Figure 1.1.** (a) *Arrow diagram for a simple epidemic model, showing the relationships between the classes of susceptible, infected, and recovered individuals.* (b) *Subgraph of the arrow diagram in* (a) *representing the recovery of infected individuals, with probability p or rate α.*

cellular automata, and parameter estimation techniques. The second and third parts of the book apply the modeling tools to biological problems.

## 1.2  Probabilities and Rates

We start with the derivation of a simple epidemic model for the spread of an infectious disease, such as influenza, through a population of healthy individuals. Assume that one infected individual is introduced into the population. In addition, assume that the infection is spread from individual to individual through contact, and that the infected recover after a certain period of time (two weeks for influenza). Recovered individuals are not available to catch the disease again.

Thus, after some time, the population consists of three types of individuals, namely, susceptible (healthy), infected, and recovered individuals. The relationships between these three classes are shown in Figure 1.1 (a). Note that in the diagram, recovered individuals can become susceptible again. In this case, we can think of recovered individuals being temporarily immune to the disease. Individuals return to the susceptible class when the immunity wears off.

In order to create a model for this situation, we need to quantify this diagram. To do that, we follow these three steps:

1. First, we identify the important quantities (the *dependent variables*) to keep track of. In our example, there are three classes of individuals. Let $S$ be the number of susceptibles in the population, $I$ the number infected, and $R$ the number recovered.

2. Second, we identify the *independent variables*, such as time $t$, space $x$, or age $a$, and so on. For our example, we write $S(x, t)$, $I(x, t)$, and $R(x, t)$ if we wish to include time and space dependence, but not age dependence.

3. Finally, we quantify the transitions and/or interactions between these classes, as indicated by the arrows in Figure 1.1 (a). To do this, we use either *probabilities* or *rates*, as explained below.

To explain the use of probabilities versus rates, we consider a subgraph of Figure 1.1 (a), concerning only the recovery of infected individuals, shown in Figure 1.1 (b). In the discussion below, note that we ignore the generation of infected individuals through contact between infected and susceptible individuals (the full epidemic model will be treated in Section 3.3.3). In order to create a model representing this particular process, we apply the three steps outlined above:

1. The dependent variable is the number of infected individuals, $I$.

2. As time progresses, infected individuals recover. Thus, the independent variable is time, $t$.

3. If we assume that 2 out of every 100 infected individuals recover per day, then the probability of recovery in a single day is $p_1 = \frac{2}{100}$. The corresponding rate, $\alpha_1$, is defined as the probability per unit of time, that is,

$$\alpha_1 = \frac{p_1}{\text{unit of time}} = \frac{2}{100} \cdot \frac{1}{\text{day}} = \frac{1}{50}\frac{1}{\text{day}}.$$

Similarly, the probability of recovery in two days is $p_2 = \frac{4}{100}$ (we use $p_n$ to denote the probability of recovering in $n$ days). The corresponding rate, $\alpha_2$, is then

$$\alpha_2 = \frac{4}{100} \cdot \frac{1}{2 \text{ days}} = \frac{1}{50}\frac{1}{\text{day}}.$$

For a time unit of $\frac{1}{2}$ of a day, we get $p_{\frac{1}{2}} = \frac{1}{100}$ and

$$\alpha_{\frac{1}{2}} = \frac{1}{100} \cdot 2\frac{1}{\text{day}} = \frac{1}{50}\frac{1}{\text{day}}.$$

We find that the rate $\alpha$ is independent of the time unit chosen, whereas the probability depends on the chosen time unit. Since the rate is independent of the chosen time unit, we can generalize. Let $\Delta t$ denote a general unit of time, and let $p_{\Delta t}$ be the probability of recovering in $\Delta t$. Then the number of infectives after one unit of time is given as

$$I(t + \Delta t) = I(t) - p_{\Delta t} I(t).$$

With some rearrangements, we get

$$I(t + \Delta t) - I(t) = -p_{\Delta t} I(t),$$
$$\frac{I(t + \Delta t) - I(t)}{\Delta t} = -\frac{p_{\Delta t}}{\Delta t} I(t),$$
$$\frac{I(t + \Delta t) - I(t)}{\Delta t} = -\alpha I(t),$$

where now the rate $\alpha = \frac{p_{\Delta t}}{\Delta t}$ appears.

Since $\alpha$ is constant for all values of $\Delta t$, we can take the limit as $\Delta t \to 0$. On the left, we obtain the differential quotient, and we obtain the following equation governing the dynamics of $I(t)$:

$$\frac{d}{dt} I(t) = -\alpha I(t).$$

To summarize, for the simple subgraph shown in Figure 1.1 (b), we found two models, namely, a *discrete-time model* with probabilities,

$$I(t + \Delta t) = I(t) - p_{\Delta t} I(t), \tag{1.1}$$

and a *continuous-time model* with rates (a *differential equation*),

$$\frac{d}{dt} I(t) = -\alpha I(t). \tag{1.2}$$

Both models can be solved, analyzed, and simulated. For the discrete-time model, (1.1), we have to specify a time unit, say $\Delta t = \frac{1}{2}$ day. Then $p_{\Delta t} = p_{\frac{1}{2}} = \alpha \cdot \frac{1}{2}$ day. If we define $I_n := I(n \cdot \Delta t)$, then we obtain the simple *difference equation*

$$I_{n+1} = \left(1 - P_{\frac{1}{2}}\right) I_n,$$

which has the solution

$$I_n = \left(1 - P_{\frac{1}{2}}\right)^n I_0, \qquad n \geq 1,$$

where $I_0$ denoted the initial number of infected individuals. The differential equation, (1.2), is solved by an exponential, $I(t) = I(0) e^{-\alpha t}$. The latter solution indicates that the number of infected individuals decreases with time, as expected intuitively (recall that the generation of new infected individuals has been ignored).

In Figure 1.2, we compare the solutions of the discrete-time and continuous-time models over a time period of 15 days, starting with 100 infected individuals ($I(0) = I_0 =$

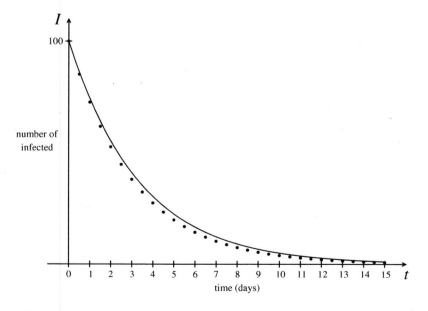

**Figure 1.2.** *Comparison of the solutions to the discrete-time model, (1.1), and the continuous-time model, (1.2), starting with 100 infected individuals ($I(0) = I_0 = 100$), and using a recovery rate of $\alpha = 0.3$.*

100) and using a recovery rate of $\alpha = 0.3$ per day and a time increment of $\Delta t = 1/2$ day. The full epidemic model corresponding to the arrow diagram shown in Figure 1.1 (a) will be discussed in detail in Section 3.3.3.

In Figure 1.2, the agreement of the discrete and the continuous models is quite convincing. However, this is not always the case. In Exercise 1.4.2, the reader is asked to vary $\Delta t$ and to investigate if the agreement is still good.

## 1.3  Model Classes

In the previous section, we derived two models for the recovery of infected individuals, namely, a discrete-time or difference equation, (1.1), and a differential equation, (1.2). The difference between these models is that the time variable is discrete for the difference equation, whereas it is continuous for the differential equation. So far, both models appear suitable. The final choice of model depends on the scientific question asked, the purpose of the model, the available data, etc.

The independent or state variables also can be chosen to be either discrete or continuous. For example, a discrete state variable may represent the number of individuals in a population, whereas a continuous variable may represent a density or a concentration.

Both of the above models are called *deterministic*. That means that if you know the state of the system at a certain point in time $t$, you can *determine* all future states by solving the corresponding model. Sometimes, however, *stochastic* effects play a dominant role. For example, in a laboratory setting you can predict that a pair of healthy rabbits will produce offspring. Outside the laboratory, life is less predictable, and the same pair of rabbits may not reproduce. In general, stochastic variations are more important for small population sizes. A model for small populations and unpredictable environments should include the uncertainty via a stochastic formulation. Large populations in constant environments (such as an aggregate of cellular slime molds, which contains about 100,000 cells) usually are modeled by deterministic models.

The number of choices presented above generates many types of models. A discussion of all types of models is beyond the scope of this book. We have chosen to restrict the material in this book to the most common model classes, summarized in the following list:

*Difference Equations:* The state (or dependent variable) can be discrete or continuous but the time is always discrete. Discrete models are suitable for seasonal events. We treat deterministic difference equations in Chapter 2 and stochastic difference equations in Chapter 5.

*Ordinary Differential Equations (ODEs):* ODEs are used to describe population evolution over a continuous time period. Deterministic ODEs are one of the major modeling tools and are discussed in detail in Chapter 3. The theory of stochastic differential equations is quite involved and is not covered in this book.

*Partial Differential Equations (PDEs):* PDEs are used if two or more continuous independent variables are used, for example, time and space, or time and age. We discuss age-structured models and reaction-diffusion equations for spatial spread in Chapter 4. Historically, stochastic PDEs were used primarily in the context of statistical physics.  Only recently have such models been considered to describe population dynamics (see, e.g., [76]).

*Stochastic Processes:* Stochastic processes and *Markov chains* are completely stochastic model classes. They are particularly useful for small populations. We treat them in detail in Chapter 5.

*Cellular Automata:* Cellular automata and related models are fully discrete models. All independent variables (such as time and space) and all dependent variables (such as population sizes) are discrete. The analysis of cellular automata is mainly restricted to computer analysis and numerical simulation. We give an introduction in Chapter 6. Cellular automata can be either deterministic or stochastic, using a random number generator.

## 1.4   Exercises for Modeling

**Exercise 1.4.1: Discrete-time versus continuous-time models.** *Assume you have a culture of bacteria growing in a petri dish, and each cell divides into two identical copies of itself every* 10 *minutes.*

(a) *Choose a unit of time, and find the corresponding probability of cell division.*

(b) *Write down a discrete-time model which balances the amount of cells at time t and at time* $t + \Delta t$.

(c) *Define the growth rate, and derive the corresponding continuous-time model.*

(d) *Solve both the discrete-time and continuous-time models, and compare the solutions.*

(e) *When is a discrete-time model appropriate? When is a continuous-time model appropriate?*

**Exercise 1.4.2: Comparison of discrete and continuous models.** *Study the two models* (1.2), (1.1) *which lead to Figure 1.2 and vary the time increment* $\Delta t$ *(e.g., try* $\Delta t = \frac{1}{4}$ *day,* $\frac{1}{8}$ *day, 1 day, 2 days, 10 days). What do you observe? Which choice of* $\Delta t$ *gives the best, and which gives the worst agreement? Can you explain why?*

**Exercise 1.4.3: Structured populations.**

(a) *Give examples of spatially structured problems. What kind of effects cannot be understood without spatial structure?*

(b) *Give an example of an age-structured problem.*

(c) *Give an example of a size-structured problem.*

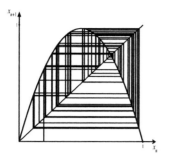

# Chapter 2
# Discrete-Time Models

## 2.1  Introduction to Discrete-Time Models

In this chapter, we use *discrete-time models* to describe dynamical phenomena in biology. Discrete-time models are appropriate when one can think about the phenomenon in terms of discrete time steps or when one wishes to describe experimental measurements that have been collected at fixed time intervals.

In general, we are concerned with a sequence of quantities,

$$x_0, x_1, x_2, x_3, x_4, \ldots,$$

where $x_i$ denotes the quantity at the $i$th measurement or after $i$ time steps. For example, $x_i$ may represent

- the size of a population of mosquitoes in year $i$;

- the proportion of individuals in a population carrying a particular allele of a gene in the $i$th generation;

- the number of cells in a bacterial culture on day $i$;

- the concentration of oxygen in the lung after the $i$th breath;

- the concentration in the blood of a drug after the $i$th dose.

You can undoubtedly think of many more such examples. Note that the time step may or may not be constant. In the example of the bacterial culture, the time step is fixed to be a day, but in the example of the oxygen concentration in the lung, the time step is variable from breath to breath. Also, time steps can be anywhere from milliseconds to years, depending on the biological problem at hand.

We can now ask ourselves, what does it mean to build a discrete-time model? In the context of our sequence of quantities $x_i$, a discrete model is a rule describing how the quantities change. In particular, a discrete model describes how $x_{n+1}$ depends on $x_n$ (and

9

perhaps $x_{n-1}, x_{n-2}, \ldots$). Restricting ourselves to the case where $x_{n+1}$ depends on $x_n$ alone, a model can then be thought of as an updating function (Adler [1]) of the form

$$x_{n+1} = f(x_n). \tag{2.1}$$

Equation (2.1) is often referred to as a discrete-time equation or difference equation, and $f$ is called a *map*.

Given some initial condition $x_0$, the updating function can be *iterated* to give $x_1 = f(x_0), x_2 = f(x_1), x_3 = f(x_2)$, and so on. The resulting simulated sequence $x_0, x_1, x_2, \ldots$ is called an *orbit* of the map. A good model should be able to produce orbits that are in close agreement with observed experimental data.

Finding the precise function $f$ that describes experimental data well or that gives a certain desired type of behavior is not always straightforward. It is often said that modeling (here, finding the right function $f$) is more of an art than a science. One starts with a particular function $f$, and then makes adjustments. Insight into how a function $f$ should be adjusted to get a better model can often be obtained from knowledge of the behavior of the current model.

Simple but powerful analytical tools are available to help determine possible types of behavior of a given model. In this chapter, we will give an introduction to some commonly used tools. We divide the chapter into two main sections. Section 2.2 deals with scalar discrete-time equations of the form (2.1), while Section 2.3 deals with systems of discrete-time equations. Throughout the chapter, applications of discrete-time equations to real biological systems, such as population growth and genetics, are discussed. More applications can be found in the exercises at the end of this chapter.

## 2.2   Scalar Discrete-Time Models

### 2.2.1   Growth of a Population and the Discrete Logistic Equation

In this section, we build a simple model describing the growth of a population of *Paramecium aurelia*. A paramecium is a unicellular organism found in large numbers in freshwater ponds. It is a member of the group of organisms called protozoa and feeds on small organisms such as bacteria and other protozoa.

We will build the model based on a classic data set collected by Gause [63]. In Table 2.1, the mean density of *Paramecium aurelia*, measured in individuals per 0.5 cm$^3$, is tabulated as a function of time, measured in days. The corresponding graph of the data is shown in Figure 2.1. The population was grown in isolation and provided with a constant level of nutrients.

Let $p_n$ be the mean density of this population on day $n$. A good starting point for building a model for $p_n$ is to think of the word equation

$$\text{future value} = \text{present value} + \text{change},$$

which readily translates to the following mathematical equation:

$$p_{n+1} = p_n + \Delta p_n. \tag{2.2}$$

The goal of the modeling process, then, is to find a reasonable approximation for $\Delta p_n$ that more or less reproduces the given set of data.

**Table 2.1.** *Growth of* Paramecium aurelia *in isolation. Here, density is the number of individuals per* 0.5 *cm*$^3$. *Data taken from Gause* [63].

| Day | Mean density of P. Aurelia | Change in density |
|-----|----------------------------|-------------------|
| $(n)$ | $(p_n)$ | $(\Delta p_n = p_{n+1} - p_n)$ |
| 0 | 2 | — |
| 1 | — | — |
| 2 | 14 | 20 |
| 3 | 34 | 32 |
| 4 | 56 | 38 |
| 5 | 94 | 95 |
| 6 | 189 | 77 |
| 7 | 266 | 64 |
| 8 | 330 | 86 |
| 9 | 416 | 91 |
| 10 | 507 | 73 |
| 11 | 580 | 30 |
| 12 | 610 | −97 |
| 13 | 513 | 80 |
| 14 | 593 | −36 |
| 15 | 557 | 3 |
| 16 | 560 | −38 |
| 17 | 522 | 43 |
| 18 | 565 | −48 |
| 19 | 517 | −17 |
| 20 | 500 | 85 |
| 21 | 585 | −85 |
| 22 | 500 | −5 |
| 23 | 495 | 30 |
| 24 | 525 | −15 |
| 25 | 510 | — |

Finding a suitable form for $\Delta p_n$ is not always easy. But let's examine the data more closely. Initially, the population increases slowly; values of $\Delta p_n$, tabulated in the third column of Table 2.1, are relatively small. As time progresses, values of $\Delta p_n$ increase and reach a maximum approximately halfway through the experiment. After that, they decrease again. We can attribute the decrease in the growth rate to intraspecific competition for nutrients and space. At the end of the experiment, the population appears to be leveling off when it reaches a mean density of approximately 540 individuals per 0.5 cm$^3$ (roughly, $\Delta p_n$ is negative when $p_n > 540$ and $\Delta p_n$ is positive when $p_n < 540$). To keep things relatively simple, we will ignore the fluctuations in the population. Note that the choice of using 540

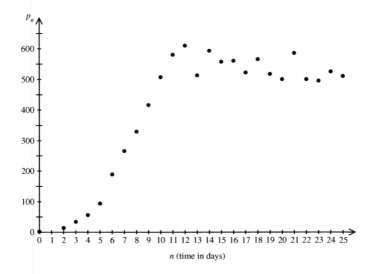

**Figure 2.1.** *Graph of the data shown in Table* 2.1.

for the limiting density may not be the best. However, it will do for a first approximative model. For clarity, then, we continue to use this number in the following discussion, but keep in mind that there may be room for improvement.

Based on the above observations, we must find a suitable form for $\Delta p_n$ that is small when $p_n$ is close to 0 and 540, and positive for intermediate values of $p_n$. The following quadratic expression fits the bill:

$$\Delta p_n = k(540 - p_n)p_n. \tag{2.3}$$

Note that this expression ensures that $\Delta p_n < 0$ when $p_n > 540$ and $\Delta p_n > 0$ when $p_n < 540$. Substituting (2.3) into (2.2), we obtain the following model for the population:

$$p_{n+1} = p_n + k(540 - p_n)p_n, \tag{2.4}$$

where the value of the parameter $k$ remains to be determined.

The experimental data contains enough information to allow us to obtain an estimate for $k$ from the data set. In particular, note that we have hypothesized that $\Delta p_n = p_{n+1} - p_n$ is proportional to the product $(540 - p_n)p_n$, with the parameter $k$ being the constant of proportionality. To test our hypothesis, we plot $\Delta p_n = p_{n+1} - p_n$ versus $p_n(540 - p_n)$ and check whether there is reasonable proportionality. The graph is shown in Figure 2.2.

Although the data looks scattered, we can fit it nicely with a straight line passing through the origin, consistent with our hypothesis. The line of best fit has slope approximately 0.00145 (in Chapter 8, you will learn how to obtain lines of best fit using Maple). Thus, setting $k = 0.00145$, we obtain the following model for the growth of the population:

$$p_{n+1} = p_n + 0.00145(540 - p_n)p_n. \tag{2.5}$$

Last but not least, we compare the behavior of our model with the observed initial data. We start our simulation with $p_2 = 14$ since there is no data point for $n = 1$. As we

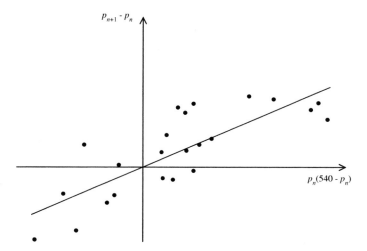

**Figure 2.2.** *Testing* (2.4) *against the data shown in Table* 2.1. *The slope of the line of best fit is approximately* 0.00145.

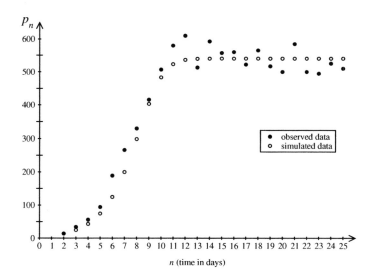

**Figure 2.3.** *Comparison of the data simulated with* (2.5) *and the data observed by Gause, from Table* 2.1.

iterate (2.5), we obtain the sequence $\{p_2, p_3, p_4, \ldots\}$. The simulated data and the observed data are shown together in Figure 2.3. We see that the agreement looks good. Recall that the choice to use 540 in the model was rather arbitrary, and improvement in the fit may be possible by adjusting this number (see the exercises).

In general, we can write the model just developed as

$$x_{n+1} = x_n + k(N - x_n)x_n, \tag{2.6}$$

where $N$ is the maximum population that can be sustained by the environment. $N$ often is referred to as the *carrying capacity* of the population. As we have just seen, this model can be used to describe the growth of a population in an environment with limited resources. The model can be used for other purposes as well. In particular, it can be used to describe the spread of an infectious disease, such as the flu or the common cold, through a small, closed population of size $N$. Here, $x_n$ is the number of infected individuals after $n$ time steps (e.g., days). Then $(N - x_n)$ is the number of individuals who have not yet become ill. The parameter $k$ is a measure of the infectivity of the disease, as well as the contact rate between healthy and infected individuals. Similarly, the model can be used to describe the spread of a rumor through a population of size $N$. In this case, $x_n$ is the number of individuals who have heard the rumor, and $N - x_n$ is the number of individuals who have not yet heard the rumor. The parameter $k$ measures how juicy the rumor is. The larger $k$, the juicier the rumor, and the faster its spread through the population.

Equation (2.6) generally is rewritten as follows:

$$\begin{aligned}
x_{n+1} &= x_n + k(N - x_n)x_n \\
&= (1 + kN)x_n - kx_n^2 \\
&= (1 + kN)\left(1 - \frac{k}{1 + kN}x_n\right)x_n \\
&= (1 + kN)\left(1 - \frac{x_n}{(1 + kN)/k}\right)x_n \\
&= r\left(1 - \frac{x_n}{K}\right)x_n, \tag{2.7}
\end{aligned}$$

where

$$r = 1 + kN, \tag{2.8}$$

$$K = \frac{1 + kN}{k}. \tag{2.9}$$

Since this model is similar in appearance to the continuous-time model known as the logistic model or the Verhulst model (you will encounter this model in Section 3.1), the model here is known as the *discrete logistic model*.

Although the discrete logistic model provides a nice fit to Gause's data, it has the unfortunate (but mathematically interesting) property that it does not exhibit logistic growth (exponential growth initially, after which growth levels off until the population's carrying capacity is reached) for all choices of the model parameters.

In Section 2.2.4, we will discuss possible alternatives to the discrete logistic model. We first make a mathematical detour. In Section 2.2.2, we introduce techniques that can be used to analyze discrete-time equations of the form

$$x_{n+1} = f(x_n).$$

In Section 2.2.3, we use these techniques to explore in some detail the dynamical behavior of the discrete logistic equation in various parameter regimes.

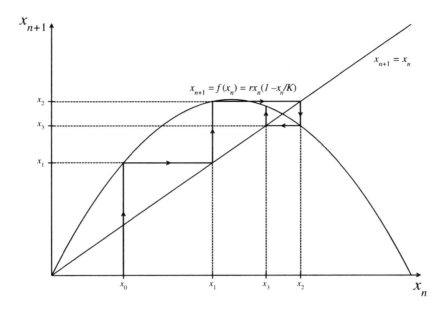

**Figure 2.4.** *Cobwebbing for the discrete logistic model* (2.7). *Parameter values used are* $r = 2.8$ *and* $K = 1$.

### 2.2.2   Cobwebbing, Fixed Points, and Linear Stability Analysis

With the fast computers of today, it is easy to generate many orbits by varying initial conditions and model parameters, and get a feel for the dynamics of the model. However, it is easy to miss some subtle behavior. We often can gain valuable insight into the model dynamics from sophisticated, but easy-to-learn, mathematical techniques. We will examine a few of these techniques in this section.

We begin with *cobwebbing*, which is a graphical solution method allowing one to quickly visualize the orbits and their long-term behavior without explicity calculating each and every iterate along the way.

We demonstrate the cobwebbing technique in Figure 2.4, which shows the graphs of the function $x_{n+1} = f(x_n) = rx_n(1 - \frac{x_n}{K})$, using $r = 2.8$ and $K = 1$, and the straight line $x_{n+1} = x_n$. We choose our first iterate, $x_0$, on the horizontal axis. The next iterate is $x_1 = f(x_0)$, which we can just read off the parabola. Visually, this is shown by a vertical line from $x_0$ on the horizontal axis to the point $(x_0, x_1)$ on the parabola. The next iterate, $x_2$, can be obtained in a similar way from $x_1$. We first need to locate $x_1$ on the horizontal axis. We already have $x_1$ on the vertical axis, and the easiest way to get it onto the horizontal axis is to reflect it through the diagonal line $x_{n+1} = x_n$. Visually, this is shown by a horizontal line from $x_1$ on the vertical axis to point $(x_1, x_1)$ on the diagonal line, and then a vertical line from point $(x_1, x_1)$ on the diagonal line to $x_1$ on the horizontal axis. This process is repeated for subsequent iterates.

In summary, one starts by traveling from $x_0$ vertically to the parabola, then horizontally to the diagonal line, vertically to the parabola, and so on, as indicated by the solid portion of

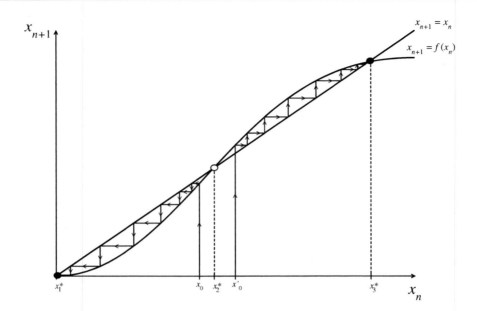

**Figure 2.5.** *Illustration of stable and unstable fixed points of the difference equation* $x_{n+1} = f(x_n)$. *The fixed points* $x_1^*$ *and* $x_3^*$ *are stable (indicated by a filled circle), and the fixed point* $x_2^*$ *is unstable (indicated by an open circle).*

the vertical and horizontal lines on the cobwebbing diagram in Figure 2.4. In this particular case, the orbit converges to the rightmost intersection of the parabola and the diagonal line.

Any intersection of the parabola and the diagonal line represents a special point. Let $x^*$ be such a point. Then $f(x^*) = x^*$. We call any such point a *fixed point* (or an *equilibrium point* or a *steady state*) of the model. If any iterate is $x^*$, then all subsequent iterates also are $x^*$. A question of interest is, what happens when an iterate is close to, but not exactly at, a fixed point? Do subsequent iterates move closer to the fixed point or further away? In the former case, the fixed point is said to be *stable*, whereas in the latter case, the fixed point is said to be *unstable*.

Examples of both stable and unstable fixed points are shown in Figure 2.5. The three fixed points shown are $x_1^*, x_2^*$, and $x_3^*$. Choosing an initial condition $x_0$ just to the left of $x_2^*$, we see that the orbit moves away from $x_2^*$, and towards $x_1^*$. Similarly, choosing the initial condition $x_0$ just to the right of $x_2^*$, we see that the orbit again moves away from $x_2^*$, but now towards $x_3^*$. Choosing the initial condition $x_0$ near $x_1^*$ or $x_3^*$ results in the orbit moving towards $x_1^*$ or $x_3^*$, respectively. We say that $x_2^*$ is an unstable fixed point of the model $x_{n+1} = f(x_n)$, and $x_1^*$ and $x_3^*$ are stable fixed points.

From Figure 2.5, note that the slope of $f$ at the stable fixed points $x_1^*$ and $x_3^*$ is less than 1 (the slope of the straight diagonal line), whereas the slope of $f$ at the unstable fixed point $x_2^*$ is greater than 1. We can formalize these ideas via a *linear stability analysis*.

We choose the $n$th iterate to be close to a fixed point $x^*$ of (2.1),

$$x_n = x^* + y_n, \tag{2.10}$$

with $y_n$ small, so that $x_n$ can be thought of as a perturbation of $x^*$. The question of interest now is, what happens to $y_n$, the deviation of $x_n$ from $x^*$, as the map is iterated? If the deviation grows, then the fixed point $x^*$ is unstable, and if the deviation decays, then it is stable. We can find the equation for the deviation by substituting (2.10) into (2.1) to obtain

$$x^* + y_{n+1} = f(x^* + y_n). \tag{2.11}$$

We expand the right-hand side using a Taylor series about $x^*$, with a remainder term of $R_2(y_n)$, to obtain

$$x^* + y_{n+1} = f(x^*) + f'(x^*)y_n + R_2(y_n). \tag{2.12}$$

Since $x^*$ is a fixed point, we can replace $f(x^*)$ on the right-hand side by $x^*$. If, in addition, we neglect all the terms in the Taylor series that have been collected in the term $R_2(y_n)$, then we are left with the following equation for the deviation:

$$y_{n+1} = f'(x^*)y_n. \tag{2.13}$$

We recognize that $f'(x^*)$ is some constant, say $\lambda$. The equation for the deviation is thus the linear difference equation

$$y_{n+1} = \lambda y_n. \tag{2.14}$$

We can write $y_{n+1}$ explicitly in terms of $\lambda$ and the initial condition $y_0$:

$$y_1 = \lambda y_0,$$
$$y_2 = \lambda y_1 = \lambda(\lambda y_0) = \lambda^2 y_0,$$
$$\vdots$$
$$y_n = \lambda^n y_0.$$

The behavior of the deviation $y_n$, and the subsequent conclusion regarding the stability of the fixed point $x^*$, can be summarized as follows:

$$\begin{array}{rl} \lambda > 1: & \text{geometric growth; fixed point } x^* \text{ is unstable;} \\ 0 < \lambda < 1: & \text{geometric decay; fixed point } x^* \text{ is stable;} \\ -1 < \lambda < 0: & \text{geometric decay with sign switch; fixed point } x^* \text{ is stable;} \\ \lambda < -1: & \text{geometric growth with sign switch; fixed point } x^* \text{ is unstable.} \end{array}$$

The four cases are illustrated in Figure 2.6. Note that no conclusion can be reached about the stability of the fixed point $x^*$ when $\lambda = \pm 1$. These two cases require advanced treatment, involving a careful examination of the neglected terms that were collected in the term $R_2(y_n)$ in (2.12), which is beyond the scope of this book. For treatment of these cases, the reader is referred to Kuznetsov [104].

More generally, we can summarize the results of the analysis in the following theorem.

**Theorem 2.1.** *Let $x^*$ be a fixed point of $x_{n+1} = f(x_n)$. Then,*

- *$x^*$ is stable when $|f'(x^*)| < 1$;*

- *$x^*$ is unstable when $|f'(x^*)| > 1$;*

- *there is no conclusion about the stability of $x^*$ when $|f'(x^*)| = 1$.*

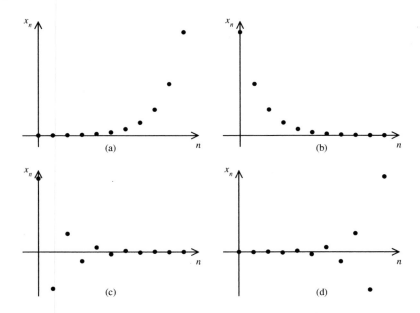

**Figure 2.6.** *Behavior of the general linear difference equation,* (2.14), *as a function of the iterates n, for the cases* (a) $\lambda > 1$; (b) $0 < \lambda < 1$; (c) $-1 < \lambda < 0$; (d) $\lambda < -1$.

That is, the linear stability of a fixed point $x^*$ is determined by the slope of the map $f(x)$ at the fixed point, as intuited earlier. The parameter $\lambda = f(x^*)$ generally is referred to as the *eigenvalue* of the map at $x^*$.

## 2.2.3   Analysis of the Discrete Logistic Equation

We now return to the discrete logistic equation, (2.7), and apply the tools discussed in the previous section. We begin with eliminating the parameter $K$ by using the transformation $\bar{x}_n = \frac{x_n}{K}$ to obtain, after dropping the overbars,

$$x_{n+1} = f(x_n) = r x_n (1 - x_n). \tag{2.15}$$

Note that if we have $x_n > 1$, then $x_{n+1} < 0$. To avoid such situations, we impose the restriction $0 \le r \le 4$ (can you think of the reason why this should be so?), so that $x_n \in [0, 1]$ for all $n$ provided $x_0 \in [0, 1]$.

The fixed points of the map can be found exactly by setting $f(x^*) = x^*$ and solving for $x^*$. There are two fixed points. The trivial fixed point, $x^* = 0$, always exists, while the nontrivial fixed point, $x^* = \frac{r-1}{r}$, is positive only when $r > 1$.

To determine the stability of the fixed points, we need $f'(x)$, which is

$$f'(x) = r(1 - 2x). \tag{2.16}$$

At the trivial fixed point, $x^* = 0$, the eigenvalue is $f'(0) = r$. That is, the trivial fixed point is stable for $0 \le r < 1$ and unstable for $1 < r \le 4$. At the nontrivial fixed point, $x^* = \frac{r-1}{r}$,

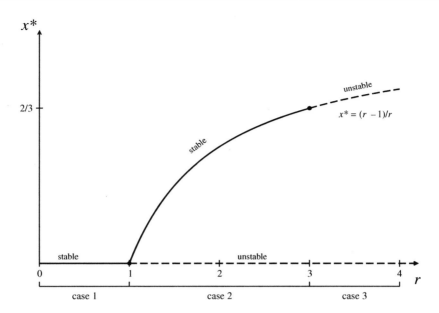

**Figure 2.7.** *Partial bifurcation diagram for the rescaled discrete logistic equation, (2.15). Shown are the fixed points and their stability as a function of the model parameter r. Solid lines indicate stability of the fixed point, and dashed lines indicate instability. The filled circles represent bifurcation points.*

the eigenvalue is $f'(\frac{r-1}{r}) = 2 - r$. That is, the nontrivial fixed point is stable for $1 < r < 3$, and unstable for $3 < r \leq 4$.

The existence and stability of the fixed points is summarized in the *bifurcation diagram* of the fixed points versus the parameter $r$, shown in Figure 2.7. Reading the diagram from left to right, note that the trivial fixed point becomes unstable as soon as the nontrivial fixed points come onto the scene at $r = 1$, when the eigenvalue moves through $+1$. The nontrivial fixed point is stable initially, but loses its stability at $r = 3$, when the eigenvalue moves through $-1$.

The two points $r = 1$ and $r = 3$ are known as *bifurcation points*. A bifurcation point is a parameter value at which there is a qualitative change in the dynamics of the map. The bifurcation at $r = 1$ is called a *transcritical bifurcation*, referring to an exchange of stability when two branches of fixed points meet (the two branches meeting here are $x^* = 0$ and $x^* = \frac{r-1}{r}$). The bifurcation at $r = 3$ is called a *flip bifurcation* or a *period-doubling bifurcation*. We will see shortly how the dynamics of the map changes at this flip bifurcation.

There are many other types of bifurcations. A detailed discussion of bifurcation theory is beyond the scope of this book, and the interested reader is referred to Alligood, Sauer, and Yorke [4], Kuznetsov [104], and Strogatz [152].

We can easily read the long-term behavior of the logistic map from the bifurcation diagram. As before, let us think of $x_n$ as the size of a population (now scaled by the factor $K$). We can distinguish three cases (indicated along the bottom of Figure 2.7). In the first case, for $0 \leq r < 1$, the population goes extinct, no matter what the size of the initial population,

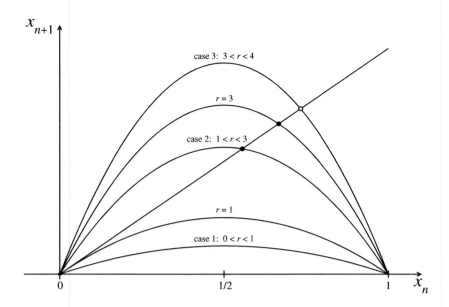

**Figure 2.8.** *Dependence of the shape of the parabola $f(x_n) = rx_n(1 - x_n)$ on the value of the model parameter $r$.*

$x_0$, is. In the second case, for values of $r$ between 1 and 3, the population reaches a nonzero steady state. The larger the value of $r$, the larger the steady-state population. What happens when the parameter exceeds 3 is not clear. Before investigating this third case, however, it pays to perform a graphical analysis complementing the results from the linear stability analysis.

Figure 2.8 shows how the shape of the parabola $f(x_n) = rx_n(1 - x_n)$ depends on the value of the model parameter $r$. Note that the roots remain fixed at $x_n = 0$ and $x_n = 1$. However, the maximum of the parabola is $\frac{r}{4}$, and thus increases with $r$.

For the first case, $0 < r < 1$, the parabola lies entirely below the diagonal line $x_{n+1} = x_n$, and the only point of intersection is at the origin. That is, the only fixed point is the trivial fixed point. Since the slope of $f$ at the origin clearly is positive but less than 1, the trivial fixed point is stable. Any population will go extinct, eventually. This situation is illustrated in Figures 2.9 (a) and (b).

When $r = 1$, the parabola is tangent to the diagonal line $x_{n+1} = x_n$ at the origin. This event marks the transition to the second case. As soon as $r > 1$, the slope of $f$ at the origin exceeds 1 (i.e., the fixed point at the origin has switched from being stable to unstable) and there is an additional point of intersection, namely, the nontrivial fixed point, $x^* = \frac{r-1}{r}$. The slope of $f$ at the nontrivial fixed point is always less than 1. Initially, for $1 < r < 3$ (case 2), the slope of $f$ at the nontrivial fixed point is greater than $-1$, and so the fixed point is stable. Any population will eventually reach a steady-state size. This situation is illustrated in Figures 2.9 (c) and (d).

When $r = 3$, the slope of $f$ at the nontrivial fixed point is $-1$, and this event marks the transition to the third case. When $3 < r \leq 4$ (case 3), the slope of $f$ at the nontrivial

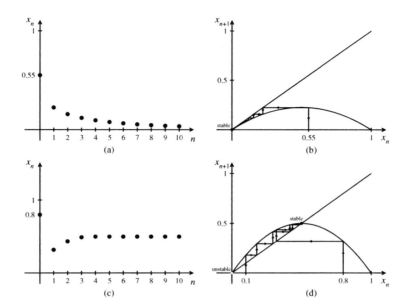

**Figure 2.9.** (a) *and* (b) *Case* 1 $(0 < r < 1)$, *for* $r = 0.9$. *The only fixed point* $\bar{x} = 0$ *is stable, and the population goes extinct.* (c) *and* (d) *Case* 2 $(1 < r < 3)$, *for* $r = 2$. *The fixed point* $\bar{x} = 0$ *is unstable, the nontrivial fixed point is stable, and the population size stabilizes.*

fixed point is less than $-1$, and so the fixed point now is unstable, as we had inferred earlier from linear stability analysis (Theorem 2.1).

We now continue with the graphical analysis and cobwebbing to determine what happens in the third case. In Figure 2.10, we show the dynamics of the discrete logistic equation for three values of $r$ between 3 and 4. The plots in the left column of Figure 2.10 show values of the iterates $x_n$ as a function of $n$ for various values of $r$. Corresponding cobwebbing diagrams are shown in the right column of Figure 2.10 (to clarify the cobwebbing diagrams shown in (b) and (d), only the last few iterates are used).

In Figures 2.10 (a) and (b), for $r = 3.2$, we observe that the population eventually oscillates between two values. We refer to the oscillation as a 2-*cycle*. In Figures 2.10 (c) and (d), for $r = 3.55$, we eventually observe a 4-*cycle*, or an oscillation between four population sizes. Values of $r$ can be found at which the discrete logistic equation exhibits an 8-*cycle*, a 16-*cycle*, and so on. But not all values of $r > 3$ give periodic oscillations. An example of an aperiodic oscillation is shown in Figures 2.10 (e) and (f), for $r = 3.88$. The orbit appears chaotic, and indeed, it can be shown that the discrete logistic equation exhibits *chaos* in the mathematical sense. A careful mathematical definition of chaos is beyond the scope of this book, and the interested reader is referred to Alligood, Sauer, and Yorke [4] and Strogatz [152] for more information. For the purposes of our discussion, it suffices to observe that the simple model under investigation can exhibit some very complicated dynamics.

We can broaden our analysis to determine the origin of the 2-cycle. When an orbit converges to a 2-cycle, it oscillates between two values, say $u$ and $v$ (see Figure 2.10 (b)),

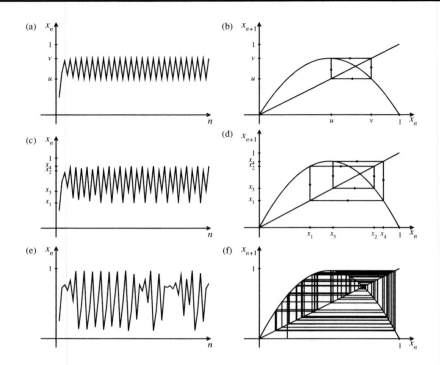

**Figure 2.10.** *Illustration of the various types of dynamical behavior of the discrete logistic equation in case* 3, *when* $3 < r \leq 4$. *(a) and (b) Two-cycle with* $r = 3.2$. *(c) and (d) Four-cycle with* $r = 3.55$. *(e) and (f) Chaos with* $r = 3.88$.

with

$$f(u) = v, \tag{2.17}$$
$$f(v) = u, \tag{2.18}$$

or, equivalently,

$$f(f(u)) = u, \tag{2.19}$$
$$f(f(v)) = v. \tag{2.20}$$

Recalling the definition of a fixed point ($x$ is a fixed point of $f(x)$ if $f(x) = x$), we see that the above equations imply that $u$ and $v$ are fixed points of the *second-iterate map*, $f(f(x)) = f^2(x)$.

The graph of the second-iterate map $f^2$ is shown in Figure 2.11 for various values of the parameter $r$. For values of $r < 3$ (Figure 2.11 (a)), the second-iterate map has two fixed points, namely, the origin, which is unstable, and the nontrivial fixed point, $x^* = \frac{r-1}{r}$, of the original logistic map, which is stable (note that any fixed point of the logistic map automatically also is a fixed point of the second-iterate map). That is, no interesting 2-cycles exist for these values of $r$. As $r$ increases, the maxima of the second-iterate map rise and

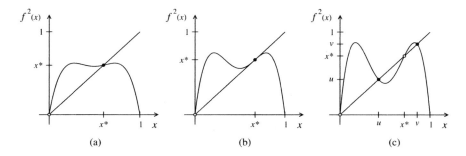

**Figure 2.11.** *The second-iterate function, $x_{n+2} = f^2(x_n)$, for the logistic map, for various values of r.* (a) $r < 3$; (b) $r = 3$; (c) $r > 3$.

the local minimum descends, until at $r = 3$ (Figure 2.11 (b)), the local minimum is tangent to the diagonal line $x_{n+2} = x_n$. At this point, two new stable fixed points of the second-iterate map emerge, namely, $u$ and $v$, corresponding to the 2-cycle (Figure 2.11 (c)). At the same time, the nontrivial fixed point $x^*$ becomes unstable (this is in accordance with our findings from the analysis of the original logistic map). The bifurcation at $r = 3$ is called a *period-doubling* or *flip bifurcation*.

Initially, $u$ and $v$ are close together, so the 2-cycle is barely noticeable. But as $r$ increases, $u$ and $v$ move away from each other, and the 2-cycle becomes more pronounced. The stability of $u$ and $v$ corresponds to the stability of the 2-cycle. That is, the 2-cycle is stable initially, since the graph of $f^2$ at $u$ and $v$ is shallow. As $r$ increases beyond $1 + \sqrt{6}$ (see Exercise 2.4.6), the slope of $f^2$ at $u$ and $v$ becomes less than $-1$, indicating that the 2-cycle becomes unstable. At this point (another flip bifurcation), the 4-cycle arises. We could continue the analysis by graphing $f^4$ for various values of $r$, but this is left as an exercise for the reader.

We can update the bifurcation diagram shown in Figure 2.7 by including information about the 2-cycle, as shown in Figure 2.12.

Ideally, we should also include information about the 4-cycle, the 8-cycle, and so on. The algebra to do so becomes unwieldy rather quickly. However, we can use the computer to create a similar diagram. The idea is to let the computer program determine the long-term behavior of the map for many values of the parameter $r$. For example, for $r = 2$, the iterates converge to $\bar{x} = \frac{1}{2}$, the stable fixed point of the map for this value of $r$. If we had computed 2000 iterates, say, from an arbitrary initial condition, then the last 100 or so iterates will all have a value virtually indistinguishable from $\frac{1}{2}$. So, plotting these last 100 iterates above $r = 2$ on a diagram of $\bar{x}$ versus $r$ just gives a point, $(r, \bar{x}) = (2, \frac{1}{2})$. If we choose $r = 3.2$ (cf. Figure 2.10 (a)), the last 100 iterates or so will jump back and forth between the values of the corresponding 2-cycle. Plotting these iterates above $r = 3.2$ gives two points, and so on. A lot of computation, using many values of $r$ close together, eventually leads to the *orbital bifurcation diagram*, also known as the *Feigenbaum diagram*, shown in Figure 2.13. Note that since the computations only detect stable behavior (stable fixed points, stable 2-cycles, and so on), the orbital bifurcation diagram differs from the bifurcation diagrams shown in Figures 2.7 and 2.12 in that the branches of unstable behavior, indicated by dashed lines, no longer are shown.

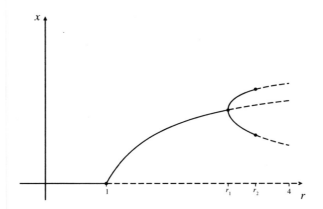

**Figure 2.12.** *Updated bifurcation diagram for the discrete logistic equation shown earlier in Figure 2.7. Shown are the fixed points, as well as the 2-cycle for values of $r > r_1 = 3$. The 2-cycle is stable up to $r_2 = 1 + \sqrt{6}$, and unstable thereafter.*

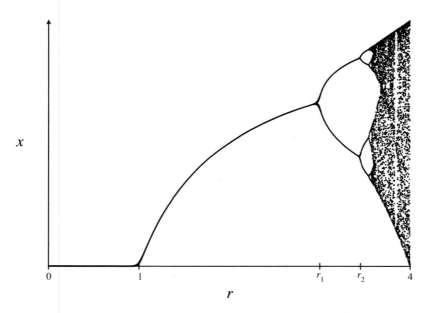

**Figure 2.13.** *Orbital bifurcation diagram for the discrete logistic equation.*

By examining the orbital bifurcation diagram, it can be seen that the 4-cycle exists only over a small range of $r$, the 8-cycle over an even smaller range of $r$, etc. It can be shown (see, e.g., Holmgren [89]) that the bifurcation points leading to higher-order cycles converge at $r \approx 3.57$. Beyond $r \approx 3.57$, the logistic map becomes chaotic, that is, the iterates no longer appear to follow a predictable pattern, although they are confined to take on only certain values (e.g., when $r = 3.6$, the iterates never take on values below 0.324

or above 0.900, but they can take on any intermediate value). We say that the attractor of the map is a *strange attractor*. By computing the orbital bifurcation diagram over a smaller range of $r$, say for $3 < r < 4$, with higher resolution, many interesting features of the map can be observed, such as *periodic windows* surrounded by chaos. A periodic window is a small range of $r$ where the attractor is periodic again. For example, near $r = 3.83$, one can find 3-cycles. From this periodic window, the transition back to chaos occurs through a series of period-doubling or flip bifurcations, leading to 6-cycles, 12-cycles, and so on.

The discrete logistic equation is a well-studied difference equation, and there are many interesting mathematical investigations that can be pursued. We will stop here and refer the interested reader to Devaney [43] and Strogatz [152].

In the 1970s, May [113] noticed that simple difference equations can give rise to very complicated dynamics. He hypothesized that the wild fluctuations observed in some natural populations might reflect chaotic orbits of low-dimensional systems of difference equations. J. M. Cushing and his colleagues [41] have followed up on this hypothesis. They conducted controlled experiments on laboratory populations of flour beetles living under constant environmental conditions. They showed that the population dynamics can be described and predicted accurately by a relatively simple model of difference equations that reflect well-understood facts about the life cycle of the flour beetle. In addition, they were able to explain observed dynamics of the population and demonstrate nonlinear phenomena such as bifurcations, periodic orbits, and chaos in real biological populations. Thus, their work has lent credibility to the use of models such as those described in this chapter. It should be no surprise, then, that these models continue to be used on a regular basis.

### 2.2.4 Alternatives to the Discrete Logistic Equation

In the previous section, we saw that the behavior of the discrete logistic equation, (2.7), is quite complex. For many choices of the model parameters, the solution does not exhibit logistic growth (exponential growth initially, followed by a leveling off of the growth rate, until the population reaches a steady state). Also, for some choices of parameter values, the model gives unrealistic results. For example, if the population $x_n > K$ in any year, then the population is extinct (negative) the next year. For these reasons, it is worthwhile to examine alternative models that do not have these problems and that are widely used.

In particular, we examine the Beverton–Holt and Ricker models. Before discussing these models in detail, we note that these two models and the discrete logistic equation belong to a class of models that can be written in the following general form:

$$x_{n+1} = f(x_n) = g(x_n)x_n. \tag{2.21}$$

Many other population models can be written in this form. Of course, the simplest model belonging to this class is the *geometric growth model*, using $g(x_n) = r$, so that

$$x_{n+1} = rx_n, \tag{2.22}$$

where $r > 0$. We have encountered this model previously, as (2.14) in the discussion of linear stability analysis. When $r > 1$, $x_n \to \infty$ as $n \to \infty$; when $r < 1$, $x_n \to 0$ as $n \to \infty$. In this case, the growth rate $g(x_n) = r$ is constant; that is, the number of offspring

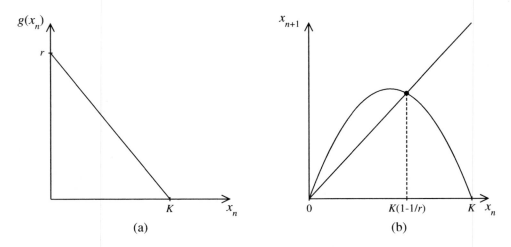

(a)                                            (b)

**Figure 2.14.** *The discrete logistic model,* (2.7). (a) *Graph of* $g(x_n) = r(1 - x_n/K)$. (b) *Graphs of* $x_{n+1} = f(x_n) = g(x_n)x_n$ *and* $x_{n+1} = x_n$.

per adult does not depend on the current population. We say that growth in the geometric model is *density independent*.

Regulatory mechanisms that control the growth of populations need to be included in any realistic model. There are many hypotheses regarding the mechanisms at play in regulating the size of populations. For example, populations are influenced by changes in the weather, a limited food supply, competition for resources such as nutrients and space, territoriality, predation, diseases, etc. The discrete logistic equation as well as the Beverton–Holt and Ricker models contain self-regulatory mechanisms that are *density dependent*; that is, the growth rate $g(x_n)$ depends nontrivially on the current population $x_n$. The models differ in their form of density dependence.

In the case of the discrete logistic model, we have $g(x_n) = r(1 - x_n/K)$. Thus, the growth rate decreases linearly, as shown in Figure 2.14 (a). It is because $g(x_n) < 0$ when $x_n > K$ that the model predicts extinction within a year whenever $x_n > K$ (see Figure 2.14 (b)). Thus, any good alternative to the discrete logistic model should have $g(x_n) > 0$. Both the Beverton–Holt and Ricker models satisfy $g(x_n) > 0$. We now discuss these models in some detail.

## The Beverton–Holt Model

The *Beverton–Holt model* was derived in the context of fisheries [22]. The growth rate is given by $g(x_n) = \frac{r}{1 + \frac{r-1}{K} x_n}$, with $r > 0$ and $K > 0$, giving

$$x_{n+1} = f(x_n) = \frac{r}{1 + \frac{r-1}{K} x_n} x_n. \tag{2.23}$$

The graph of $g(x)$, for $r > 1$, is shown in Figure 2.15 (a), and the resulting Beverton–Holt model is shown in Figure 2.15 (b). We see that the Beverton–Holt map increases monotonically, approaching the asymptote $x_{n+1} = rK/(r - 1)$.

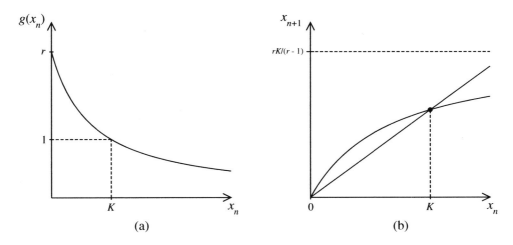

**Figure 2.15.** *The Beverton–Holt model,* (2.23). *(a) Graph of* $g(x_n) = r/(1 + \frac{r-1}{K}x_n)$. *(b) Graphs of* $x_{n+1} = f(x_n) = g(x_n)x_n$ *and* $x_{n+1} = x_n$.

Fixed points of the Beverton–Holt map are the nontrivial fixed point at the origin, $x_1^* = 0$, and the nontrivial fixed point at the carrying capacity, $x_2^* = K$. To determine the stability of the fixed points, we need

$$f'(x) = \frac{r}{\left(1 + \frac{r-1}{K}x\right)}.$$

Then

$$f'(x_1^*) = f'(0) = r,$$

$$f'(x_2^*) = f'(K) = \frac{1}{r}.$$

Thus, when $r > 1$, the trivial fixed point, $x_1^* = 0$, is unstable, and the nontrivial fixed point, $x_2^* = K$, is stable (when $0 < r < 1$, the stability reverses). Cobwebbing confirms our analytical result. In fact, it is easy to verify with cobwebbing that convergence to $x_2^* = K$ is monotonic. That is, starting from a small initial condition $0 < x_0 \ll K$, the population initially increases quickly. Growth slows down when the population approaches the carrying capacity $K$. Similarly, when the initial condition $x_0 > K$, the population decreases smoothly to $K$. Complex behavior such as cycles and chaos is not possible.

The Beverton–Holt model is one of the few nonlinear models for which a solution in closed form can be written down (see the exercises). It can be shown, then, that the solution behavior of the Beverton–Holt model is precisely that of the continuous version of the logistic model (equation (3.6), discussed in detail in Section 3.1). In fact, the Beverton–Holt model is the time-one map of the continuous logistic equation, and as such, we can consider it to be another discrete analog of the continous logistic equation. We defer derivation of the Beverton–Holt equation via the time-one map to Section 3.6.2. In the meantime, we reiterate that we now have two discrete-time models that can be considered analogous to

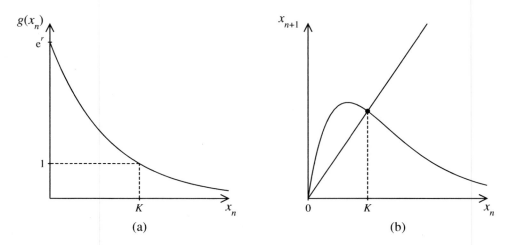

**Figure 2.16.** *The Ricker model,* (2.24). (a) *Graph of* $g(x_n) = \exp\left[r\left(1 - \frac{x_n}{K}\right)\right]$. (b) *Graphs of* $x_{n+1} = f(x_n) = g(x_n)x_n$ *and* $x_{n+1} = x_n$.

the continuous logistic equation. In particular, we have the discrete logistic equation, which is analogous in the sense that the form of the equation is the same, and the Beverton–Holt model, which is analogous in the sense that the form of the solution is the same.

In Section 2.2.1, we fit the discrete logistic equation, (2.7), to Gause's classic data for the growth of *Paramecium aurelia* (cf. Figure 2.1). As we have just seen, it may be more appropriate to fit the Beverton–Holt model to the data instead. This is left as an exercise for the reader.

### The Ricker Model

The *Ricker model* also has its roots in fisheries [137]. In this case, $g(x) = \exp\left[r\left(1 - \frac{x_n}{K}\right)\right]$, with $r > 0$ and $K > 0$, giving

$$x_{n+1} = f(x_n) = \exp\left[r\left(1 - \frac{x_n}{K}\right)\right]x_n. \tag{2.24}$$

We can think of the factor $\exp(r)$ as a constant reproduction factor, and of the factor $\exp\left(-rx_n/K\right)$ as a density-dependent mortality factor. The larger the population $x_n$, the more severe the mortality factor.

The graph of $g(x)$ is shown in Figure 2.16 (a), and the resulting Ricker map is shown in Figure 2.16 (b). Comparing Figures 2.15 (a) and 2.16 (a), we see that the shape of the graph of $g(x)$ is similar to that for the Beverton–Holt model. However, the exponential function decreases more quickly than the inverse function, and as a result, the Ricker map has a single local maximum (it sometimes is referred to as a single-hump map), as is the case in the discrete logistic map (see Figure 2.14 (b)). However, the Ricker map remains positive for all values of $x_n$. Thus, the model can exhibit complex dynamics such as cycles and chaos, but it never yields unrealistic (negative) populations.

Fixed points of the Ricker map are given by the trivial fixed point at the origin, $x_1^* = 0$, and the nontrivial fixed point at the carrying capacity, $x_2^* = K$. To determine the stability of the fixed points, we need

$$f'(x) = e^{r(1-\frac{x}{K})} \left[ 1 - \frac{rx}{K} \right]. \tag{2.25}$$

Then

$$f'(x_1^*) = f'(0) = e^r > 1, \tag{2.26}$$
$$f'(x_2^*) = f'(K) = 1 - r. \tag{2.27}$$

Thus, the trivial fixed point $x_1^* = 0$ is always unstable. Since $|1-r| < 1$ when $0 < r < 2$, the nontrivial fixed point is stable for $0 < r < 2$ and unstable for $r > 2$. Numerical simulation of the model shows that the nontrivial fixed point is reached from any initial condition $x_0 > 0$ when $0 < r < 2$; that is, the nontrivial fixed point is globally asymptotically stable. When $r > 2$, cycles and chaos are observed.

Further detailed analysis of the Ricker model mirrors the investigation of the discrete logistic equation in the previous section and is the subject of Section 8.2 in the chapter on Maple.

### 2.2.5 Models in Population Genetics

We now move away from population biology and show another application of discrete-time equations, namely, in the study of population genetics. Population genetics concerns itself with the genetic basis for evolution in a population.

We begin with a review of some terminology in the study of genetics. We consider *diploid* organisms, whose genetic material rests on two sets of *chromosomes*, one obtained from each parent. Chromosomes contain genes, which are the fundamental units of heredity, carrying information from one generation to the next. Due to mutations, a gene can exist in different forms, or alleles. Two homologous alleles, one originating from each parent, interact to produce a trait, such as eye color in humans or wing color in moths.

Suppose we are interested in a trait determined by one gene for which there are two alleles. For example, consider wing coloration in moths. Let the two alleles be denoted by W and w. That is, individual moths may have one of three *genotypes* (allelic composition): WW, Ww, and ww. Individuals with WW or ww are called *homozygous*; those with Ww are called *heterozygous*. Let's further suppose that individuals have one of two *phenotypes* (outward expression of the genetic code): individuals with genotype WW and Ww develop white wings, whereas individuals with genotype ww develop black wings. In this case, the W allele is capable of expressing the color trait at the expense of the w allele; we say it is *dominant*. Similarly, the w allele fails to have an impact when paired with the W allele; we say it is *recessive*.

A question of interest in population genetics is how the genetic make-up of a population changes over time. In particular, how do allele frequencies change, if at all, across the generations? Do recessive alleles disappear gradually? What happens when there is selection?

**Table 2.2.** *Punnett square summarizing how the frequencies of the W and w alleles in the current generation give rise to three different genotypes in the next generation.*

|        |   |         | Mother |               |
|--------|---|---------|--------|---------------|
|        |   |         | W      | w             |
|        |   |         | $p_n$  | $1 - p_n$     |
|        | W | $p_n$   | $p_n^2$ | $p_n(1 - p_n)$ |
| Father |   |         |        |               |
|        | w | $1 - p_n$ | $p_n(1 - p_n)$ | $(1 - p_n)^2$ |

We divide this section into three parts. First, we develop a model to track the frequency of the W allele in the population and derive a well-known result in population genetics known as the Hardy–Weinberg law. We then introduce selection and examine the effect of selection in a population with two phenotypes and a population with three phenotypes. The model for the latter case exhibits a type of behavior not encountered previously, namely, bistability. Throughout this section, we restrict ourselves to the study of organisms with discrete generations, so that discrete-time equations are appropriate. The development of this section was inspired by [154].

**The Hardy–Weinberg Law**

Let $p_n$ be the frequency of the W allele in the population, that is, the number of alleles W divided by the total number of alleles in the population, during the $n$th generation. Similarly, let $q_n$ be the frequency of the w allele during the $n$th generation. Of course, since $p_n + q_n = 1$, it is sufficient to track only $p_n$, since $q_n$ can always be recovered via $q_n = 1 - p_n$.

To derive a model for $p_n$, we need to make a number of assumptions. To begin, we assume the following:

- mating is completely random (white moths don't preferentially mate with other white moths nor with black moths);

- all genotypes are equally fit; that is, all genotypes are equally likely to survive to breed;

- there is an absence of mutation;

- the frequency of the allele in either sex is the same as in the entire population.

To compute $p_{n+1}$, it helps to construct a *Punnett square*, as shown in Table 2.2, which summarizes the frequencies of the alleles in the current generation and the resulting frequencies of the three different genotypes in the next generation. Thus, the frequencies of the next generation with genotypes WW, Ww (= wW), and ww are $p_n^2$, $2p_n(1 - p_n)$, and $(1 - p_n)^2$, respectively. The frequency of the W allele in the next generation is equivalent to the probability of obtaining a W allele by randomly choosing one allele from a random individual. The probability of obtaining a W allele from an individual with genotype WW is 1, from an individual with genotype Ww, it is $\frac{1}{2}$, and from an individual with genotype ww, it is 0. Computing a weighted average of these probabilities thus yields the following expression for the frequency of the W allele:

$$p_{n+1} = \frac{1 \cdot p_n^2 + \frac{1}{2} \cdot 2p_n(1 - p_n) + 0 \cdot (1 - p_n)^2}{p_n^2 + 2p_n(1 - p_n) + (1 - p_n)^2} \tag{2.28}$$

$$= \frac{p_n}{(p_n + 1 - p_n)^2} \tag{2.29}$$

$$= p_n. \tag{2.30}$$

We see that allele frequencies do not change from generation to generation, provided the assumptions stated above hold, of course. This conclusion is known as the *Hardy–Weinberg law*.

### Selection in a Population with Two Phenotypes

The assumptions stated in the previous section imply that there is no selection. What happens when there is selection? For example, suppose that white-winged moths are more conspicous and therefore more likely to be eaten by birds than black-winged moths. Will white-winged moths become extinct? What if white-winged moths have the selective advantage instead? Will black-winged moths become extinct?

Let $\alpha$ be the fraction of white-winged moths surviving to produce the next generation, with $0 \le \alpha \le 1$. Similarly, let $\gamma$ be the fraction of black-winged moths surviving, with $0 \le \gamma \le 1$. Choosing $\alpha > \gamma$ gives white-winged moths a selective advantage, while $\alpha < \gamma$ gives black-winged moths a selective advantage.

Just before reproduction, the genotype ratio WW : Ww : ww is $\alpha p_n^2 : 2\alpha p_n(1 - p_n) : \gamma(1 - p_n)^2$. The resulting frequency of W alleles in the next generation is then

$$p_{n+1} = \frac{1 \cdot \alpha p_n^2 + \frac{1}{2} \cdot 2\alpha p_n(1 - p_n) + 0 \cdot \gamma(1 - p_n)^2}{\alpha p_n^2 + 2\alpha p_n(1 - p_n) + \gamma(1 - p_n)^2} \tag{2.31}$$

$$= \frac{\alpha p_n}{(\gamma - \alpha)p_n^2 - 2(\gamma - \alpha)p_n + \gamma}. \tag{2.32}$$

Note that when $\alpha = \gamma$, that is, when there is no selective advantage, we recover the Hardy–Weinberg equilibrium, namely $p_{n+1} = p_n$. But when $\alpha \ne \gamma$, we have a nonlinear equation that warrants further investigation.

Let's begin by finding the fixed points $p^*$ of this equation and determining their stability with a linear stability analysis. First, we look for values of $p^*$ such that $f(p^*) = p^*$, where

$$f(p) = \frac{\alpha p}{(\gamma - \alpha)p^2 - 2(\gamma - \alpha)p + \gamma}.$$

It is easy to show that this equation has two distinct roots, corresponding to two fixed points, namely, $p_1^* = 0$ and $p_2^* = 1$. In terms of the genetic problem at hand, $p_1^* = 0$ means that the W allele has become extinct and all moths have black wings, whereas $p_2^* = 1$ means that the w allele has become extinct and all moths have white wings. Intuitively, we expect

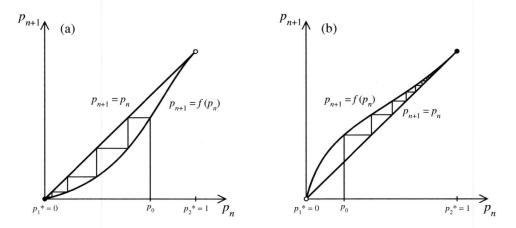

**Figure 2.17.** *Cobwebbing diagrams for allele frequency equation* (2.32).
(a) $\alpha < \gamma$ ($\alpha = 0.2$ *and* $\gamma = 0.8$); (b) $\alpha > \gamma$ ($\alpha = 0.8$ *and* $\gamma = 0.2$).

$p_1^* = 0$ to be stable when black-winged moths have the selective advantage ($\alpha < \gamma$), and
$p_2^* = 1$ to be stable when white-winged moths have the selective advantage ($\alpha > \gamma$).

To determine the stability of these fixed points with linear stability analysis, we find

$$f'(p) = \frac{-\alpha\left((\gamma - \alpha)p^2 - \gamma\right)}{\left((\gamma - \alpha)p^2 - 2(\gamma - \alpha)p + \gamma\right)^2},$$

so that $f'(p_1^*) = f'(0) = \frac{\alpha}{\gamma}$ and $f'(p_2^*) = f'(1) = 1$. The appearance of the ratio $\alpha/\gamma$
looks promising in light of our intuition discussed above. Let's check the details. When
$\alpha < \gamma$, we have $0 < \frac{\alpha}{\gamma} < 1$, and so we conclude that the fixed point $p_1^* = 0$ is stable.
That is, when black-winged moths have the selective advantage, the W allele can indeed
become extinct if its frequency becomes sufficiently small (since the linear stability analysis
only is local, we cannot conclude that the W allele will become extinct per se). Similarly,
when $\alpha > \gamma$, the fixed point $p_1^* = 0$ is unstable. Since $f'(p_2^*) = 1$, the linear stability
analysis does not yield a conclusion about the stability of the other fixed point, $p_2^* = 1$ (see
Theorem 2.1). A graphical stability analysis will be helpful.

Cobwebbing diagrams for the allele frequency equation (2.32) are shown in Fig-
ure 2.17. In Figure 2.17 (a), the case $\alpha < \gamma$ is shown. Indeed, any initial condition $p_0$ with
$0 \le p_0 < 1$ leads to the fixed point $p_1^* = 0$. We conclude that $p_1^* = 0$ is stable, and $p_2^* = 1$
is unstable (for any $0 < p_0 \le 1$, which are the only biologically sensible initial conditions).
In Figure 2.17 (b), the case $\alpha > \gamma$ is shown. Here, the situation is reversed. The fixed point
$p_1^* = 0$ is unstable, and $p_2^* = 1$ is stable (again, for any $0 < p_0 \le 1$).

In summary, selection on the level of phenotype (white-winged versus black-winged),
when one allele is dominant and the other recessive, eventually leads to extinction of one
of the alleles. How fast the allele is driven to extinction depends on the relative strength of
the model parameters $\alpha$ and $\gamma$. The larger the difference between $\alpha$ and $\gamma$, the faster the
approach to extinction.

### Selection in a Population with Three Phenotypes

Let's generalize this investigation one more step into the effect of selection. Suppose that the W allele is no longer dominant and that the three genotypes, WW, Ww, and ww, give rise to three distinct phenotypes. For ease of discussion, we will assume that individuals with genotypes WW and ww will develop white and black wings, respectively, as before, and that individuals with genotype Ww will develop gray wings. Furthermore, introduce the parameter $\beta$ to represent the selective pressure on gray-winged moths, with $0 \le \beta \le 1$. Just before reproduction, the genotype ratio WW : Ww : ww is $\alpha p_n^2 : 2\beta p_n(1 - p_n) : \gamma(1 - p_n)^2$. Questions of interest now include, under which conditions will all three phenotypes co-exist? Similarly, can gray-winged moths be driven extinct? If so, will white-winged moths or black-winged moths survive?

As before, it suffices to study the iterative map for the frequency of W alleles, which is

$$p_{n+1} = \frac{1 \cdot \alpha p_n^2 + \frac{1}{2} \cdot 2\beta p_n(1 - p_n) + 0 \cdot \gamma(1 - p_n)^2}{\alpha p_n^2 + 2\beta p_n(1 - p_n) + \gamma(1 - p_n)^2} \tag{2.33}$$

$$= \frac{(\alpha - \beta)p_n^2 + \beta p_n}{(\alpha - 2\beta + \gamma)p_n^2 + 2(\beta - \gamma)p_n + \gamma}. \tag{2.34}$$

This more general allele frequency equation looks a bit more intimidating than the previous one, but doing the analysis is still quite reasonable. In particular, fixed points $p$ satisfy $p^* = f(p^*)$, where

$$f(p) = \frac{(\alpha - \beta)p^2 + \beta p}{(\alpha - 2\beta + \gamma)p^2 + 2(\beta - \gamma)p + \gamma},$$

yielding a cubic equation in $p$. One fixed point can be found by inspection, namely, $p_1^* = 0$. We are then left with a quadratic equation, and its roots are $p_2^* = 1$ and

$$p_3^* = \frac{\gamma - \beta}{\alpha - 2\beta + \gamma}.$$

It is easy to check that $p_3^* \in (0, 1)$ only when $\beta < \alpha, \gamma$ or when $\beta > \alpha, \gamma$. Otherwise, $p_1^* = 0$ and $p_2^* = 1$ are the only biologically relevant fixed points. We can now proceed to determine the stability of the fixed points. As before, we require $f'(p)$, which is

$$f'(p) = \frac{(\alpha\beta - 2\alpha\gamma + \beta\gamma)p^2 + 2\gamma(\alpha - \beta)p + \beta\gamma}{\left((\alpha - 2\beta + \gamma)p^2 + 2(\beta - \gamma)p + \gamma\right)^2}.$$

Then (with a little help from Maple to simplify the algebra)

$$f'(p_1^*) = f'(0) = \frac{\beta}{\gamma}, \tag{2.35}$$

$$f'(p_2^*) = f'(1) = \frac{\beta}{\alpha}, \tag{2.36}$$

$$f'(p_3^*) = -\frac{\alpha\beta - 2\alpha\gamma + \beta\gamma}{\alpha\gamma - \beta^2}. \tag{2.37}$$

**Table 2.3.** *Summary of the outcome of the more general allele frequency equation, (2.34).*

| Case | Model parameter | Fixed points and stability | Long-term behavior | Biological interpretation |
|------|----------------|---------------------------|-------------------|--------------------------|
| I | $\alpha > \beta > \gamma$ | 0 is unstable<br>1 is stable | $p \to 1$ | White-winged moths have the selective advantage and black-winged moths have the selective disadvantage; w allele becomes extinct; all moths will have genotype WW (white-winged). |
| II | $\alpha < \beta < \gamma$ | 0 is stable<br>1 is unstable | $p \to 0$ | Black-winged moths have the selective advantage and white-winged moths have the selective disadvantage; W allele becomes extinct; all moths will have genotype ww (black-winged). |
| III | $\beta > \gamma > \alpha$<br>or<br>$\beta > \alpha > \gamma$ | 0 is unstable<br>1 is unstable<br>$p_3^*$ is stable | $p \to p_3^*$ | Gray-winged moths have the selective advantage; both W and w alleles remain in the population, and their frequencies reach an equilibrium; all genotypes coexist. |
| IV | $\beta < \gamma < \alpha$<br>or<br>$\beta < \alpha < \gamma$ | 0 is stable<br>1 is stable<br>$p_3^*$ is unstable | $p \to 0$<br>or<br>$p \to 1$ | Gray-winged moths have the selective disadvantage; either the W or the w allele becomes extinct; all moths will have either genotype WW or ww (white-winged or black-winged). |

We can now determine the fate of the moth population under different conditions by studying the outcome of the model with different parameter sets. It can be shown that there are four fundamentally different cases, as summarized in Table 2.3 (the reader is asked to work out the details in the exercises). Representative cobweb diagrams for each case are shown in Figure 2.18.

Cases I and II are straighforward, and the results are rather intuitive. In case I, white-winged moths (genotype WW) have the selective advantage and black-winged moths (genotype ww) have the selective disadvantage, so that the w allele is driven to extinction. Consequently, both gray-winged and black-winged moths are driven to extinction, and the only moths remaining are the white-winged moths (Figure 2.18 (a)). Case II is just the opposite, with black-winged moths having the selective advantage and white-winged moths the selective disadvantage, so that the only moths remaining are black-winged moths (Figure 2.18 (b)).

A bit more interesting are cases III and IV. In case III (shown in Figure 2.18 (c)), gray-winged moths have the selective advantage. Consequently, both W and w alleles remain

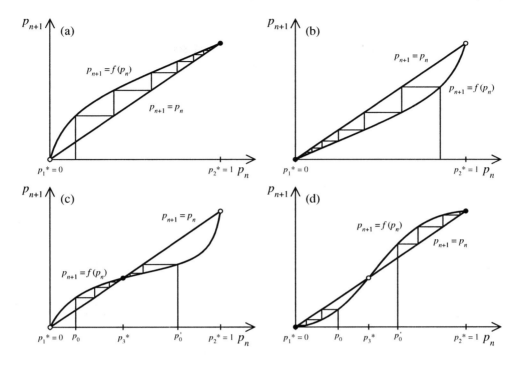

**Figure 2.18.** *Cobweb diagrams for the more general allele frequency equation,* (2.34), *corresponding to the four cases summarized in Table 2.3.* (a) *Case* I ($\alpha > \beta > \gamma$), *with* $\alpha = 0.8$, $\beta = 0.5$, *and* $\gamma = 0.1$. (b) *Case* II ($\alpha < \beta < \gamma$), *with* $\alpha = 0.1$, $\beta = 0.5$, *and* $\gamma = 0.8$. (c) *Case* III ($\beta > \alpha, \gamma$), *with* $\alpha = 0.1$, $\beta = 0.9$, *and* $\gamma = 0.3$. (d) *Case* IV ($\beta < \alpha, \gamma$), *with* $\alpha = 0.9$, $\beta = 0.1$, *and* $\gamma = 0.7$.

in the population ($p \longrightarrow p_3^*$, which lies between 0 and 1), and all three genotypes coexist. The equilibrium frequency for the W allele, $p_3^*$, depends on the relative strength of the three selective pressure parameters. The larger the value of $\alpha$, the larger $p_3^*$, that is, the higher the equilibrium frequency of the W allele, as might be expected biologically.

Finally, in case IV (shown in Figure 2.18 (d)), where gray-winged moths have the selective disadvantage, we see a new and interesting dynamical behavior, known as *bistability*. There are two stable fixed points, separated by an unstable fixed point. As time progresses, either one of the stable fixed points is approached, depending on the initial condition. If the initial frequency of the W allele is greater than $p_3^*$, then $p \longrightarrow p_2^* = 1$ is approached. That is, provided the frequency of the W allele is sufficiently large initially, it will become dominant. If it is less than $p_3^*$, then $p \longrightarrow p_1^* = 0$ is approached, and the W allele becomes extinct.

It turns out that the ideas presented here find application in a real-life situation. The peppered moth (*Biston betularia*) is common in both Europe and North America. Normally, the moth has a "peppered" appearance, but sometimes it is completely black (melanic). On lichen-covered tree trunks, the normal form is camouflaged, while the melanic form is rather conspicuous and at a selective disadvantage. During the industrial revolution in England, lichen were killed by pollution, resulting in much darker tree trunks. Consequently, the selective pressure on the two forms of the moths reversed, and the frequencies of the allele for the gene responsible for wing coloration adapted quickly [101].

In this section, we have only scratched the surface of the types of problems in the area of population genetics that can be studied with discrete-time equations. For further exploration, the reader is referred to Chapter 4 in the text by Britton [29] and to Section 3.6 in the text by Edelstein-Keshet [51].

## 2.3   Systems of Discrete-Time Equations

### 2.3.1   Love Affairs: Introduction

Consider the relationship between two lovers, say Romeo and Juliet (with apologies to Shakespeare). It is not unreasonable to think that their feelings for each other are dynamic. In [151, 152], Strogatz developed a simple model, consisting of a system of ODEs, describing the dynamic love affair. Here, we will consider a discrete-time version of the model.

Let $R_n$ be Romeo's love/hate for Juliet on day $n$, and let $J_n$ be Juliet's love/hate for Romeo on day $n$. We will agree upon the following interpretation of the values of $R_n$ (similarly for $J_n$): when $R_n > 0$, Romeo loves Juliet; when $R_n < 0$, Romeo hates Juliet; and when $R_n = 0$, Romeo is neutral towards Juliet. The larger the $|R_n|$, the stronger the feeling of love/hate.

Next, let's assume that Romeo and Juliet respond to their own feelings in a linear fashion. In particular, assume

$$R_{n+1} = a_R R_n, \tag{2.38}$$

$$J_{n+1} = a_J J_n. \tag{2.39}$$

It seems reasonable to take $a_R, a_J > 0$ so that we're not dealing with wild mood swings (love one day, hate the next, and so on). Depending on the magnitude of the $a$ parameter, there are two romantic styles. If $0 < a_R, a_J < 1$, then the initial feeling becomes neutral as time progresses. If $a_R, a_J > 1$, then the initial feeling intensifies.

Now we add simple linear terms that represent the response of Romeo and Juliet to the feelings of the other, to get the following system of equations:

$$R_{n+1} = a_R R_n + p_R J_n, \tag{2.40}$$

$$J_{n+1} = a_J J_n + p_J R_n. \tag{2.41}$$

The $p$ parameters describe how their love/hate changes in response to the current feeling of the other. We allow $p_R, p_J \in \mathbb{R}$. In this case, the sign of the $p$ parameter determines a particular romantic style. For example, if $p_R > 0$, then Romeo gets excited by Juliet's love for him, while he gets discouraged by Juliet's hate for him. In contrast, if $p_R < 0$, then Juliet's hate for him contributes to his love for her, while Juliet's love for him contributes to his hate for her.

Both Romeo and Juliet thus have four romantic styles. The outcome of their love affair depends on the particular combination of romantic styles, the relative size of the $a$ and $p$ parameters, and the initial feelings for each other. It is easy (and instructive too) to

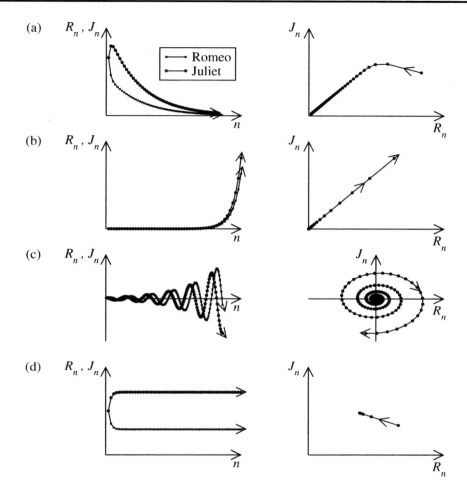

**Figure 2.19.** *Four case studies for Romeo and Juliet. Graphs in the left column show $R_n$ and $J_n$ as functions of $n$. Graphs in the right column show the orbits in the $(R_n, J_n)$ phase plane. In all cases, initial conditions used are $R_0 = J_0 = 1$. (a) $a_R = 0.5$, $a_J = 0.7$, $p_R = 0.2$, $p_J = 0.5$. (b) $a_R = 0.5$, $a_J = 0.7$, $p_R = 0.7$, $p_J = 0.9$. (c) $a_R = 1.0$, $a_J = 1.0$, $p_R = 0.2$, $p_J = -0.2$. (d) $a_R = 0.5$, $a_J = 0.8$, $p_R = 0.2$, $p_J = 0.5$.*

simulate this model, for example, on a spreadsheet, and experiment with different sets of parameters. Here, we present four case studies, illustrating typical types of behavior of the system.

In Figure 2.19, we show two side-by-side graphs for each of the four case studies. Graphs in the left column show $R_n$ and $J_n$ as functions of $n$. Graphs in the right column show the orbits in the $(R_n, J_n)$ *phase plane.* In these graphs, time $n$ is suppressed, and successive coordinates $(R_n, J_n)$ are plotted. The direction of the orbit as $n$ increases is indicated with arrows.

In Figure 2.19 (a), both Romeo and Juliet can be considered cautious lovers [152]. If they were to respond to their own feelings only, their feelings for each other would become neutral ($0 < a_R, a_J < 1$). Although they respond in kind to each other, they only do so tentatively ($p_R$ and $p_J$ are both positive, but relatively small). Different initial conditions all lead to the same outcome: as $n \to \infty$, $R_n, J_n \to 0$; that is, the love affair fizzles, and both Romeo and Juliet become neutral to each other.

In Figure 2.19 (b), the $p$ parameters have been increased slightly, so that Romeo and Juliet both respond more decisively to the feelings of the other. In the case shown, the love affair results in a love fest, with both $R_n, J_n \to +\infty$ as $n \to \infty$. With different initial conditions, the love affair may result in war instead, with both $R_n, J_n \to -\infty$ as $n \to \infty$.

In Figure 2.19 (c), we consider a case where both Romeo and Juliet remain true to their initial feelings ($a_R, a_J = 1$), but their $p$ parameters have opposite sign (do opposites attract?). Here, the love affair exhibits growing oscillations; that is, Romeo and Juliet experience a perpetual cycle of love and hate, with their feelings ever intensifying as time progresses.

Finally, in Figure 2.19 (d), we achieve an equilibrium of perpetual love, albeit one in which Juliet loves Romeo more than Romeo loves Juliet.

One can continue to vary the model parameters and initial conditions to investigate the outcome of the love affair. This becomes tiring quickly, and unsatisfying. Instead, it would be nice to be able to predict the outcome of the love affair given a set of model parameters. We can do so by extending the concept of fixed points and their stability from scalar equations to systems of equations, which we do in the next section. We return to Romeo and Juliet in Section 2.3.3, where we apply the results of linear stability analysis.

## 2.3.2   Fixed Points and Linear Stability Analysis for Systems of Discrete-Time Equations

In this section, we extend the concept of fixed points and their stability from scalar equations to systems of equations. For ease of notation, we present the material for a two-dimensional system, but the results are generalized readily to higher-dimensional systems, as we will see shortly.

Consider the following two-dimensional discrete-time system:

$$x_{n+1} = f(x_n, y_n), \tag{2.42}$$

$$y_{n+1} = g(x_n, y_n). \tag{2.43}$$

Fixed points of this system are all points $(x^*, y^*)$ satisfying $f(x^*, y^*) = x^*$ and $g(x^*, y^*) = y^*$.

To determine the stability of a fixed point, consider a small perturbation from the fixed point by letting

$$x_n = x^* + u_n, \tag{2.44}$$

$$y_n = y^* + v_n, \tag{2.45}$$

where both $u_n$ and $v_n$ are understood to be small, so that $x_n$ and $y_n$ can be thought of as perturbations of $x^*$ and $y^*$, respectively. Similar to the situation for one-dimensional discrete-time equations, discussed earlier, the question of interest is what happens to $u_n$ and $v_n$, the deviations of $x_n$ and $y_n$ from $x^*$ and $y^*$, respectively, as the map is iterated.

We can find the map for the deviation $(u_n, v_n)$ by substituting (2.44)–(2.45) into (2.42)–(2.43) to obtain

$$x^* + u_{n+1} = f(x^* + u_n, y^* + v_n), \tag{2.46}$$

$$y^* + v_{n+1} = g(x^* + v_n, y^* + v_n). \tag{2.47}$$

We expand the right-hand side using a Taylor series about $(x^*, y^*)$, with remainder terms $R_{x,2}$ and $R_{y,2}$, respectively, to obtain

$$x^* + u_{n+1} = f(x^*, y^*) + \frac{\partial f}{\partial x}(x^*, y^*)u_n + \frac{\partial f}{\partial y}(x^*, y^*)v_n + R_{x,2}(u_n, v_n), \tag{2.48}$$

$$y^* + v_{n+1} = g(x^*, y^*) + \frac{\partial g}{\partial x}(x^*, y^*)u_n + \frac{\partial g}{\partial y}(x^*, y^*)v_n + R_{y,2}(u_n, v_n). \tag{2.49}$$

Since $(x^*, y^*)$ is a fixed point, we can replace $f(x^*, y^*)$ and $g(x^*, y^*)$ on the right-hand side by $x^*$ and $y^*$, respectively. If, in addition, we neglect all the terms in the Taylor series that have been collected in $R_{x,2}$ and $R_{y,2}$, then we are left with the following map for the deviation:

$$u_{n+1} = \frac{\partial f}{\partial x}(x^*, y^*)u_n + \frac{\partial f}{\partial y}(x^*, y^*)v_n, \tag{2.50}$$

$$v_{n+1} = \frac{\partial g}{\partial x}(x^*, y^*)u_n + \frac{\partial g}{\partial y}(x^*, y^*)v_n. \tag{2.51}$$

As before, we recognize that the partial derivatives appearing here are evaluated at the fixed point $(x^*, y^*)$, and so they are all constants. We thus have a linear map, which can be rewritten in matrix form:

$$\mathbf{w}_{n+1} = J\mathbf{w}_n, \tag{2.52}$$

where

$$\mathbf{w}_n = \begin{pmatrix} u_n \\ v_n \end{pmatrix}, \tag{2.53}$$

and

$$J = \begin{bmatrix} \frac{\partial f}{\partial x}(x^*, y^*) & \frac{\partial f}{\partial y}(x^*, y^*) \\ \frac{\partial g}{\partial x}(x^*, y^*) & \frac{\partial g}{\partial y}(x^*, y^*) \end{bmatrix} \tag{2.54}$$

is the *Jacobian matrix* of the original map, evaluated at the fixed point $(x^*, y^*)$ (earlier, we used $J$ to denote Juliet's love/hate for Romeo; however, the meaning should be clear from the context of the equation).

Since we started with a two-dimensional system of equations, (2.42)–(2.43), $\mathbf{w}$ is a 2-vector, and the Jacobian matrix is a $2 \times 2$ matrix. In general, if we start with an $m$-dimensional system, $\mathbf{w}$ is an $m$-vector, and the Jacobian matrix has dimension $m \times m$. Thus, now that we have switched to matrix notation, the results that follow apply not only to two-dimensional systems, but also to $m$-dimensional systems in general.

Motivated by the form of the solution for scalar equations, we look for solutions of the form

$$\mathbf{w}_n = \lambda^n \mathbf{c}, \tag{2.55}$$

where $\mathbf{c}$ is a constant vector. Substituting (2.55) into (2.52) gives

$$\lambda^{n+1}\mathbf{c} = \lambda^n J\mathbf{c}. \tag{2.56}$$

Cancelling $\lambda^n$ and rearranging gives

$$(J - \lambda I)\mathbf{c} = \mathbf{0}, \tag{2.57}$$

where $I$ denotes the identity matrix and $\mathbf{0}$ denotes the zero vector.

We recognize this last equation as the eigenvalue problem from linear algebra for the matrix $J$: a nonzero vector $\mathbf{c}$ satisfying the equation is an *eigenvector* corresponding to the *eigenvalue* $\lambda$.

In order to obtain a nonzero vector $\mathbf{c}$, we need

$$\det(J - \lambda I) = 0, \tag{2.58}$$

known as the *characteristic equation* of the matrix $J$. Since $J$ is an $m \times m$ matrix, setting $\det(J - \lambda I) = 0$ gives a polynomial equation of degree $m$ for $\lambda$. This polynomial is known as the *characteristic polynomial* of the matrix. In general, the characteristic polynomial has $m$ distinct roots, $\lambda_1, \lambda_2, \ldots, \lambda_m$. The superposition principle then yields the following general solution of (2.52) for the deviation $\mathbf{w}_n$:

$$\mathbf{w}_n = \sum_{i=1}^{m} A_i \lambda_i^n \mathbf{c}_i, \tag{2.59}$$

where the $A_i$'s are arbitrary constants (determined by initial conditions), and $\mathbf{c}_i$ is the eigenvector corresponding to the eigenvalue $\lambda_i$.

We are now in a position to evaluate the dynamics of the deviation $\mathbf{w}_n$. As was the case for scalar equations, the dynamics are determined by the size of the eigenvalues $\lambda_i$ relative to 1. In general, if all eigenvalues $|\lambda_i| < 1$, then $|\mathbf{w}_n| \to 0$ as $n \to \infty$. If at least one of the eigenvalues $|\lambda_i| > 1$, then $|\mathbf{w}_n| \to \infty$ as $n \to \infty$.

The implication for the stability of the fixed points of the original map thus can be summarized in the following theorem.

**Theorem 2.2.** *Let $\mathbf{x}^*$ be a fixed point of the m-dimensional difference equation $\mathbf{x}_{n+1} = \mathbf{f}(\mathbf{x}_n)$, where $\mathbf{x} \in \mathbb{R}^m$, $\mathbf{f} : \mathbb{R}^m \to \mathbb{R}^m$, and $\mathbf{f}$ is at least twice continuously differentiable. Let $J$ be the Jacobian matrix of $\mathbf{f}$, evaluated at $\mathbf{x}^*$. Then*

- $\mathbf{x}^*$ *is stable if all eigenvalues of the Jacobian matrix $J$ have magnitude less than 1;*

- $\mathbf{x}^*$ *is unstable if at least one of the eigenvalues has magnitude greater than 1.*

It is not always necessary to calculate the eigenvalues of the Jacobian matrix $J$ explicitly. In particular, Jury [96] derived necessary and sufficient conditions for all eigenvalues of the Jacobian matrix to have magnitude less than 1. The so-called Jury conditions can be written down in terms of the coefficients of the characteristic polynomial. They are easy to write down and apply for two- and three-dimensional systems, but quickly become unwieldy for high-dimensional systems.

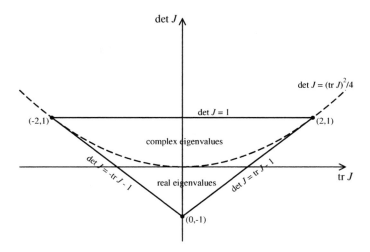

**Figure 2.20.** *The stability triangle for two-dimensional discrete-time systems in* (tr $J$, det $J$) *space. Any fixed point for which* (tr $J$, det $J$) *lies within the triangle is stable. Any fixed point for which* (tr $J$, det $J$) *lies outside the triangle is unstable. Eigenvalues are complex if* (tr $J$, det $J$) *lies above the dashed line, and real if* (tr $J$, det $J$) *lies below the dashed line.*

For two-dimensional systems, the characteristic polynomial can be written as

$$\lambda^2 - \text{tr } J\lambda + \det J = 0. \tag{2.60}$$

In the exercises, the reader is asked to verify that the following Jury conditions,

$$|\text{tr } J| < 1 + \det J < 2, \tag{2.61}$$

are necessary and sufficient conditions for all eigenvalues of $J$ to have magnitude less than 1, that is, for the fixed point in question to be stable. The Jury conditions for systems of higher dimension can be found in the paper by Jury [96] and also in the text by Edelstein-Keshet [51].

For two-dimensional systems, the Jury conditions can be visualized in (tr $J$, det $J$) space, as shown in Figure 2.20. In particular, the Jury conditions hold on the interior of the stability triangle, determined by the intersection of the three following regions:

$$\det J < 1, \tag{2.62}$$
$$\det J > \text{tr } J - 1, \tag{2.63}$$
$$\det J > -\text{tr } J - 1. \tag{2.64}$$

The system undergoes a bifurcation as parameter values are varied and (tr $J$, det $J$) crosses one of the boundaries of this triangle. On the boundary of the triangle, one of the Jury conditions is violated through equality, giving rise to bifurcations. A treatment of bifurcations is beyond the scope of this book, but the reader is referred to [104] for details.

### 2.3.3   Love Affairs: Model Analysis

We now return to the love affair of Romeo and Juliet and attempt to understand the outcome of their affair as observed in the four case studies discussed in Section 2.3.1 in terms of the stability of any fixed points.

Fixed points $(R^*, J^*)$ of the system for Romeo and Juliet, (2.40)–(2.41), must satisfy

$$R^* = a_R R^* + p_R J^*, \tag{2.65}$$

$$J^* = a_J J^* + p_J R^*. \tag{2.66}$$

Rearranging, we obtain the following linear system for $R^*$ and $J^*$:

$$(a_R - 1)R^* + p_R J^* = 0, \tag{2.67}$$

$$p_J R^* + (a_J - 1)J^* = 0, \tag{2.68}$$

which can be written as

$$\begin{bmatrix} a_R - 1 & p_R \\ p_J & a_J - 1 \end{bmatrix} \begin{bmatrix} R^* \\ J^* \end{bmatrix} = \begin{bmatrix} 0 \\ 0 \end{bmatrix}. \tag{2.69}$$

We have a homogeneous linear system. Thus, this system has a unique solution, namely, $(R^*, J^*) = (0, 0)$, provided that $\det(A) \neq 0$, where

$$A = \begin{bmatrix} a_R - 1 & p_R \\ p_J & a_J - 1 \end{bmatrix}. \tag{2.70}$$

It can be verified easily that this condition holds for the three case studies shown in Figures 2.19 (a)–(c). When $\det(A) = 0$, as is the case for the fourth case study, shown in Figure 2.19 (d), there is an infinite number of solutions or fixed points. We begin by discussing the results of the first three case studies and defer discussion of the fourth case study to later.

Orbits shown in the phase planes in Figures 2.19 (a)–(c) suggest that the fixed point $(R^*, J^*) = (0, 0)$ is stable in the first case study, while it is unstable in the second and third case studies. We can easily verify this with a linear stability analysis. Since the original system, (2.40)–(2.41), is already linear, there is little work to be done.

The Jacobian matrix is simply

$$J = \begin{bmatrix} a_R & p_R \\ p_J & a_J \end{bmatrix}, \tag{2.71}$$

with

$$\operatorname{tr} J = a_R + a_J, \tag{2.72}$$

$$\det J = a_R a_J - p_R p_J. \tag{2.73}$$

Applying the Jury conditions, (2.61), we require

$$|a_R + a_J| < 1 + a_R a_J - p_R p_J < 2 \tag{2.74}$$

for the fixed point $(R^*, J^*) = (0, 0)$ to be stable.

Indeed, for the first case study ($a_R = 0.5$, $a_J = 0.7$, $p_R = 0.2$, $p_J = 0.5$, shown in Figure 2.19 (a)), both Jury conditions are satisfied, and hence the fixed point is stable, as we had guessed. For the second case study ($a_R = 0.5$, $a_J = 0.7$, $p_R = 0.7$, $p_J = 0.9$, shown in Figure 2.19 (b)), the first Jury condition is violated, and hence the fixed point is unstable, also as we had guessed. For the third case study ($a_R = a_J = 1.0$, $p_R = 0.2$, $p_J = -0.2$, shown in Figure 2.19 (c)), it is the second Jury condition that is violated, verifying that indeed the fixed point is unstable in this case as well. The oscillatory nature of the love affair results from the fact that the eigenvalues no longer are real in this case, but are complex conjugates.

Last but not least, we examine the fourth case study in more detail. Recall that in this case, the fixed point $(R^*, J^*) = (0, 0)$ no longer is unique, since $\det(A) = 0$, where $A$ is the matrix defined in (2.70). Instead, from (2.69), we note that there is a continuum of fixed points satisfying $(a_R - 1)R^* + p_R J^* = 0$ or, equivalently, $p_J R^* + (a_J - 1)J^* = 0$ (since $\det(A) = 0$, one of the equations in (2.69) is redundant; we can choose either one of the two to work with). Choosing to work with the first equation, we obtain

$$J^* = \frac{1 - a_R}{p_R} R^*. \tag{2.75}$$

Thus, all points of the form

$$(R^*, J^*) = \left( R^*, \frac{1 - a_R}{p_R} R^* \right) \tag{2.76}$$

are fixed points. For example, in our case ($a_R = 0.5$, $a_J = 0.8$, $p_R = 0.2$, $p_J = 0.5$), all fixed points are of the form

$$(R^*, J^*) = \left( R^*, \frac{1 - 0.5}{0.2} R^* \right) = (R^*, 2.5R^*); \tag{2.77}$$

that is, if the fixed points are stable, Romeo and Juliet will either both love or hate each other, with Juliet's feeling always 2.5 times as strong as Romeo's. We will verify shortly that the fixed points indeed are stable with our choice of parameter values. For now, we investigate which of the infinite number of fixed points is approached as $n \to \infty$.

Consider the original system, using the specific model parameters from the fourth case study:

$$R_{n+1} = 0.5R_n + 0.2J_n, \tag{2.78}$$
$$J_{n+1} = 0.5R_n + 0.8J_n. \tag{2.79}$$

Note that $a_R + p_J = 0.5 + 0.5 = 1$ and $p_R + a_J = 0.2 + 0.8 = 1$. In terms of the love affair, we can interpret these conditions as follows: the total amount of love/hate that Romeo and Juliet feel for each other initially is preserved on all subsequent days. Each day, Romeo's love/hate for Juliet is split 50/50 between Romeo and Juliet. Similarly, Juliet's love/hate for Romeo is split unequally, with 20% transferred to Romeo and the remaining 80% retained by Juliet herself.

The reader is asked in the exercises to show that $\det(A) = 0$ whenever Romeo and Juliet preserve their love/hate from day to day, that is, whenever $a_R + p_J = 1$ and

$p_R + a_J = 1$. In what follows, we will restrict ourselves to the situation in which Romeo and Juliet are in love/hate-preserving mode.

Although the insight just obtained is perhaps somewhat unromantic, it does allow us to determine the final outcome of Romeo and Juliet's relationship. Since the total amount of love/hate between Romeo and Juliet is initially $R_0 + J_0$ and it is preserved, we must have $R^* + J^* = R_0 + J_0$, provided that the orbit converges to $(R^*, J^*)$. Thus,

$$R^* + J^* = R^* + \frac{1 - a_R}{p_R} R^* = R^* + \frac{p_J}{p_R} R^* = \frac{p_R + p_J}{p_R} R^* = R_0 + J_0, \qquad (2.80)$$

yielding the following solution for the fixed point that is approached:

$$R^* = \frac{p_R}{p_R + p_J}(R_0 + J_0), \qquad (2.81)$$

$$J^* = R_0 + J_0 - R^*. \qquad (2.82)$$

For our choice of parameter values and initial conditions ($R_0 = J_0 = 1$), we obtain $R^* \approx 0.571429$ and $J^* \approx 1.428571$.

We now investigate the stability of the fixed points. In the exercises, the reader is asked to show that the two eigenvalues of the Jacobian matrix are

$$\lambda_1 = 1, \qquad (2.83)$$

$$\lambda_2 = a_R + a_J - 1 > -1. \qquad (2.84)$$

The first eigenvalue, $\lambda_1$, is precisely equal to 1, reflecting the fact that the first Jury condition is just violated through equality. Because the original system is linear, the stability of the fixed points is determined by the magnitude of the second eigenvalue, $\lambda_2$ (remember, this is not the case for nonlinear systems!). The fixed points are stable provided $|\lambda_2| < 1$, that is, provided $a_R + a_J < 2$, which simply is the second Jury condition. For our choice of parameter values, $\lambda_2 = 0.5 + 0.8 - 1 = 0.3$; that is, the fixed points are stable (as expected from the solution shown in Figures 2.19).

### 2.3.4  Host–Parasitoid Models

*Host–parasitoid models* are a classic example of the use of discrete-time systems in population dynamics. These types of models address the life cycles of two interacting species of insects, one a host and the other a parasitoid.

Parasitoids are insects whose females lay their eggs in or on the bodies of the host insects. Parasitoid eggs develop into parasitoid larvae at the expense of their host. Hosts that have been parasitized thus give rise to the next generation of parasitoids, while only hosts that are *not* parasitized will give rise to the next generation of hosts.

We will limit our attention to hosts and parasitoids with one nonoverlapping generation per year so that discrete-time equations are appropriate.

Let $H_n$ and $P_n$ be the number of the hosts and parasitoids, respectively, at generation $n$. Further, let $f(H_n, P_n)$ be the fraction of hosts that are *not* parasitized. This fraction is a function of the rate of encounter of the two insect species and will be specified shortly. We thus have the following:

$$f(H_n, P_n)H_n = \text{number of hosts } not \text{ parasitized,}$$
$$[1 - f(H_n, P_n)] H_n = \text{number of hosts parasitized.}$$

The following two assumptions allow us to complete the basic host–parasitoid model:

1. The host population grows geometrically in the absence of the parasitoids, with reproductive rate $k > 1$.

2. The average number of eggs laid in a single host that give rise to adult parasitoids is $c$.

We obtain

$$H_{n+1} = kf(H_n, P_n)H_n, \tag{2.85}$$
$$P_{n+1} = c[1 - f(H_n, P_n)] H_n. \tag{2.86}$$

We now develop the functional form of $f(H_n, P_n)$. We assume that encounters between hosts and parasitoids occur at random and are independent (the latter means that parasitoids do not distinguish between hosts that have been parasitized and hosts that have not yet been parasitized). The *Law of Mass Action*, which will be treated in depth in Section 3.3.1, states that the number of encounters is proportional to the product $H_n P_n$, that is, $a H_n P_n$, where $a$ is the constant of proportionality representing the searching efficiency of the parasitoids. The *average* number of encounters per host is thus

$$v = \frac{a H_n P_n}{H_n} = a P_n. \tag{2.87}$$

Of course, not all hosts experience this many encounters. Some will experience more, others less. Let

$$p(i) = \text{the probability that a host experiences } i \text{ encounters.} \tag{2.88}$$

Since we assumed that encounters are random and independent, they are said to follow a *Poisson process*, and we can use the *Poisson distribution* for $p(i)$ [1]. In particular,

$$p(i) = \frac{v^i e^{-v}}{i!}. \tag{2.89}$$

Recalling that we defined $f(H_n, P_n)$ to be the fraction of hosts *not* parasitized, we have

$$f(H_n, P_n) = p(0) = \frac{v^0 e^{-v}}{0!} = e^{-v} = e^{-a P_n}. \tag{2.90}$$

Substituting (2.90) into (2.85)–(2.86), we obtain Nicholson and Bailey's classic model [126],

$$H_{n+1} = k H_n e^{-a P_n}, \tag{2.91}$$
$$P_{n+1} = c H_n [1 - e^{-a P_n}]. \tag{2.92}$$

It can be shown (see the exercises) that the Nicholson–Bailey model has two fixed points, namely, the trivial fixed point, $(H_1^*, P_1^*) = (0, 0)$, and the following nontrivial fixed point:

$$(H_2^*, P_2^*) = \left( \frac{k \ln k}{ac(k - 1)}, \frac{\ln k}{a} \right), \qquad (2.93)$$

provided $k > 1$. The trivial fixed point represents the situation in which both the host and the parasitoid are extinct. We are interested in situations where there is coexistence of the two insect species. Hence, of interest is the stability of the nontrivial fixed point.

The Jacobian matrix, evaluated at the nontrivial fixed point, is

$$J(H_2^*, P_2^*) = \begin{bmatrix} 1 & -\frac{k \ln k}{c(k-1)} \\ \frac{c(k-1)}{k} & \frac{\ln k}{k-1} \end{bmatrix}, \qquad (2.94)$$

so that

$$\operatorname{tr} J = 1 + \frac{\ln k}{k - 1}, \qquad (2.95)$$

$$\det J = \ln k + \frac{\ln k}{k - 1}. \qquad (2.96)$$

Since $k > 1$, the first of the Jury conditions, (2.61), always is satisfied. Since $\det J > 1$ for all $k > 1$ (see the exercises), the second Jury condition can never be satisfied. We conclude that the nontrivial fixed point, $(H_2^*, P_2^*)$, is always unstable.

Instability of the nontrivial steady state in itself does not preclude coexistence of the two insect species. For example, coexistence could come in the form of a stable cycle. However, for the Nicholson–Bailey model, no choice of parameter values leads to coexistence. Instead, the model exhibits growing oscillations, an example of which is shown in Figure 2.21. We observe that parasitoid levels can become extremely low. That is, the model predicts near-extinction of the parasitoids. Of course, as soon as the parasitoids have gone extinct, the hosts grow geometrically.

It appears that the model is not very realistic. Indeed, it cannot be used to predict long-term dynamics of a host–parasitoid interaction. However, the model has been used successfully to describe short-term oscillations in host–parasitoid systems. For example, Burnett [33] used the model to fit data for approximately two dozen generations of populations of the greenhouse whitefly *Trialeurodes vaporairorum* and the parasitoid *Encarsia formosa* grown under laboratory conditions.

The work of Nicholson and Bailey [126] was an important milestone in the modeling of host–parasitoid systems, showing that host–parasitoid interactions can result in large-amplitude oscillations when host density is limited solely by the parasitoid. Further, the Nicholson–Bailey model is used as a starting point for many contemporary models. These models all include features that have a stabilizing effect on the nontrivial fixed point.

For example, Beddington, Free, and Lawton [16] modified the equation for the host population, so that its growth is density-dependent instead of geometric in the absence of parasitoids. In particular, they replaced

$$H_{n+1} = k H_n$$

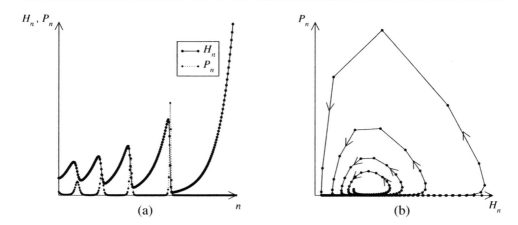

**Figure 2.21.** *The Nicholson–Bailey model, (2.91)–(2.92), exhibits growing oscillations. The left panel shows a typical solution for $H_n$ and $P_n$ as functions of $n$. The right panel shows the orbit in the $(H_n, P_n)$ phase plane. Model parameters used are $k = 1.05$, $a = 0.005$, and $c = 3$, and initial conditions are $H_0 = 50$ and $P_0 = 10$.*

in (2.91) with

$$H_{n+1} = e^{r(1-H_n/K)} H_n,$$

where $K$ is the carrying capacity of the host insect population, and $r$ determines the rate of approach to the carrying capacity. Their full host–parasitoid model thus reads

$$H_{n+1} = e^{r(1-H_n/K)} H_n e^{-aP_n}, \tag{2.97}$$
$$P_{n+1} = cH_n[1 - e^{-aP_n}]. \tag{2.98}$$

Two simulations of this model are shown in Figure 2.22. The simulation shown in Figure 2.22 (a) shows co-existence at a stable fixed point, and the one in Figure 2.22 (b) shows coexistence in a stable cycle. The determination of fixed points and their stability is tedious, and the reader is referred to [16] for details.

Ecological processes other than intraspecific competition in the host population also can stabilize the system. Examples are intraspecific competition in the parasitoid population, spatial heterogeneity of the environment, parasitoid dispersal among host patches, and so forth. It has proven extremely difficult to ascertain which, if any, of these mechanisms operate in nature, and research continues in this fascinating area of mathematical biology. One reason why host–parasitoid systems continue to receive much attention is their potential for biological control, where parasitoids are introduced to reduce the host population of a pest on agricultural crops. Questions of interest are what the qualities of a parasitoid should be, what can go wrong, and so on. Readers interested in learning more about host–parasitoid systems and biological control are referred to the books by Godfray [68] and Hassell [80] and the article by Murdoch [119].

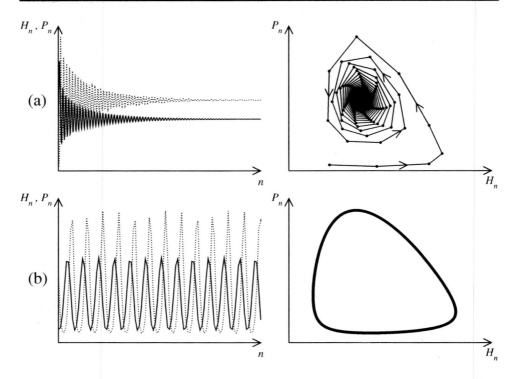

**Figure 2.22.** *Two types of behavior exhibited by the Beddington model,* (2.97)–(2.98). *Graphs in the left column show* $H_n$ *and* $P_n$ *as functions of n; graphs in the right column show corresponding orbits in the* $(H_n, P_n)$ *phase plane.* (a) *The host and parasitoid coexist at a stable fixed point* ($K = 200$). (b) *The host and parasitoid coexist in a stable cycle* ($K = 250$). *Other model parameters are* $r = 1.1$, $a = 0.005$, *and* $c = 3$.

## 2.4   Exercises for Discrete-Time Models

**Exercise 2.4.1: German population.** *Write down a simple discrete birth-death model describing the following situation.  Individuals die at rate* $\delta$ *and are born at rate* $\mu$*.  On December* 31, 1998, *Germany had a population of* 82,037,000. *In* 1999, *there were* 770,744 *live births and* 846,330 *deaths (source:* Statistisches Bundesamt). *Find* $\delta$ *and* $\mu$*. What will happen to the German population in the future?  How should the model be altered to be more realistic?*

**Exercise 2.4.2: Drug prescriptions.** *Consider the following model for a drug prescription:*

$$a_{n+1} = a_n - ka_n + b,$$

*where* $a_n$ *is the amount of a drug (in mg, say) in the bloodstream after administration of n dosages at regular intervals (hourly, say).*

  (a) *Discuss the meaning of the model parameters k and b.  What can you say about their size and sign?*

(b) *Find the fixed points of the model and their stability via linearization.*

(c) *Perform a cobwebbing analysis for this model. What happens to the amount of drug in the bloodstream in the long run? How does the result depend on the model parameters?*

(d) *How should b be chosen to ensure that the drug is effective, but not toxic?*

**Exercise 2.4.3: Improving the fit of the logistic model to the data.** Note to the instructor: This question requires nonlinear fitting techniques, which are not treated in this chapter, nor in the chapter on Maple. However, students may be asked to attempt this question after studying the project on cell competition in Section 10.1.

*In Section 2.2.1, we fit (2.4) to Gause's data. Recall that the choice to use the number 540 in this equation was rather arbitrary. Consider the more general model,*

$$p_{n+1} = p_n + k(N - p_n)p_n.$$

(a) *Use nonlinear fitting techniques to determine the best fit of both model parameters, k and N.*

(b) *Simulate the model with the best fit values for k and N, and make a plot to compare the model results with the data observed by Gause. Were you able to improve upon the comparison shown in Figure 2.3?*

**Exercise 2.4.4: Fluctuations in the population of P. aurelia.** *In Section 2.2.1, we ignored the fluctuations in the population of* P. aurelia *at carrying capacity. Discuss possible reasons for the appearance of the fluctuations.*

**Exercise 2.4.5: Whale population.** *Consider the survival of a population of whales, and assume that if the number of whales falls below a minimum survival level m, then the species will become extinct. In addition, assume that the population is limited by the carrying capacity M of the environment. That is, if the whale population is above M, then it will experience a decline because the environment cannot sustain that large a population level.*

(a) *Let $a_n$ represent the whale population after n years. Discuss the model*

$$a_{n+1} = a_n + k(M - a_n)(a_n - m),$$

*where $k > 0$. Does it make sense in terms of the description above?*

(b) *Find the fixed points of the model, and determine their stability via linearization. You may assume that $M = 5000$, $m = 100$, and $k = 0.0001$.*

(c) *Perform a graphical stability analysis. Are your results consistent with the results from (b)?*

(d) *Sketch the graphs of $a_n$ versus n for various initial conditions.*

(e) *The model has two serious shortcomings. What are they? Hint: Consider what happens when $a_0 < m$, and when $a_0 \gg M$.*

(f) *Think of a possible way to fix the model so as to overcome the shortcomings. You are encouraged to be creative, innovative—you do not need to write down the equation of an improved model; it is sufficient to describe your ideas with words and/or sketches of graphs.*

**Exercise 2.4.6: Second-iterate map.** *This exercise deals with the second-iterate map, $f^2(x)$, for the logistic map, $f(x) = rx(1 - x)$.*

(a) *Compute $f^2(x)$.*

(b) *Find the fixed points of $f^2(x)$. Verify that a nontrivial 2-cycle exists only for $r > 3$.*

(c) *Compute $\frac{d}{dx} f^2(x)$.*

(d) *Verify that the nontrivial 2-cycle is stable for $3 < r < 1 + \sqrt{6}$, and unstable for $r > 1 + \sqrt{6}$.*

**Exercise 2.4.7: Fourth-iterate map.** *This exercise deals with the fourth-iterate map, $f^4(x)$, for the logistic map, $f(x) = rx(1 - x)$.*

(a) *Graph $f^4(x)$ for various values of the model parameter $r$. Compare to the graphs of $f(x)$ and $f^2(x)$.*

(b) *At which value of $r$ does a 4-cycle appear?*

(c) *At which value of $r$ does the 4-cycle become unstable?*

**Exercise 2.4.8: Exact solution for the Beverton–Holt model.** *The Beverton–Holt model, (2.23), is one of the few nonlinear models which has a solution in closed form, that is, $x_n$ in terms of the model parameters and the initial condition $x_0$. Use the transformation $u_n = \frac{1}{x_n}$ to show that the solution can be written as*

$$x_n = \frac{r x_0}{1 + \frac{r^n - 1}{K} x_0}.$$

**Exercise 2.4.9: Fitting the Beverton–Holt model to Gause's data.** *In Section 2.2.1, we fit Gause's data for P. aurelia with the discrete logistic equation. In Section 2.2.4, we learned about alternatives to the discrete logistic equation. In particular, we saw that the Beverton–Holt model would be a suitable alternative model to describe populations undergoing logistic growth. Fit the Beverton–Holt model to the data in Table 2.1.*

*Hint: The use of line-fitting techniques with Maple will be helpful (see Chapter 8).*

**Exercise 2.4.10: The tent map.** *The tent map is an approximation to the discrete logistic equation: $x_{n+1} = f(x_n)$ with*

$$f(x) = \begin{cases} \mu x & for \quad 0 \le x \le 0.5, \\ \mu(1 - x) & for \ 0.5 < x \le 1. \end{cases}$$

(a) *Sketch the graph of $f$ for $\mu > 0$.*

(b) *Find the steady states and their stability.*

(c) *Find orbits of period 2.*

(d) *Plot f for $\mu = 2$. Carefully try to find an orbit of period 3.*

**Exercise 2.4.11: Blood cell population.** *In this exercise, we will investigate a model for the size of the red blood cell population in the human body (see also [65]). Let $x_n$ be the number of red blood cells in the human body on day n. We wish to write down an updating function for the number of red blood cells on day $n + 1$. We will think of the updating function in terms of destruction and production of red blood cells. If we let $d(x_n)$ represent the number of red blood cells lost due to cell death on day n, and $p(x_n)$ the number of red blood cells gained due to production by the bone marrow on day n, then we can write*

$$x_{n+1} = x_n - d(x_n) + p(x_n);$$

*that is, the number of red blood cells tomorrow is the number of red blood cells today minus those destroyed plus those produced.*

*It is widely accepted that a constant fraction c of cells is destroyed each day, that is, $d(x_n) = cx_n$. There is less information on the production of red blood cells, but the qualitative features of $p(x_n)$ are generally assumed to be as for the Ricker curve. That is, if there are not many red blood cells, then the bone marrow is rather productive, whereas if there are already many red blood cells, the bone marrow is less productive. Two possible forms for $p(x_n)$ are*

$$p_1(x) = axe^{-bx},$$

*with $a > 0$ and $b > 0$ (see [105]), and*

$$p_2(x) = \frac{b\theta^m x}{\theta^m + x^m},$$

*with $b > 0$, $\theta > 0$, and $m > 0$ (see [112]).*

(a) *Sketch a graph of $p_2(x_n)$ for different values of $\theta$ and m. What is the significance of b, $\theta$, and m?*

(b) *It is known that the production of red blood cells involves a delay of several days. How would you modify the above model to take account of the delay?*

**Exercise 2.4.12: Population genetics.** *The general allele frequency equation, (2.34), exhibits four fundamentally different outcomes, summarized in Table 2.3. Prove that indeed there are four fundamentally different cases.*

**Exercise 2.4.13: Competition.** *Consider the following simple competition model:*

$$A_{n+1} = \mu_1 A_n - \mu_3 A_n B_n,$$
$$B_{n+1} = \mu_2 B_n - \mu_4 A_n B_n,$$

*where $\mu_1$, $\mu_2$, $\mu_3$, $\mu_4$ are positive constants.*

(a) *Find all fixed points.*

(b) *Determine the stability of the fixed points for the specific case $\mu_1 = 1.2$, $\mu_2 = 1.3$, $\mu_3 = 0.001$, and $\mu_4 = 0.002$.*

**Exercise 2.4.14: Spread of infectious disease.** *Consider the following model for the spread of an infectious disease (such as the flu or the common cold) through a population of size $N$:*

$$I_{n+1} = I_n + kI_n(N - I_n),$$
$$I_0 = 1,$$

*where $I_n$ is the number of infected (and infectious) individuals on day n, and k is a measure of the infectivity and how well the population mixes.*

(a) *What does the model predict? You may assume that $kN < 2$.*

*The above model does not take into account recovery of individuals. Consider recovery with immunity (i.e., once a person recovers, (s)he cannot get sick a second time), and assume that an individual recovers in exactly d days.*

(b) *Modify the model to incorporate immunity. Explain (justify) your model. What additional assumptions have you made?*

**Exercise 2.4.15: Jury conditions.** *Let $J$ be the Jacobian matrix, (2.54), corresponding to the general two-dimensional discrete-time system, (2.42)–(2.43).*

(a) *Show that the characteristic polynomial for $J$ can be written as*

$$\lambda^2 - \operatorname{tr} J\lambda + \det J = 0.$$

(b) *Show that necessary and sufficient conditions for both eigenvalues of $J$ to have magnitude less than 1 are the following Jury conditions:*

$$|\operatorname{tr} J| < 1 + \det J < 2.$$

**Exercise 2.4.16: Romeo and Juliet in love/hate-preserving mode.** *Consider the discrete-time model developed for the relationship between Romeo and Juliet, (2.40)–(2.41), and assume that the amount of love/hate that Romeo and Juliet feel for each other initially is preserved on all subsequent days, that is, $a_R + p_J = 1$ and $a_J + p_R = 1$.*

(a) *Show that $\det(A) = 0$, where the matrix A is defined in (2.70).*

(b) *Show that the two eigenvalues of the Jacobian matrix are*

$$\lambda_1 = 1,$$
$$\lambda_2 = a_R + a_J - 1.$$

**Exercise 2.4.17: Host–parasitoid systems: The Poisson distribution.** *Assuming that the average number of encounters with a parasitoid per host is $v$, the Poisson distribution states that*

$$P(i) = \text{the probability that a host experiences } i \text{ encounters}$$
$$= \frac{v^i e^{-v}}{i!}.$$

*Show that*

$$\sum_{i=0}^{\infty} P(i) = 1.$$

**Exercise 2.4.18: Host–parasitoid systems: The Nicholson–Bailey model.** *Consider the Nicholson–Bailey model, (2.91)–(2.92).*

(a) *Show that fixed points of the Nicholson–Bailey model are the trivial fixed point, $(H_1^*, P_1^*) = (0, 0)$, and the following nontrivial fixed point:*

$$(H_2^*, P_2^*) = \left( \frac{k \ln k}{ac(k-1)}, \frac{\ln k}{a} \right),$$

*provided $k > 1$. Why is the restriction $k > 1$ necessary?*

(b) *Determine the stability of the trivial fixed point, $(H_1^*, P_1^*)$.*

(c) *In the text, we investigated the stability of the nontrivial fixed point, $(H_2^*, P_2^*)$, and stated that*

$$\ln k + \frac{\ln k}{k-1} > 1$$

*for all $k > 1$. The inequality implies that the second Jury condition cannot be satisfied; that is, the nontrivial fixed point is always unstable. Prove the inequality.*

*Hint: Consider $f(k) = k \ln k - k + 1$, and show $f(k) > f(1) = 0$ for $k > 1$.*

**Exercise 2.4.19: Host–parasitoid systems: The Beddington model.** *Consider the Beddington model, (2.97)–(2.98).*

(a) *Determine all fixed points.*

(b) *Determine the stability of the fixed points. Under which conditions on the model parameters are fixed points stable? Unstable?*

(c) *Use Maple to iterate the model, and confirm the results of the stability analysis.*

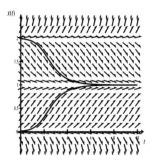

# Chapter 3

# Ordinary Differential Equations

## 3.1  Introduction to ODEs

We have seen in the introductory section (Section 1.2) that the recovery of infected individuals can be modeled by a differential equation (equation (1.2)). In general, differential equations are extremely useful in modeling biological processes. On the one hand, modeling with differential equations is quite transparent, even for complicated biological systems. On the other hand, there is an enormously powerful mathematical theory available, which includes explicit solutions, approximate solutions, numerical solutions, qualitative behavior, and the theory of dynamical systems. As soon as a model has been derived, it can be treated as a mathematical entity. General theorems and analytical methods can be applied to generate useful results. These results can then be interpreted in biological terms.

The possibility of abstraction is the essential advantage of mathematical modeling. For example, to understand the behavior of solutions of the equation

$$x'(t) = 2x(t),$$

it does not matter whether $x(t)$ describes a growing fish population, a growing tumor, or the increase in infected individuals. Mathematically, it is just the equation for exponential growth which can be treated and solved without referring to the interpretation at hand. Once the results are established, they need to be understood in biological terms.

An *ordinary differential equation* (ODE) is an equation for an unknown function of one variable, say $x(t)$, which involves the function and some of its derivatives. For example,

$$x'(t) = 2, \quad y'(t) = 3t, \quad z'(t) = \frac{1}{2}z(t) \tag{3.1}$$

are three differential equations. A *solution* is a function which satisfies the differential equation. For the above examples, (3.1), it is easy to check that the solutions are of the form

$$x(t) = 2t + c_1, \quad y(t) = \frac{3}{2}t^2 + c_2, \quad z(t) = c_3 e^{\frac{1}{2}t}, \tag{3.2}$$

respectively, with constants of integration $c_1$, $c_2$, and $c_3$. The solutions as given in (3.2) are called *general solutions*. If we specify one value for $x$, $y$, or $z$, then the value of $c_1$, $c_2$, or $c_3$ is fixed and we obtain a unique solution. In many cases, we specify an *initial condition*, for example,

$$x(0) = 1, \quad y(0) = 2, \quad \text{and} \quad z(0) = 1.$$

With use of the above general solutions, we find $c_1 = 1$, $c_2 = 2$, and $c_3 = 1$.

We say that $x(t) = 2t + 1$ solves the *initial value problem*

$$x'(t) = 2t, \qquad x(0) = 1.$$

Similarly, $y(t) = \frac{3}{2}t^2 + 2$ solves the initial value problem $y'(t) = 3t$, $y(0) = 2$, and $z(t) = e^{\frac{1}{2}t}$ solves the initial value problem $z'(t) = \frac{1}{2}z(t)$, $z(0) = 1$.

In general, an ODE for an unknown function, $x(t)$

$$x'(t) = f(x(t), t),$$

has the following interpretation. The left-hand side, $x'(t)$, describes the *rate of change* of the quantity $x(t)$ over time. The right-hand side, $f(x(t), t)$, describes all sources of change in $x(t)$. For the recovery from a disease (equation (1.2)), the change in the amount of infected individuals, $\frac{d}{dt}I(t)$, is given by the recovery rate $-\alpha$ times the number of infected $I(t)$.

To solve a differential equation means to use *local information* ("What happens next?") to deduce *long-time behavior* ("What happens in the future?").

This interpretation makes ODEs useful for modeling biological processes. If we know all factors for the process at hand, and if we know the rates of change these factors invoke, then we can write down a differential equation. We analyze and solve it and find explanations and predictions for our biological question.

Before we come to modeling, we will introduce some of the wonderful analytical methods for ODEs, which can be explained using elementary calculus.

## 3.2   Scalar Equations

We first study *scalar equations of the first order*, that is, equations of the form

$$x'(t) = f(x(t), t),$$

where $x(t)$ is a scalar function and the equation involves first-order derivatives. If the function $f(x, t)$ does not depend on $t$, we call the equation *autonomous*. For first-order autonomous scalar ODEs,

$$x' = f(x), \tag{3.3}$$

the *phase-line analysis* explains the qualitative behavior of solutions without even solving the equation. We consider $f(x) = x(1 - x)(2 - x)$, for which the graph is shown in Figure 3.1. The function $f(x)$ has zeros at 0, 1, and 2. We can easily check that $x(t) \equiv 0$, $x(t) \equiv 1$, and $x(t) \equiv 2$ are three constant solutions to the differential equation

$$x' = x(1 - x)(2 - x). \tag{3.4}$$

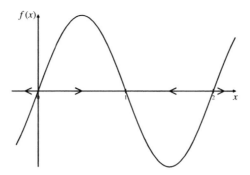

**Figure 3.1.** *Phase-line analysis of* $f(x) = x(1 - x)(2 - x)$. *The arrowheads indicate whether the solution of the corresponding ODE is increasing or decreasing.*

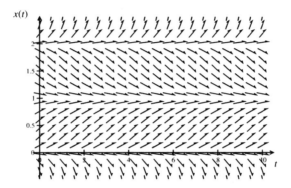

**Figure 3.2.** *Vector field of* (3.4).

These special solutions are called *equilibria* or *steady states* of (3.4). If initially, at $t = 0$, a solution has value 0 (or 1 or 2), it remains so for all times $t > 0$. The left-hand side of (3.4) describes the change of $x(t)$ over time: the solution $x(t)$ is increasing whenever $f(x) > 0$ and decreasing whenever $f(x) < 0$. In our example, $x(t)$ is increasing in the intervals $(0, 1)$ and $(2, \infty)$; it is decreasing in the intervals $(-\infty, 0)$ and $(1, 2)$. We indicate this behavior by adding arrowheads to the $x$-axis in Figure 3.1. If, for instance, the initial condition $x(0)$ is in $(0, 1)$, then the solution will grow and converge to $x = 1$ for $t \rightarrow \infty$.

To get an even better qualitative understanding of the behavior of the solutions of (3.4), we plot the corresponding *vector field*. For that, we evaluate the *slope* of the solution $x(t)$ for many points $(t, x)$ and draw a short arrow indicating the slope in the $(t, x)$ diagram. Since $x' = f(x)$, the slope is given by $f(x)$. In Figure 3.2, we show the time interval of $[0, 10]$ and the $x$-interval of $[0, 2.5]$. At approximately 200 points, we have indicated the slope of the solution with a short arrow. Now, solutions $x(t)$ must have slope $x'(t)$, hence solution curves are *tangential* to these short arrows. In Figure 3.3, we show two typical solutions. Note how nicely they follow the vector field. The steady states 0, 1, 2, which we discussed earlier, appear as lines with horizontal arrows, which means the slope

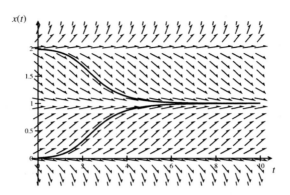

**Figure 3.3.** *Two typical solutions in the vector field of* (3.4).

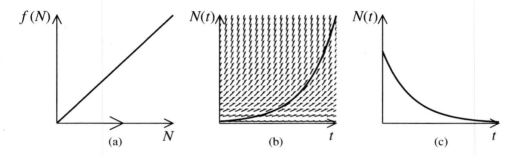

**Figure 3.4.** (a) *Phase-line analysis for Example* 3.2.1, *for r* = 2; (b) *vector field and typical solution for r* = 2; (c) *typical solution for r* = −2.

is zero ($x'(t) = 0$). Solutions which do not start at 0, 1, or 2 tend to get away from the two equilibria $x = 0$ and $x = 2$, while they converge to $x = 1$. We say that $x = 1$ is a *stable equilibrium*, and $x = 0$ and $x = 2$ are *unstable equilibria*.

With the above *phase-line analysis* and *vector-field analysis* we get a very good understanding of the qualitative properties of the solutions *without* solving (3.4). Equation (3.4) can be solved explicitly as well, using separation and partial fractions (see the exercises). These classical solution techniques can be found in most introductory ODE textbooks (such as Boyce and DiPrima [25]).

**Example 3.2.1:** *Exponential Growth and Exponential Functions.* Solutions to the exponential growth equation,

$$N' = rN, \tag{3.5}$$

have the form $N(t) = N_0 e^{rt}$, where $N_0 = N(0)$ is the initial condition. The phase line, the vector field, and a typical solution for $r > 0$ are shown in Figures 3.4 (a) and (b). Figure 3.4 (c) shows a solution for $r < 0$. For $r > 0$, equation (3.5) describes exponential growth, which can be applied to population growth. In the case of $r < 0$, equation (3.5) describes exponential decay, which can be applied to radioactive decay or to the decay of a drug in the blood circulation.

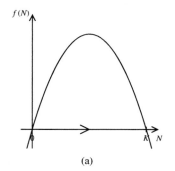
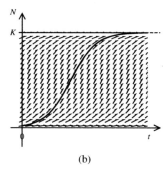

(a)                                     (b)

**Figure 3.5.** (a) *Phase-line analysis for Example 3.2.1;* (b) *vector field and a typical solution.*

### Logistic Growth

In many cases, exponential growth is not an appropriate model. At a certain size, a growing population will reach a limit where all available resources are used to sustain the high population level, but the habitat will not support any more individuals. Hence, we alter the previous model and consider the logistic equation, also known as the Verhulst equation,

$$N' = rN\left(1 - \frac{N}{K}\right), \tag{3.6}$$

where $r > 0$ is the intrinsic growth rate and $K$ is the carrying capacity. Compared to the exponential growth model (3.5), the logistic equation contains the additional term $-(r/K)N^2$. This term can be understood as a competition term from individuals of the same species who compete for the same resources. We use phase-line analysis in Figure 3.5 to obtain the qualitative behavior of the solution $N(t)$. The population grows and converges to the equilibrium solution $N \equiv K$ for $t \to \infty$. Note that it will not reach $K$ in finite time, since solutions of ODEs do not intersect (see Theorem 3.2). In Section 10.1, we use a logistic equation to model growth of cell populations.

## 3.2.1  The Picard–Lindelöf Theorem

Let us return to the theory of differential equations. There is a general result which states that, under reasonable assumptions, solutions of differential equations do not intersect. To formulate the corresponding theorem, we need the notion of *Lipschitz-continuity*. This means that the function $f(x)$ is continuous and, in addition, it satisfies a growth inequality.

**Definition 3.1.** *A function $f : D \to \mathbb{R}$ with domain $D \subset \mathbb{R}$ is called* Lipschitz continuous *if there is a constant $L > 0$ such that $|f(x) - f(y)| \leq L|x - y|$ for all $x, y \in D$.*

**Theorem 3.2 (Picard–Lindelöf).**  *Assume the function $f : D \to \mathbb{R}$ is Lipschitz continuous. Let the initial condition $x_0$ lie in $D$. Then there is an $\varepsilon > 0$ such that the initial* *value problem*

$$\frac{d}{dt}x = f(x), \quad x(0) = x_0$$

*has a unique solution $x(t)$ for $0 \le t \le \varepsilon$.*

**Remark 3.2.1.**

1. Although the theorem is formulated for a small time interval $[0, \varepsilon]$, it implies that solutions to different initial data *never* intersect. Why?

2. The same result holds true for systems of differential equations which can be written in vector notation as

$$\frac{d}{dt}\begin{pmatrix} x_1(t) \\ \vdots \\ x_n(t) \end{pmatrix} = \begin{pmatrix} f_1(x_1, \ldots, x_n) \\ \vdots \\ f_n(x_1, \ldots, x_n) \end{pmatrix}.$$

   In this case, we require that each of the functions $f_1, \ldots, f_n$ is Lipschitz continuous in all of its arguments.

3. A continuously differentiable function is always Lipschitz continuous (on a bounded domain $D \subset \mathbb{R}$). Hence as soon as $f_1 \ldots, f_n$ are differentiable, solutions will never intersect. This includes all differential equations which are built from polynomials, exponentials, sine functions or cosine functions, etc.

## 3.3 Systems of Equations

For many biological processes, it is necessary to follow the time evolution of more than one factor or more than one species. For example, a predator–prey system needs two dependent variables: predator and prey. Similarly, an epidemic SIR model needs three variables: the susceptibles, $S$, the infected, $I$, and the recovered, $R$. We are automatically led to systems of differential equations. To introduce modeling with systems of ordinary differential equations, we study chemical networks, interacting populations, and the spread of an infectious disease in the following sections.

### 3.3.1 Reaction Kinetics

In this section, we introduce the *Law of Mass Action*, and we show how to use it to model chemical reactions. This method is certainly important for the modeling of many physiological processes. It also serves as a good tool for modeling populations. In fact, as we will see shortly, the epidemic model of Figure 1.1 can be understood in terms of reaction kinetics.

**The Law of Mass Action**

First, we consider an irreversible reaction process in which reactants $A$ and $B$ produce $C$:

$$A + B \xrightarrow{k} C,$$

where $k$ is the reaction constant. We are interested in the product $C$ and we argue as follows:

$$\begin{bmatrix} \text{change of} \\ \text{the product} \\ \text{over time} \end{bmatrix} = \begin{bmatrix} \text{number of} \\ \text{collisions of} \\ \text{molecules } A \\ \text{and } B \end{bmatrix} \cdot \begin{bmatrix} \text{probability that a} \\ \text{collision has enough} \\ \text{kinetic energy to} \\ \text{initiate a reaction} \end{bmatrix} .$$

Let $a = [A]$, $b = [B]$, and $c = [C]$ denote the concentrations of the reactants $A$, $B$, and $C$, respectively. The product $r_1 ab \Delta t$ is a good approximation to the number of collisions in time $\Delta t$. The probability that a collision has enough energy to overcome the activation energy of this reaction is denoted by a constant $r_2$. If we let $\Delta C$ denote the change of the product $C$ over time, then the above work equation can be written as

$$\Delta C = abk \Delta t,$$

where $k = r_1 r_2$. Dividing both sides by $\Delta t$, we obtain

$$\frac{\Delta C}{\Delta t} = k \cdot a \cdot b.$$

In the limit of $\Delta t \to 0$, we get

$$\frac{dc}{dt} = k \cdot a \cdot b,$$

which is called the *Law of Mass Action*. Please note that although it is called a *Law* of Mass Action, it is indeed a mathematical *model*. It is no longer valid if the concentration of one participating is many orders of magnitude larger than the other.

**Reversible Reactions**

For a reversible reaction,

$$A + B \underset{k_-}{\overset{k_+}{\rightleftharpoons}} C,$$

we assume that the molecules of $C$ break apart at a rate that is in proportion to the concentration of $C$ molecules. If we balance all production and consumption terms for each participating chemical, then we obtain the following system of differential equations:

$$\frac{dc}{dt} = k_+ \, a \, b - k_- \, c,$$

$$\frac{da}{dt} = k_- \, c - k_+ \, a \, b,$$

$$\frac{db}{dt} = k_- \, c - k_+ \, a \, b.$$

**Michaelis–Menten Kinetics**

One situation where the Law of Mass Action is not directly applicable is the enzymatic reaction

$$E + S \overset{k_+}{\underset{k_-}{\rightleftharpoons}} E + P,$$

involving the reaction of a substrate $S$ with an enzyme $E$ to gain a product $P$ and the enzyme. Note that enzymes are catalyzers. A better description of an enzymatic reaction is the *Michaelis–Menten kinetic*. We assume that substrate $S$ and enzyme $E$ form an intermediate complex, $C$, which then decays into $P$ and $E$:

$$E + S \overset{k_1}{\underset{k_{-1}}{\rightleftharpoons}} C \overset{k_2}{\rightarrow} E + P,$$

where $k_1$, $k_{-1}$, and $k_2$ are rate constants. Let $s = [S]$, $e = [E]$, $c = [C]$, $p = [P]$. As before, we can describe the process using a system of differential equations, as follows:

$$\begin{aligned}
\frac{ds}{dt} &= -k_1 \, s \, e + k_{-1} \, c, \\
\frac{de}{dt} &= -k_1 \, s \, e + k_{-1} \, c + k_2 \, c, \\
\frac{dc}{dt} &= k_1 \, s \, e - k_{-1} \, c - k_2 \, c, \\
\frac{dp}{dt} &= k_2 \, c.
\end{aligned} \tag{3.7}$$

The Michaelis–Menten kinetics model serves as a well-established model for enzyme kinetics and is used widely in the area of mathematical physiology. For more details on the Michaelis–Menten kinetics and on related models, we refer to Keener and Sneyd [99].

### 3.3.2   A General Interaction Model for Two Populations

To further explain modeling with systems of differential equations we investigate the following general two-species interaction model:

$$\begin{aligned}
\dot{x} &= \alpha x + \beta xy, \\
\dot{y} &= \gamma y + \delta xy,
\end{aligned} \tag{3.8}$$

where $x(t)$ and $y(t)$ denote the concentrations (or numbers) of two populations and $\alpha$, $\beta$, $\gamma$, and $\delta$ are constant real parameters. In what follows, we will treat all terms of this model and explain the situations they describe. This approach leads to a systematic understanding of typical model ingredients which can later be used and/or modified for specific applications.

The linear terms $\alpha x$ and $\gamma y$ describe the growth or decay of the corresponding population $x$ or $y$ in isolation. For example, if $\alpha > 0$ and $\beta = 0$, then population $x$ will grow like $e^{\alpha t}$; if $\alpha < 0$, it will decay exponentially. Similarly, if $\delta = 0$, then the sign of $\gamma$ decides whether $y(t)$ is exponentially growing or decaying.

**Table 3.1.** *Classification of the general two-species interaction model,* (3.8).

| $\alpha$ | $\beta$ | $\gamma$ | $\delta$ | |
|---|---|---|---|---|
| $+$ | $+$ | $+$ | $-$ | Predator $(x)$ – prey $(y)$ models |
| $+$ | $+$ | $-$ | $-$ | |
| $-$ | $+$ | $+$ | $-$ | |
| $-$ | $+$ | $-$ | $-$ | |
| $+$ | $+$ | $+$ | $+$ | Mutualism or symbiosis models |
| $+$ | $+$ | $-$ | $+$ | |
| $-$ | $+$ | $-$ | $+$ | |
| $+$ | $-$ | $+$ | $-$ | Competition models |
| $+$ | $-$ | $-$ | $-$ | |
| $-$ | $-$ | $-$ | $-$ | |

Interaction of the two populations is modeled by the nonlinear terms $\beta xy$ and $\delta xy$. These terms look like "Law of Mass Action" terms. This reflects that two individuals of $x$ and $y$ have to meet before they can interact. If they meet, then $\beta$ or $\delta$ describes the likelihood that an interaction indeed occurs. Alternatively, the term $\beta y$ can be interpreted as a rate of change of $x$ due to interaction with $y$. It has turned out that the application of the Law of Mass Action to populations can be very useful.

Now, assume $\beta > 0$ and $\delta < 0$. This means that whenever $x$ and $y$ meet, population $x$ grows, while population $y$ declines. Be reminded of the fact that the left-hand side of the second equation of (3.8), $\dot{y}$, denotes the change of $y$ with respect to $t$. The negative term $-|\delta|xy$ on the right-hand side contributes negatively to the growth of $y$. In the case of $\beta > 0$ and $\delta < 0$, population $x$ benefits from interaction and population $y$ suffers. We denote this form of interaction *predator–prey interaction*. In that case, $x$ denotes the predator density (or number) and $y$ denotes the prey.

The situation for $\beta < 0$ and $\delta > 0$ is very similar. In this case $x$ denotes the prey and $y$ the predator. We consider these two cases ($\beta < 0, \delta > 0$ and $\beta > 0, \delta < 0$) as identical. In fact, a switch of notation from $(x, y)$ to $(y, x)$ leads from one case to the other.

Considering the general model (3.8), each of the four parameters can have two signs. Thus, there are $2^4 = 16$ cases to consider. Since we consider the switch of notation as equivalent, there are only 10 qualitatively different cases. These cases are represented by the following sign patterns for $(\alpha, \beta, \gamma, \delta)$:

$$(+ + + +), (+ + + -), (+ + - +), (+ + - -), (+ - + -),$$
$$(- + - +), (- + + -), (+ - - -), (- + - -), (- - - -).$$

The reader might want to verify that the other cases, like $(+ - - +)$ or $(- - + +)$, are included via a transformation of variables.

In Table 3.1, we show a summary of these cases. We have seen that the signs of $\alpha$ and $\gamma$ describe only the growth properties of each species in isolation. We use the interaction terms to classify the cases. All cases with $\beta > 0$ and $\delta < 0$ represent *predator–prey* models. If both $\beta > 0$ and $\delta > 0$, then interactions are beneficial to both populations. We call these models *mutualistic* or *symbiotic* models. Finally, if $\beta < 0$ and $\delta < 0$, then the interaction is disadvantageous for both populations, and the models are called *competition* models.

With the general model (3.8) in hand, it is not difficult to choose the sign pattern of the model to a given experiment. For example, in Section 10.1 we model the competition between two cell populations $N_1$ and $N_2$:

$$
\begin{aligned}
\dot{N}_1 &= r_1 N_1 \left( 1 - \frac{N_1}{K_1} \right) - \frac{\beta_{12}}{K_1} N_1 N_2, \\
\dot{N}_2 &= r_2 N_2 \left( 1 - \frac{N_2}{K_2} \right) - \frac{\beta_{21}}{K_2} N_1 N_2.
\end{aligned}
\tag{3.9}
$$

The signs of the interaction terms (assuming $K_1$, $K_2 > 0$ and $\beta_{12}$, $\beta_{21} > 0$) show that this is indeed a competition model. The difference with model (3.8) is in the growth terms of the species in isolation. We use logistic growth instead of exponential growth.

In Section 3.4.3, we will analyze model (3.8) qualitatively. For further details on predator–prey, competition, and mutualism models, we refer to Murray [121, 122], Edelstein-Keshet [51], or Britton [29].

### 3.3.3   A Basic Epidemic Model

In this section, we consider the spread of an infectious disease in a host population. Let $S$, $I$, and $R$ denote the number of susceptible, infectious, and recovered individuals, respectively. The infection process can be described as shown in Figure 1.1 in Section 1.2. The parameter $\beta > 0$ is the transmission coefficient, $\alpha > 0$ is the recovery rate, and $\gamma > 0$ is the rate for the loss of immunity.

If the disease is transmitted through direct contact, then the rate of new incidences, $\beta I S$, is in proportion to the number of susceptible and to the number of infectious individuals, according to the Law of Mass Action. With these assumptions, the disease process shown in Figure 1.1 is described by the following classical SIR (susceptibles-infected-recovered) model:

$$
\begin{aligned}
S' &= -\beta I S + \gamma R, \\
I' &= \beta I S - \alpha I, \\
R' &= \alpha I - \gamma R.
\end{aligned}
\tag{3.10}
$$

For simplicity, we assume that $\gamma = 0$. This can be understood as assuming the mean immune period $\frac{1}{\gamma} = \infty$; namely, the disease incurs permanent immunity. The simplified model is known as the Kermack–McKendrick model [100]:

$$
\begin{aligned}
S' &= -\beta I S, \\
I' &= \beta I S - \alpha I.
\end{aligned}
\tag{3.11}
$$

We will study this model further in Section 3.4.4, after we learn some techniques of qualitative analysis.

### 3.3.4 Nondimensionalization

Some of the models mentioned above have a large number of free parameters. If the model is used to describe an experiment or an observation, then the parameters might be found in the literature or by fitting to experiments. Mathematical analysis, however, has the advantage of studying the qualitative behavior of a model for all possible values of the parameters, which means we try to understand questions of growth, death, extinctions, epidemic outbreaks, and so on, based on algebraic relations between the parameters. In the following sections we will develop the corresponding theory. Here we show an important technique to reduce the number of free parameters without losing any properties of the model. The method is called *nondimensionalization* and the complexity of a model can be reduced significantly. We give two examples.

**Example 3.3.1:** First we study the logistic growth model (3.6). The quantity $N$ has the dimension of a population size. We can nondimensionalize if we relate $N$ to some reference population size. In this case we can choose the carrying capacity $K$ and define $\tilde{N} = N/K$. Then the logistic model reads (divide (3.6) by $K$)

$$\tilde{N}' = r\tilde{N}(1 - \tilde{N}).$$

In a next step we transform the time variable and introduce $\tilde{t} = rt$. Then from the chain rule we obtain

$$\frac{d\tilde{N}}{dt} = \frac{d\tilde{N}}{d\tilde{t}}\frac{d\tilde{t}}{dt} = r\frac{d\tilde{N}}{d\tilde{t}}.$$

Hence the logistic equation now reads

$$\frac{d}{d\tilde{t}}\tilde{N} = \tilde{N}(1 - \tilde{N}). \tag{3.12}$$

Both parameters $r$ and $K$ have disappeared. The transformations which we used are all equivalent transformations and the original function can be generated from $\tilde{N}$:

$$N(t) = K\tilde{N}(r\tilde{t}).$$

To understand the qualitative behavior of $N(t)$, it is sufficient to study (3.12) instead. You will find in the literature that many authors nondimensionalize and then disregard the tilde. Then the nondimensionalized model that corresponds to the logistic equation reads

$$N' = N(1 - N).$$

**Example 3.3.2:** Here we reduce the number of parameters for the general two-species interaction model (3.8) for the case of $\alpha, \beta, \delta > 0$. We set

$$\tilde{x} = \frac{\delta}{\alpha}x, \quad \tilde{y} = \frac{\beta}{\alpha}y, \quad \tilde{t} = \alpha t, \quad \mu = \frac{\gamma}{\alpha}.$$

If we apply these transformations to the original equations (3.8) and remove the tilde from the resulting equations, we obtain

$$\begin{aligned} x' &= x + xy, \\ y' &= \mu y + xy. \end{aligned} \tag{3.13}$$

The resulting model depends on a single parameter $\mu$ and the qualitative behavior can be studied, depending on $\mu$. Also here the transformations are invertible and the original functions $x$ and $y$ can be generated from (3.13).

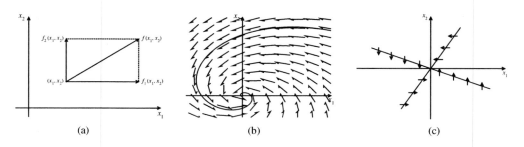

**Figure 3.6.** (a) *Construction of the vector field for* (3.14); (b) *the direction field for*
*Example* 3.4.2 *and one typical trajectory*; (c) *the nullclines with vector field at the nullclines.*

## 3.4   Qualitative Analysis of 2 × 2 Systems

In this section, we develop a qualitative theory for systems of two differential equations,
much in the spirit of Section 3.2, where we introduced phase-line and vector-field analysis.
Here, we will use *phase-plane* analysis, *vector-field* analysis, and the *phase portrait*. With
these methods, the qualitative behavior of a system of equations can be understood without
solving the equations explicitly. Explicit solution methods can be found in other textbooks
on ODEs (such as Boyce and DiPrima [25]).

Consider a system of two differential equations,

$$x_1' = f_1(x_1, x_2),$$
$$x_2' = f_2(x_1, x_2). \tag{3.14}$$

At each $x = (x_1, x_2) \in \mathbb{R}^2$, the *vector field* $f(x) = (f_1(x), f_2(x))$ represents a vector, as
shown in Figure 3.6. A solution $x(t) = (x_1(t), x_2(t))$ represents a parametric curve in the
$(x_1, x_2)$ plane, called a *trajectory* or an *orbit*, whose tangent vector $x'(t) = (x_1'(t), x_2'(t))$ is
specified by the *vector field* $f(x(t)) = (f_1(x_1(t), x_2(t)), f_2(x_1(t), x_2(t)))$. We can obtain a
good impression of the overall dynamics if we plot many vectors in the $(x_1, x_2)$ plane. For
each chosen point $(x_1, x_2)$, we calculate $(f_1(x_1, x_2), f_2(x_1, x_2))$ and sketch this vector. In
Figure 3.6 (a), we show how to calculate one such vector. We repeat this procedure at many
different points until the whole plane is filled with vectors. It is sometimes convenient to
consider only the direction of vector field and not the magnitude. This yields a *direction
field* for the system (Figure 3.6 (b)).

Since solution curves are tangential to the vector field, $f$, we often can follow trajectories just by following the arrows. In Figure 3.6 (b), a typical solution curve is shown (in
this case, we have a spiral converging to the stable origin). The vector field can be used to
sketch more than one typical solution, starting at different initial conditions. The sketch of
the $(x_1, x_2)$ plane with a number of typical solutions is called a *phase portrait*. Of course,
"typical" is a rather vague notion and you need some experience to be able to decide which
solutions represent the qualitative behavior. We will demonstrate and practice this in what
follows.

Many computer packages provide a routine to draw the vector field and the phase
portrait of an ODE system. In Chapter 8, we will learn how to do this with Maple.

Another helpful tool for obtaining insight into the phase portrait are *nullclines* (or *0-isoclines*). The $x_1$-nullcline, $n_1$, is the set of points $(x_1, x_2)$ such that $x_1' = f(x_1, x_2) = 0$, that is,

$$n_1 := \{(x_1, x_2)| f_1(x_1, x_2) = 0\}.$$

Similarly, the $x_2$-nullcline, $n_2$, is

$$n_2 := \{(x_1, x_2)| f_2(x_1, x_2) = 0\}.$$

On the $x_1$-nullcline, $n_1$, all vectors of the vector field are vertical (since $x_1' = 0$). Similarly, on $n_2$, all vectors are horizontal (since $x_2' = 0$). At intersections of $n_1$ and $n_2$, we have $x_1' = 0$ and $x_2' = 0$. Hence a *steady-state* or *equilibrium point* exists at any intersection of $n_1$ and $n_2$. In Figure 3.6 (c), we show the nullclines corresponding to the vector field of Figure 3.6 (b).

In general, *equilibria*, or *steady states* of (3.14) are solutions of

$$f_1(x_1, x_2) = 0, \qquad f_2(x_1, x_2) = 0,$$

which we denote by $(\bar{x}_1, \bar{x}_2)$. The steady states play an important role in the understanding of the whole dynamics. In many cases, if the behavior near each steady state is known, then the global behavior of solutions can be understood quite well. It turns out that we can classify all possible behaviors which can occur near a steady state. We will do so in the following two sections. In Section 3.4.1, we first treat specific linear systems. After that, we generalize to arbitrary linear systems. In Section 3.4.2, we consider nonlinear systems. Phase-plane analysis will then be applied to the population interaction model (in Section 3.4.3) and to the epidemic model (in Section 3.4.4).

## 3.4.1   Phase-Plane Analysis: Linear Systems

### Step 1: Specific Linear Systems

### (1a) Real Eigenvalues

Consider the simplest linear system,

$$\begin{aligned} x_1' &= \lambda_1 x_1, \\ x_2' &= \lambda_2 x_2, \end{aligned} \tag{3.15}$$

whose unique steady state is the origin, $(\bar{x}_1, \bar{x}_2) = (0, 0)$. In matrix form, we can write

$$\frac{d}{dt}\begin{pmatrix} x_1 \\ x_2 \end{pmatrix} = \begin{pmatrix} \lambda_1 & 0 \\ 0 & \lambda_2 \end{pmatrix} \begin{pmatrix} x_1 \\ x_2 \end{pmatrix}.$$

Note that $\lambda_1$ and $\lambda_2$ are the *eigenvalues* of the matrix

$$A = \begin{pmatrix} \lambda_1 & 0 \\ 0 & \lambda_2 \end{pmatrix}.$$

Solutions to (3.15) are

$$x_1(t) = x_1(0)e^{\lambda_1 t}, \qquad x_2(t) = x_2(0)e^{\lambda_2 t}.$$

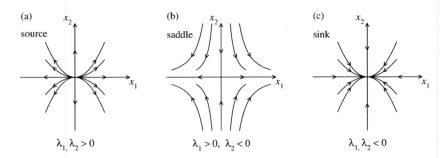

**Figure 3.7.** *Three qualitatively different phase portraits for system* (3.15) *depending on the sign pattern of* $\lambda_1$ *and* $\lambda_2$. (a) $\lambda_1, \lambda_2 > 0$; (b) $\lambda_1 > 0, \lambda_2 < 0$; (c) $\lambda_1, \lambda_2 < 0$. *Here it is assumed that* $\lambda_2$ *is the larger eigenvalue when sketching* (a) *and* (c).

Plotting the parametric curves $(x_1(t), x_2(t))$ for different initial values $(x_1(0), x_2(0))$, we arrive at three distinct phase portraits, depending on the signs of $\lambda_1$ and $\lambda_2$, as shown in Figure 3.7.

Case (a): If both eigenvalues $\lambda_1$ and $\lambda_2$ are positive, then all solutions diverge from the steady state $(0, 0)$. In Figure 3.7 (a), several trajectories are shown for positive, negative, or mixed initial conditions. In this case, the steady state $(0, 0)$ is called a *source* or an *unstable node*.

Case (b): If the eigenvalues have opposite signs, $\lambda_1 > 0$ and $\lambda_2 < 0$, say, then $x_1(t)$ is exponentially increasing, while $x_2(t)$ is decreasing. All solutions approach the $x_1$-axis, as shown in Figure 3.7 (b). In this case, the steady state $(0, 0)$ is called a *saddle*.

Case (c): If both eigenvalues are negative, then all solutions converge to the steady state $(0, 0)$, as shown in Figure 3.7 (c). The steady state is called a *sink* or *stable node*.

### (1b) Complex Eigenvalues

Consider the linear system

$$\frac{d}{dt}\begin{pmatrix} x_1 \\ x_2 \end{pmatrix} = \begin{pmatrix} \alpha & \beta \\ -\beta & \alpha \end{pmatrix}\begin{pmatrix} x_1 \\ x_2 \end{pmatrix}. \tag{3.16}$$

For $\beta \neq 0$, the system has the origin, $(0, 0)$, as its only steady state. The coefficient matrix $A = \begin{pmatrix} \alpha & \beta \\ -\beta & \alpha \end{pmatrix}$ has two complex conjugate eigenvalues

$$\lambda_1 = \alpha + \beta i \qquad \text{and} \qquad \lambda_2 = \alpha - \beta i.$$

We can verify (see the exercises) that (3.16) has two special solutions, namely,

$$x^{(1)}(t) = e^{\alpha t}\begin{pmatrix} \cos \beta t \\ -\sin \beta t \end{pmatrix}, \qquad x^{(2)}(t) = e^{\alpha t}\begin{pmatrix} \sin \beta t \\ \cos \beta t \end{pmatrix}.$$

The superposition principle of linear systems implies that all solutions to (3.16) are of the form

$$x(t) = c_1 x^{(1)}(t) + c_2 x^{(2)}(t) = a e^{\alpha t}\begin{pmatrix} \cos(\beta t + \phi) \\ -\sin(\beta t + \phi) \end{pmatrix}$$

(a)　　　$\alpha = 0$, center

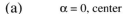
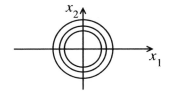
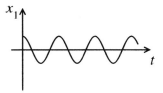

(b)　　$\alpha > 0$, unstable spiral

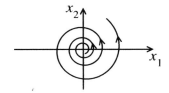

(c)　　$\alpha < 0$, stable spiral

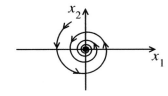

**Figure 3.8.** *Three qualitatively different cases for system (3.16), depending on the value of the parameter* $\alpha$. *(a)* $\alpha = 0$; *(b)* $\alpha > 0$; *(c)* $\alpha < 0$. *Graphs in the left column show phase portraits. Graphs in the right column show a typical solution for* $x_1(t)$. *Here it is assumed that* $\beta > 0$ *when sketching (a)–(c), so that the spirals move clockwise.*

or

$$x_1(t) = ae^{\alpha t}\cos(\beta t + \phi),$$
$$x_2(t) = -ae^{\alpha t}\sin(\beta t + \phi),$$

(3.17)

where $a$ and $\phi$ are determined by the initial conditions, $(x_1(0), x_2(0))$.

Using (3.17), we can classify three distinct cases.

Case (a): $\alpha = 0$, so that both eigenvalues are purely imaginary. All solutions are periodic, and all trajectories are closed orbits surrounding the steady state $(0, 0)$, as shown in Figure 3.8 (a). The steady state is called a *center*.

Case (b): $\alpha > 0$, so that both eigenvalues have positive real parts. The exponential function $e^{\alpha t}$ grows for $t > 0$. All trajectories spiral away from the steady state $(0, 0)$, as shown in Figure 3.8 (b). The steady state is called an *unstable spiral* or a *spiral source*.

Case (c): $\alpha < 0$, so that both eigenvalues have negative real parts. The exponential function $e^{\alpha t}$ decays for $t > 0$. All trajectories spiral towards the steady state $(0, 0)$, as shown in Figure 3.8 (c). The steady state is called a *stable spiral* or a *spiral sink*.

Corresponding solutions $x_1(t)$ for each case are shown in Figure 3.8.

**Step 2: General Linear Systems**

We now consider a general linear system,

$$\frac{d}{dt}\begin{pmatrix} x_1 \\ x_2 \end{pmatrix} = \begin{pmatrix} a & b \\ c & d \end{pmatrix}\begin{pmatrix} x_1 \\ x_2 \end{pmatrix}, \qquad A = \begin{pmatrix} a & b \\ c & d \end{pmatrix}. \tag{3.18}$$

If we make the transformation of coordinates

$$\begin{pmatrix} y_1 \\ y_2 \end{pmatrix} = P^{-1}\begin{pmatrix} x_1 \\ x_2 \end{pmatrix},$$

where $P$ is a $2 \times 2$ invertible matrix, then $y = (y_1, y_2)$ satisfies the system

$$\frac{d}{dt}\begin{pmatrix} y_1 \\ y_2 \end{pmatrix} = B\begin{pmatrix} y_1 \\ y_2 \end{pmatrix}, \tag{3.19}$$

where $B = P^{-1}AP$. The matrix $B$ is *similar* to $A$—it has identical eigenvalues (Hirsch and Smale [86]). Hence,

Systems (3.18) and (3.19) have the same phase portraits.

It is known from linear algebra (see [106]) that if $A$ has two distinct real eigenvalues $\lambda_1$ and $\lambda_2$ such that $\lambda_1 \neq \lambda_2$, then we can choose $P$ such that

$$B = \begin{pmatrix} \lambda_1 & 0 \\ 0 & \lambda_2 \end{pmatrix}.$$

If $A$ has two complex conjugate eigenvalues $\lambda_1 = \bar{\lambda}_2 = \alpha + \beta i$, then we can choose $P$ such that

$$B = \begin{pmatrix} \alpha & \beta \\ -\beta & \alpha \end{pmatrix}.$$

Thus, we conclude that the phase portraits of (3.18) will be the same as those of systems (3.15) or (3.16), studied earlier. Before presenting a theorem about the stability of the origin, we work out the details of computing the matrix $B$ for two specific examples.

**Example 3.4.1:** Consider the linear system

$$\begin{aligned} \dot{x} &= 2x - 2y, \\ \dot{y} &= 2x - 3y. \end{aligned} \tag{3.20}$$

In vector matrix notation, we have

$$\frac{d}{dt}\begin{pmatrix} x \\ y \end{pmatrix} = \begin{pmatrix} 2 & -2 \\ 2 & -3 \end{pmatrix}\begin{pmatrix} x \\ y \end{pmatrix}, \qquad A = \begin{pmatrix} 2 & -2 \\ 2 & -3 \end{pmatrix}\begin{pmatrix} x \\ y \end{pmatrix}.$$

It is straightforward to verify that the eigenvalues and corresponding eigenvectors of $A$ are $\lambda_1 = 1$, $\zeta_1 = \binom{2}{1}$ and $\lambda_2 = -2$, $\zeta_2 = \binom{1}{2}$.

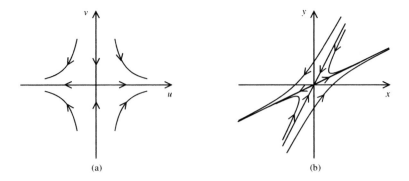

**Figure 3.9.** *The phase portraits of* (a) *(3.21) and* (b) *(3.20).*

The eigenvalues of $A$ are real and distinct. If we use the eigenvectors $\zeta_1$ and $\zeta_2$ as columns of a matrix $P$, we obtain the transformation

$$P = \begin{pmatrix} 2 & 1 \\ 1 & 2 \end{pmatrix}, \quad P^{-1} = \frac{1}{3}\begin{pmatrix} 2 & -1 \\ -1 & 2 \end{pmatrix},$$

then

$$B = P^{-1}AP = \begin{pmatrix} 1 & 0 \\ 0 & -2 \end{pmatrix}.$$

From the solution of the related linear system

$$\frac{d}{dt}\begin{pmatrix} u \\ v \end{pmatrix} = \begin{pmatrix} 1 & 0 \\ 0 & -2 \end{pmatrix}\begin{pmatrix} u \\ v \end{pmatrix}, \tag{3.21}$$

we can recover the solution of (3.20) via

$$\begin{pmatrix} x \\ y \end{pmatrix} = P\begin{pmatrix} u \\ v \end{pmatrix}.$$

The phase portrait of (3.21) is shown in Figure 3.9 (a), and the corresponding phase portrait of (3.20) is shown in Figure 3.9 (b). The transformation $P$ maps the unstable direction $\binom{1}{0}$ of (3.21) onto the unstable direction $\binom{2}{1}$ of (3.20). Similarly, the stable direction $\binom{0}{1}$ of (3.21) is mapped onto the stable direction $\binom{1}{2}$ of (3.20). Note that the phase portrait shown in Figure 3.9 (b) is a compressed and rotated version of the phase portrait shown in Figure 3.9 (a).

**Example 3.4.2:** We consider the system

$$\frac{d}{dt}\begin{pmatrix} x \\ y \end{pmatrix} = \begin{pmatrix} 1 & -4 \\ 2 & -3 \end{pmatrix}\begin{pmatrix} x \\ y \end{pmatrix}.$$

The eigenvalues of the corresponding matrix are $\lambda_1 = -1 + 2i$ and $\lambda_2 = \bar{\lambda}_1 = -1 - 2i$. The corresponding (complex) eigenvectors are

$$\zeta_1 = \begin{pmatrix} -4 \\ -2 + 2i \end{pmatrix} \quad \text{and} \quad \zeta_2 = \bar{\zeta}_1.$$

We write $\zeta_1 = \phi + i\psi$ with real vectors $\phi$ and $\psi$ and obtain the transformation matrix $P = (\phi\psi)$ (see Perko [132]), namely,

$$P = \begin{pmatrix} -4 & 0 \\ -2 & 2 \end{pmatrix},$$

with inverse

$$P^{-1} = \frac{1}{4}\begin{pmatrix} -1 & 0 \\ -1 & 2 \end{pmatrix}.$$

Using this transformation $P$, we obtain the matrix

$$B = P^{-1}AP = \begin{pmatrix} -1 & 2 \\ -2 & -1 \end{pmatrix},$$

and the transformed system has the form of (3.16):

$$\frac{d}{dt}\begin{pmatrix} u \\ v \end{pmatrix} = \begin{pmatrix} -1 & 2 \\ -2 & -1 \end{pmatrix}\begin{pmatrix} u \\ v \end{pmatrix}.$$

The solution can be written in the general form

$$\begin{pmatrix} u \\ v \end{pmatrix}(t) = e^{-t}\left(c_1\begin{pmatrix} -\sin(2t) \\ \cos(2t) \end{pmatrix} + c_2\begin{pmatrix} \cos(2t) \\ \sin(2t) \end{pmatrix}\right).$$

The solution describes oscillations around $(0, 0)$ with frequency $\pi^{-1}$, where the amplitude decays exponentially like $e^{-t}$. Hence solutions converge to $(0, 0)$ and the steady state $(0, 0)$ is a stable spiral. The vector field and one solution curve were shown in Figure 3.6 (b).

In all the cases discussed above, solutions only converge to the steady state at $(0, 0)$ when both eigenvalues $\lambda_1, \lambda_2 < 0$ (the origin is a stable node), or when the real part of the eigenvalues satisfies $\alpha < 0$ (the origin is a stable spiral). When solutions converge to the steady state, we say the steady state is *asymptotically stable*.

We have seen that we can classify the equilibria of a linear system according to the eigenvalues of the corresponding coefficient matrix,

$$A = \begin{pmatrix} a & b \\ c & d \end{pmatrix}.$$

Sometimes it is more convenient to use two other characteristic values of $A$, namely, the *trace*, tr $A = a + d$, and the *determinant*, det $A = ad - bc$. It is known that the trace is always the sum of the eigenvalues, tr $A = \lambda_1 + \lambda_2$, and the determinant is the product, det $A = \lambda_1\lambda_2$. Moreover, one can use the trace and determinant to calculate the eigenvalues. In the exercises, the reader is asked to show that

$$\lambda_{1,2} = \frac{\text{tr } A}{2} \pm \frac{1}{2}\sqrt{(\text{tr } A)^2 - 4 \det A}. \tag{3.22}$$

Note that the formula in (3.22) holds only for $2 \times 2$ matrices. For higher-order matrices, there is no simple formula of this form.

From (3.22), we see that it is necessary to have tr $A < 0$ in order to have a steady state that is asymptotically stable (otherwise at least one eigenvalue would have a positive real part). If tr $A < 0$, then the discriminant, $(\text{tr } A)^2 - 4 \det A$, is either negative or smaller than $(\text{tr } A)^2$. Hence the real part of the eigenvalues is always negative, and $(0, 0)$ is asymptotically stable. We can summarize our conclusions in the following theorem.

**Theorem 3.3.** *For a linear system, (3.18), the following are equivalent:*

- *the equilibrium $(0, 0)$ is asymptotically stable;*

- *all eigenvalues of A have negative real parts;*

- $\det A = ad - bc > 0$ *and* $\text{tr } A = a + d < 0$.

We can treat all different combinations for the sign of trace and determinant and obtain a complete picture of possible behavior near an equilibrium point. Figure 3.10 shows the "zoo" of all possible types of behavior for steady states of two-dimensional systems.

We can summarize the possible types of behavior as follows:

1. Case $\det A < 0$. Then $(\text{tr } A)^2 - 4 \det A > (\text{tr } A)^2$. From formula (3.22), it follows that there is one positive and one negative eigenvalue, $\lambda_1 > 0$ and $\lambda_2 < 0$, say. Hence, $(0,0)$ is a saddle point. Moreover, solutions grow as $e^{\lambda_1 t}$ in the direction of the eigenvector $\varphi_1$ corresponding to $\lambda_1$, and solutions decay as $e^{\lambda_2 t}$ in the direction of the eigenvector $\varphi_2$ corresponding to $\lambda_2$. In Figure 3.10, the *stable* and *unstable* eigenvectors are shown.

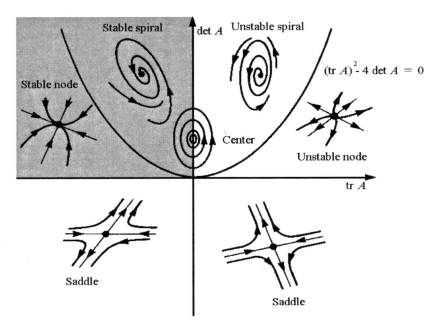

**Figure 3.10.** *The zoo for the general linear system, (3.18). This is a modified version of Figure 5.14 in Edelstein-Keshet [51].*

2. Case $\det A > 0$, $\operatorname{tr} A < 0$. If $(\operatorname{tr} A)^2 < 4 \det A$ (above the parabola in Figure 3.10), then $\lambda_1, \lambda_2$ are complex conjugate eigenvalues with real part $\frac{\operatorname{tr} A}{2} < 0$, and $(0, 0)$ is a stable spiral. If $(\operatorname{tr} A)^2 > 4 \det A$ (below the parabola), then $\lambda_1, \lambda_2$ are real, but they have the same sign, and $(0, 0)$ is a stable node.

3. Case $\det A > 0$, $\operatorname{tr} A > 0$. Depending on the sign of $(\operatorname{tr} A)^2 - 4 \det A$, we have either an unstable spiral or an unstable node.

4. Case $\det A > 0$, $\operatorname{tr} A = 0$. In this case we have a center.

5. The remaining cases ($\det A = 0$ or $(\operatorname{tr} A)^2 - 4 \det A = 0$) will not be discussed. We refer to Hirsch and Smale [86] for these cases.

## 3.4.2   Nonlinear Systems and Linearization

Consider a nonlinear system in $\mathbb{R}^2$,

$$
\begin{aligned}
x_1' &= f_1(x_1, x_2), \\
x_2' &= f_2(x_1, x_2),
\end{aligned}
\tag{3.23}
$$

where $f_1$ and $f_2$ are continuously differentiable functions.

In general, each pair $(\bar{x}_1, \bar{x}_2)$ satisfying $f_1(\bar{x}_1, \bar{x}_2) = f_2(\bar{x}_1, \bar{x}_2) = 0$ is called an equilibrium or a steady state for (3.23). We would like to understand the behavior of the solutions near equilibria.

For linear systems, we observed that solutions converge to $(0, 0)$, they diverge away from $(0, 0)$, or, in the center case, they stay close by. Before we can generalize these observations to nonlinear systems, we need some definitions from dynamical systems theory (see Perko [132]).

**Definition 3.4.**

(a) *A steady state $(\bar{x}_1, \bar{x}_2)$ is called* stable *if a solution which starts nearby stays nearby.*

   *More formally:* $(\bar{x}_1, \bar{x}_2)$ is stable if for all $\varepsilon > 0$, there exists a $\delta > 0$ such that solutions to initial data $(x_1^0, x_2^0)$ with $\|(x_1^0, x_2^0) - (\bar{x}_1, \bar{x}_2)\| < \delta$ satisfy $\|(x_1(t), x_2(t)) - (\bar{x}_1, \bar{x}_2)\| < \varepsilon$ for all time $t > 0$. Here, $\|.\|$ denotes the Euclidean vector norm.

(b) *A steady state $(\bar{x}_1, \bar{x}_2)$ which is not stable is called* unstable *(there is at least one solution which diverges from $(\bar{x}_1, \bar{x}_2)$).*

(c) *A steady state $(\bar{x}_1, \bar{x}_2)$ is called* asymptotically stable *if $(\bar{x}_1, \bar{x}_2)$ is stable and all solutions near $(\bar{x}_1, \bar{x}_2)$ converge to $(\bar{x}_1, \bar{x}_2)$.*

   *More formally:* $(\bar{x}_1, \bar{x}_2)$ is asymptotically stable if $(\bar{x}_1, \bar{x}_2)$ is stable, and there exists a $\delta > 0$ such that all solutions with initial data $(x_1^0, x_2^0)$, with $\|(x_1^0, x_2^0) - (\bar{x}_1, \bar{x}_2)\| < \delta$, satisfy $\lim_{t \to \infty} \|(x_1(t), x_2(t)) - (\bar{x}_1, \bar{x}_2)\| = 0$.

We can determine the stability of a steady state $(\bar{x}_1, \bar{x}_2)$ by linearizing (3.23). The process is similar to the linearization of discrete-time systems, treated in Section 2.3.2.

Let

$$x_1(t) = \bar{x}_1 + z_1(t),$$
$$x_2(t) = \bar{x}_2 + z_2(t),$$

where $z_1(t)$ and $z_2(t)$ are assumed to be small, so that they can be thought of as perturbations to the steady state. We denote $\bar{x} = (\bar{x}_1, \bar{x}_2)$ and $z = (z_1, z_2)$, and write the Taylor expansion of $f = (f_1, f_2)$ about $(\bar{x}_1, \bar{x}_2)$:

$$f(\bar{x} + z) = f(\bar{x}) + Df(\bar{x}) \cdot z + \text{higher-order terms},$$

where

$$Df(\bar{x}_1, \bar{x}_2) := \begin{pmatrix} a & b \\ c & d \end{pmatrix} = \begin{pmatrix} \frac{\partial f_1}{\partial x_1} & \frac{\partial f_1}{\partial x_2} \\ \frac{\partial f_2}{\partial x_1} & \frac{\partial f_2}{\partial x_2} \end{pmatrix} \Bigg|_{(x_1, x_2) = (\bar{x}_1, \bar{x}_2)}$$

contains the partial derivatives of $f$ evaluated at $(\bar{x}_1, \bar{x}_2)$ (for a reminder on partial derivatives, see Section 4.1). The matrix $Df(\bar{x}_1, \bar{x}_2)$ is called the *Jacobian matrix* of $f$ at $(\bar{x}_1, \bar{x}_2)$.

We substitute the Taylor expansion into (3.23) and we drop the higher-order terms. Since $x_1' = \frac{d}{dt}(\bar{x}_1 + z_1(t)) = z_1'$ and $x_2' = z_2'$, and since $f(\bar{x}) = 0$, we obtain a linear system governing the dynamics of the perturbation $(z_1, z_2)$:

$$\frac{d}{dt} \begin{pmatrix} z_1 \\ z_2 \end{pmatrix} = \begin{pmatrix} a & b \\ c & d \end{pmatrix} \begin{pmatrix} z_1 \\ z_2 \end{pmatrix}. \tag{3.24}$$

We know already from the previous section how to treat linear systems. For most (but not all) steady states, conclusions obtained for the linearized system indeed carry over to the original nonlinear system.

**Definition 3.5.** $(\bar{x}_1, \bar{x}_2)$ *is called* hyperbolic *if all eigenvalues of the Jacobian $Df(\bar{x}_1, \bar{x}_2)$ have nonzero real part.*

**Theorem 3.6 (Hartman–Grobman).** *Assume that $(\bar{x}_1, \bar{x}_2)$ is a hyperbolic equilibrium. Then, in a small neighborhood of $(\bar{x}_1, \bar{x}_2)$, the phase portrait of the nonlinear system, (3.23), is equivalent to that of the linearized system, (3.24).*

**Remark 3.4.1.**

1. By Theorems 3.3 and 3.6, at a hyperbolic equilibrium $\bar{x}$, stability properties are determined by the eigenvalues of the Jacobian matrix, $Df(\bar{x}_1, \bar{x}_2)$. This method of linearization may fail for nonhyperbolic equilibria.

2. The phrase "equivalent to" in the above theorem refers to *topological equivalence* of vector fields. This means that in a neighborhood of $(\bar{x}_1, \bar{x}_2)$, there is a homeomorphism (a continuous one-to-one map between open sets) which maps the vector field of the nonlinear system to the vector field of its linearization. In that case, the phase portrait near the stationary point is one of those shown in Figure 3.10. The theory behind the Hartman–Grobman theorem is given in Perko [132].

   For an example, recall Example 3.4.1. The two phase portraits in Figure 3.9 are topologically equivalent, and the homeomorphism is given by the matrix $P$.

### 3.4.3  Qualitative Analysis of the General Population Interaction Model

In this section, we use the qualitative theory developed above to re-examine the general two-species model, (3.8). From the 10 different cases as summarized in Table 3.1, we select one example for predator–prey, one example for mutualism, and one example for competition, and treat these in detail. The other cases are left as exercises. Before we consider specific cases, we determine the steady states and their linearizations.

We begin by writing (3.8) in vector notation:

$$\frac{d}{dt}\begin{pmatrix} x \\ y \end{pmatrix} = \begin{pmatrix} f_1(x, y) \\ f_2(x, y) \end{pmatrix},\tag{3.25}$$

with $f_1(x, y) = \alpha x + \beta xy$ and $f_2(x, y) = \gamma y + \delta xy$. To find the $x$-nullcline, $n_x$, we set $f_1 = 0$. Hence,

$$n_x = \left\{ (x, y)|x = 0, \text{ or } y = -\frac{\alpha}{\beta} \right\}.$$

Similarly, the $y$-nullcline is

$$n_y = \left\{ (x, y)|y = 0, \text{ or } x = -\frac{\gamma}{\delta} \right\}.$$

The steady states $(\bar{x}, \bar{y})$ are intersection points of the nullclines, and they satisfy $f_1(\bar{x}, \bar{y}) = 0$ and $f_2(\bar{x}, \bar{y}) = 0$. We find two steady states, namely,

$$P_1 = (0, 0) \quad \text{and} \quad P_2 = \left( -\frac{\gamma}{\delta}, -\frac{\alpha}{\beta} \right).$$

The linearization of (3.25) is given by

$$\frac{d}{dt}\begin{pmatrix} z_1 \\ z_2 \end{pmatrix} = Df(\bar{x}, \bar{y})\begin{pmatrix} z_1 \\ z_2 \end{pmatrix},$$

and

$$Df = \begin{pmatrix} \frac{\partial f_1}{\partial x} & \frac{\partial f_1}{\partial y} \\ \frac{\partial f_2}{\partial x} & \frac{\partial f_2}{\partial y} \end{pmatrix} = \begin{pmatrix} \alpha + \beta y & \beta x \\ \delta y & \gamma + \delta x \end{pmatrix}.$$

We evaluate this matrix at the two steady states, $P_1$ and $P_2$. For $P_1$, we find

$$Df(0, 0) = \begin{pmatrix} \alpha & 0 \\ 0 & \gamma \end{pmatrix},\tag{3.26}$$

which has the two eigenvalues $\lambda_1 = \alpha$ and $\lambda_2 = \gamma$. Similarly, for $P_2$, we find

$$Df\left( -\frac{\gamma}{\delta}, -\frac{\alpha}{\beta} \right) = \begin{pmatrix} 0 & -\frac{\beta \gamma}{\delta} \\ -\frac{\alpha \delta}{\beta} & 0 \end{pmatrix} = A.$$

Since $\operatorname{tr} A = 0$ and $\det A = -\alpha \gamma$, formula (3.22) gives that the eigenvalues are given by

$$\lambda_{1/2} = \pm\sqrt{\alpha \gamma}.\tag{3.27}$$

To identify the type of steady states, we need to have more information. In particular, we need to know the signs of the parameters $\alpha$, $\beta$, $\gamma$, and $\delta$. Analysis of three specific cases follows.

**Case $(-++-)$: A predator–prey model.** We assume that $\alpha < 0$, $\beta > 0$, $\gamma > 0$, and $\delta < 0$. From (3.26), we see that one eigenvalue is negative ($\lambda_1 = \alpha < 0$), and the other eigenvalue is positive ($\lambda_2 = \gamma > 0$). Hence, $P_1 = (0,0)$ is a saddle.

Before we study $P_2 = (-\frac{\gamma}{\delta}, -\frac{\alpha}{\beta})$, we have to ensure that it is *biologically relevant*, i.e., $-\frac{\gamma}{\delta} > 0$ and $-\frac{\alpha}{\beta} > 0$. Since $\gamma$, $\delta$ and $\alpha$, $\beta$ have opposite signs, this is indeed true.

In (3.27), the product $\alpha\gamma < 0$, so that the eigenvalues are purely imaginary, namely,

$$\lambda_{1/2} = \pm i \sqrt{|\alpha\gamma|}.$$

Hence $(-\frac{\gamma}{\delta}, -\frac{\alpha}{\beta})$ is a center.

Thus, $P_2$ is not hyperbolic, and the Hartman–Grobman theorem does not apply. We cannot decide the type of steady state: $P_2$ may be a center, a stable spiral, or an unstable spiral. We can obtain the missing information from a first integral, found by integrating

$$\frac{dy}{dx} = \frac{dy/dt}{dx/dt} = \frac{y(\gamma + \delta x)}{x(\alpha + \beta y)}$$

via separation of variables to yield $h(x, y) = \alpha \ln y + \beta y - (\gamma \ln x + \delta x) = $ const. The phase portion is comprised of closed curves, each corresponding to $h(x, y) = $ const for a fixed constant (Figure 3.11 (c)). We refer to $P_2$ as a *nonlinear center*. The vector field and the phase portrait for the case $(-++-)$ are shown in Figure 3.11. We observe predator–prey oscillations between periods of high and low population sizes.

**Case $(-+-+)$: Mutualism of two species which cannot survive alone ($\alpha < 0$ and $\gamma < 0$).** The eigenvalues of $Df(0,0)$ are $\alpha < 0$ and $\gamma < 0$; hence $(0,0)$ is a stable node. Also, $-\frac{\alpha}{\beta} > 0$ and $-\frac{\gamma}{\delta} > 0$, and hence $P_2$ is biologically relevant. The product $\alpha\gamma > 0$; hence $P_2$ is a saddle. The vector field and phase portrait are given in Figure 3.12.

From the phase portrait we see that if the initial populations for $x$ and $y$ are big enough, then both populations can benefit and grow. If one of them is too small initially, then both species go extinct (converge to zero).

**Case $(+---)$: A competition model.** In this case $(0,0)$ is a saddle and $P_2$ is not biologically relevant ($-\frac{\gamma}{\delta} < 0$). The vector field and phase portrait are given in Figure 3.13.

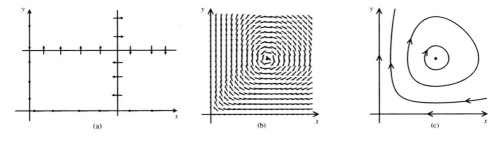

(a)          (b)          (c)

**Figure 3.11.** (a) *Nullclines; (b) direction field; and* (c) *phase portrait for the two-species model, (3.8), with sign pattern* $(-++-)$ *(predator–prey).*

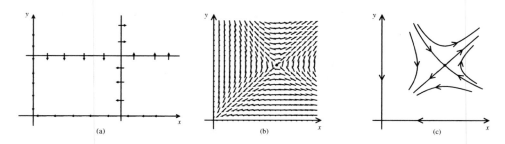

**Figure 3.12.** (a) *Nullclines;* (b) *direction field; and* (c) *phase portrait for the two-species model,* (3.8), *with sign pattern* $(-+-+)$ *(mutualism).*

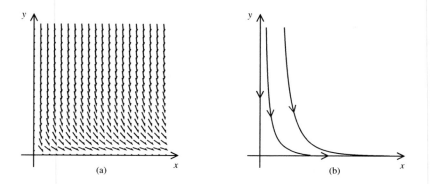

**Figure 3.13.** (a) *Direction field and* (b) *phase portrait for the two-species model,* (3.8), *with sign pattern* $(+---)$ *(competition).*

Population $y$ goes extinct while population $x$ can grow without competition. See Exercise 3.9.11 for the remaining cases.

### 3.4.4   Qualitative Analysis of the Epidemic Model

Here, we return to the epidemic model, (3.11), derived in Section 3.3.3. To find steady states, we set $S' = 0$ and $I' = 0$. If $S' = 0$, then either $S = 0$ or $I = 0$. To satisfy $I' = 0$, we must have either $I = 0$ or $S = \frac{\alpha}{\beta}$. Therefore, (3.11) has a ray of steady states along the positive $S$-axis,

$$\{(S,0) \quad | \quad S \geq 0\}.$$

To find the stability of each steady state $(\bar{S}, 0)$, we examine the Jacobian matrix,

$$\begin{pmatrix} -\beta I & -\beta S \\ \beta I & \beta S - \alpha \end{pmatrix}\Bigg|_{\substack{S=\bar{S} \\ I=0}} = \begin{pmatrix} 0 & -\beta \bar{S} \\ 0 & \beta \bar{S} - \alpha \end{pmatrix}.$$

The two eigenvalues are $\lambda_1 = 0$ and $\lambda_2 = \beta \bar{S} - \alpha$.

The eigenvalue $\lambda_1 = 0$ corresponds to the neutrally stable direction along the ray of steady states. The second eigenvalue, $\lambda_2 = \beta \bar{S} - \alpha$, is positive if $\bar{S} > \frac{\alpha}{\beta}$ and negative if

$\bar{S} < \frac{\alpha}{\beta}$. To construct the phase portrait, we write one unknown, $I$, as a function of the other, $S$. This way, we still follow the trajectory of an epidemic, but we forget about the time course for a moment. To achieve this, we use the chain rule. In particular, if $I = I(S(t))$, then

$$\frac{dI}{dt} = \frac{dI}{dS} \cdot \frac{dS}{dt}.$$

Hence,

$$\frac{dI}{dS} = \frac{I'}{S'} = \frac{\beta I S - \alpha I}{-\beta I S} = -1 + \frac{\alpha}{\beta S}.$$

If we regard $I$ as a function of $S$, and integrate the above equation from $S_0$ to $S$, then we obtain

$$I(S) - I(S_0) = -(S - S_0) + \frac{\alpha}{\beta}(\ln S - \ln S_0).$$

Therefore,

$$I(S) = \frac{\alpha}{\beta} \ln S - S + C_1,$$

where the constant $C_1$ is determined by the initial condition

$$C_1 = I(S_0) + S_0 - \frac{\alpha}{\beta} \ln S_0.$$

To obtain the phase portrait of (3.11), we shift the graph of $\frac{\alpha}{\beta} \ln S - S$ vertically, where the amount of the shift is determined by the values of $S_0$ and $C_1$. As shown in Figure 3.14, steady states to the right of $\frac{\alpha}{\beta}$, namely, $\bar{S} > \frac{\alpha}{\beta}$, are unstable in the direction away from the $S$-axis, and those to the left of $\frac{\alpha}{\beta}$ are stable. This conclusion agrees with our earlier result obtained from local stability analysis.

Biologically, the phase portrait in Figure 3.14 reveals an important fact in epidemiology: $\frac{\alpha}{\beta}$ represents the critical population size to sustain an epidemic; if the initial susceptible population is below $\frac{\alpha}{\beta}$, then no epidemic is possible and the number of infections decreases, whereas if $S_0 > \frac{\alpha}{\beta}$, then the number of infections initially increases, reaching its maximum when $S = \frac{\alpha}{\beta}$, and then declines.

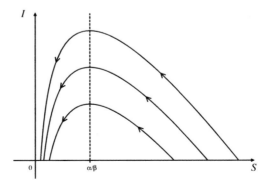

**Figure 3.14.** *Phase portrait of the epidemic model,* (3.11).

## 3.5    General Systems of Three or More Equations

The qualitative theory of linearization, vector fields, and phase portraits can be extended easily to systems of three or more differential equations, although the graphical illustration can become tricky for higher-dimensional systems. Here, we follow the steps of Section 3.4.3 and just indicate how they generalize to higher-dimensional systems. For more details, we refer the reader to texts about dynamical system theory (Perko [132] or Hirsch and Smale [86]).

We write a general system for the $n$ unknowns, $x_1(t), \ldots, x_n(t), n \geq 3$, as

$$\dot{x} = f(x), \tag{3.28}$$

where $x(t) = (x_1(t), \ldots, x_n(t))^T$, and $f(x) = (f_1(x_1, \ldots, x_n), \ldots, f_n(x_1, \ldots, x_n))^T$.

We can define nullclines in the same way as before. In particular, for $j = 1, \ldots, n$, let

$$n_j = \{x \,|\, f_j(x) = 0\}.$$

Steady states $\bar{x}$ of (3.28) are solutions of $f(\bar{x}) = 0$ or, equivalently,

$$f_1(\bar{x}_1, \ldots, \bar{x}_n) = 0, \ldots, f_n(\bar{x}_1, \ldots, \bar{x}_n) = 0.$$

The linearization of (3.28) at $\bar{x} = (\bar{x}_1, \ldots, \bar{x}_n)$ is given by the Jacobian matrix,

$$Df(\bar{x}) = \begin{pmatrix} \frac{\partial f_1}{\partial x_1}(\bar{x}) & \cdots & \frac{\partial f_1}{\partial x_n}(\bar{x}) \\ \vdots & & \vdots \\ \frac{\partial f_n}{\partial x_1}(\bar{x}) & \cdots & \frac{\partial f_n}{\partial x_n}(\bar{x}) \end{pmatrix}.$$

As before, the eigenvalues of $Df(\bar{x})$ determine the stability of $\bar{x}$. Suppose, for example, that we have a six-dimensional problem and that we find two positive real eigenvalues ($\lambda_1, \lambda_2 > 0$), one negative real eigenvalue ($\lambda_3 < 0$), one zero eigenvalue ($\lambda_4 = 0$), and a pair of complex conjugate eigenvalues ($\lambda_5 = \bar{\lambda}_6$) with negative real part. Let $\zeta_1, \ldots, \zeta_6$ denote the corresponding eigenvectors. Solutions which start close to $\bar{x}$ will grow in the $\zeta_1$ and $\zeta_2$ direction, and they will decay along the $\zeta_3$ direction. Since $\lambda_4 = 0$, we cannot conclude from linearization whether the solutions grow or decay in the $\zeta_4$ direction. The two complex eigenvalues $\lambda_5$ and $\lambda_6$ indicate rotations in the plane which is spanned by Re $\zeta_5$ and Im $\zeta_5$. Since the real parts of $\lambda_5$ and $\lambda_6$ are negative, solutions will spiral towards $\bar{x}$ in this plane.

We call the plane spanned by $\zeta_1$ and $\zeta_2$ which goes through $\bar{x}$ the (local) *unstable manifold*, $M^u(\bar{x})$. The three-dimensional space spanned by $(\zeta_2, \text{Re } \zeta_5, \text{Im } \zeta_5)$ is called the (local) *stable manifold*, $M^s(\bar{x})$. The line through $\bar{x}$ in direction $\zeta_4$ is called the (local) *center manifold*.

Visualizing the phase portrait in six-dimensional space is difficult, if not impossible. Even though the graphical analysis is limited, it is important to know the dimensions of the stable, unstable, and center manifolds. This can be used to understand the stability of steady states, to prove or disprove the existence of limit cycles, or to predict the existence of complicated dynamics. A detailed study of stable/unstable/center manifolds is out of reach of this textbook. For further treatment of manifolds, we refer to Perko [132] or Hirsch and Smale [86].

# 3.6   Discrete-Time Models from Continuous-Time Models

There are important connections between discrete-time and continuous-time models. We have seen already that linearization and linear stability analysis play a dominant role for ODEs as well as for discrete models. There are other connections. In this section, we will discuss two approaches by which to derive discrete-time equations corresponding to a given ODE. In Section 3.6.1, we discuss *numerical methods*, and in Section 3.6.2, we discuss the derivation of *time-one maps*.

## 3.6.1   Numerical Methods

In many cases, differential equations cannot be solved explicitly. We have learned how to gain useful information from qualitative analysis and phase portraits without solving the equations. However, sometimes it is desirable or necessary to find a quantitatively accurate solution as well, which can be accomplished using computational approaches. Computational approaches mostly use numerical methods to approximate the solution of an ODE. We do not go into great detail about numerical methods here, but we can explain the principle via the easiest numerical method, namely, the *Euler method*.

Assume we are interested in solving the differential equation

$$x' = f(x).$$

We choose a small time step, $\Delta t$, $\Delta t > 0$, and we approximate the derivative by its differential quotient, that is,

$$x'(t) \approx \frac{x(t + \Delta t) - x(t)}{\Delta t}.$$

We use this differential quotient in the differential equation and obtain, after some rearrangements,

$$x(t + \Delta t) = x(t) + \Delta t f(x(t)).$$

If we let $x_n = x(t)$ and $x_{n+1} = x(t + \Delta t)$, then we obtain

$$x_{n+1} = x_n + f(x_n)\Delta t,$$

which we recognize as a discrete-time dynamical system. All methods, such as cobwebbing, linear stability analysis, or bifurcation analysis, from Chapter 2 can be applied to the numerical scheme. Details of more sophisticated numerical schemes can be found in Burden and Faires [31].

## 3.6.2   The Time-One Map

Another connection between discrete-time equations and ODEs comes from the explicit solution to an ODE. As seen in Section 3.1, the general solution of a differential equation has an unknown constant of integration. This constant can be found if one value of the solution is specified. In many cases, this is the initial condition.

Consider the following initial value problem:

$$x' = f(x), \qquad x(0) = x_0.$$

Given $x_0$, let the solution be denoted by $x(t, x_0)$. In this notation, we treat the initial condition as a parameter. Of course, for different initial values we find different solutions.

Now assume that an experiment is described by the differential equation. The experiment runs for one time unit, after which it is stopped. The result is $x(1, x_0)$. When the experiment is continued, the value $x(1, x_0)$ serves as initial value for the same differential equation. Thus, after two time units, the result is

$$x(2, x_0) = x(1, x(1, x_0));$$

see Perko [132]. We can continue in this fashion. Mathematically, we define a map between the time steps $n$ and $n + 1$. If we let $x_n = x(n, x_0)$, then we can write

$$x_{n+1} = x(1, x_n),$$

or

$$x_{n+1} = f(x_n), \qquad \text{with} \quad f(z) = x(1, z),$$

where now the initial condition, $z$, becomes the argument of $f$. Hence we find a discrete-time dynamical system, also known as the *time-one map* of the ODE.

To illustrate this method further we discuss two examples.

**Example 3.6.1:** *Exponential Growth.* The initial value problem for exponential growth

$$x' = rx, \qquad x(0) = x_0,$$

is solved by

$$x(t, x_0) = x_0 e^{rt}.$$

Thus, after one unit of time, we find

$$x(1, x_0) = x_0 e^r.$$

After two units of time, we find

$$x(2, x_0) = x_0 e^{2r} = x(1, x(1, x_0)).$$

In this case the time-one map reads

$$x_{n+1} = \alpha x_n, \qquad \alpha = e^r.$$

By this method, the linear discrete-time equation, $x_{n+1} = \alpha x_n$, and the linear differential equation, $x' = rx$, are directly related.

The next question is, What is the corresponding difference equation which appears as a time-one map of the logistic equation?

**Example 3.6.2:** *Logistic Growth.* It can be shown (see the exercises) that the initial value problem for the logistic equation

$$N' = \mu N \left( 1 - \frac{N}{K} \right), \qquad N(0) = N_0,$$

is solved by

$$N(t, N_0) = \frac{e^{\mu t} N_0}{1 + \frac{e^{\mu t} - 1}{K} N_0}.$$

If we let $r = e^\mu$, then the time-one map can be written as

$$N_{n+1} = \frac{r N_n}{1 + \frac{r-1}{K} N_n},$$

which we recognize as the Beverton–Holt model (equation (2.23) in Section 2.2.4).

**Remark 3.6.1.**

1. Note that we can also choose other time increments, such as $t = \frac{1}{2}$ or $t = 2$, and use the above procedure. This would give the time-half map or time-two map, respectively, and so on.

2. If $x(t, x_0)$ is a periodic orbit of period $T$, then the time-$T$ map for initial values close to $x_0$ is called a *Poincaré map*.

## 3.7 Elementary Bifurcations

Mathematical models often give rise to differential equations that have many parameters, such as $\alpha$ in the recovery model (1.2) or the growth rate $r$ and the carrying capacity $K$ in the logistic model (3.6). When parameter values are changed, we may expect a change in the behavior of the solution of the differential equation. If variation of a parameter changes the qualitative behavior of the solution, we call it a *bifurcation*.

For example, consider the equation for linear growth or linear decay,

$$x' = \mu x.$$

If $\mu > 0$, solutions grow exponentially; if $\mu < 0$, all solutions tend to zero. The qualitative behavior of solutions for $\mu < 0$ and $\mu > 0$ are quite different, whereas the behavior of the solution for $\mu = 1$ and $\mu = 2$ are very similar. For this example, $\mu = 0$ is a bifurcation value.

To understand a mathematical model properly it is important to know when and how a bifurcation occurs. In this section, we introduce four common bifurcations, namely, bifurcations that occur at equilibria.

We consider a scalar differential equation depending on a scalar parameter,

$$x' = f(x, \mu), \qquad x \in \mathbb{R}, \ \mu \in \mathbb{R}, \tag{3.29}$$

where $\mu$ is the parameter, and $f : \mathbb{R}^2 \to \mathbb{R}$ is continuously differentiable.

**Definition 3.7.** *We say that $\bar{x}$ is a* bifurcation point *and $\bar{\mu}$ is a* bifurcation value *if*

$$f(\bar{x}, \bar{\mu}) = 0 \quad and \quad \frac{\partial}{\partial x} f(\bar{x}, \bar{\mu}) = 0,$$

*where $\frac{\partial}{\partial x} f$ denotes the partial derivative with respect to $x$.*

Note that $f(\bar{x}, \bar{\mu}) = 0$ implies that $\bar{x}$ is a steady state of the differential equation $x' = f(x, \bar{\mu})$. Recall that $\bar{x}$ is a hyperbolic steady state if $f_x(\bar{x}, \bar{\mu}) \neq 0$. Thus, the second

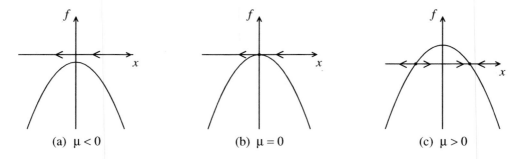

(a) $\mu < 0$                     (b) $\mu = 0$                     (c) $\mu > 0$

**Figure 3.15.** *Phase-line analysis of a saddle-node bifurcation.*

equation in the definition implies that bifurcation points must be nonhyperbolic steady states. In the following sections, we give the *normal forms* of the four most common bifurcations. The first three (saddle-node, transcritical, and pitchfork) can be exhibited in scalar equations. The Hopf bifurcation can occur in systems having dimension at least 2.

### 3.7.1  Saddle-Node Bifurcation

Consider
$$x' = f(x, \mu) = \mu - x^2, \qquad x \in \mathbb{R}, \ \mu \in \mathbb{R}. \tag{3.30}$$
We have $f(x, \mu) = 0$ if and only if $\bar{x} = \pm\sqrt{\mu}$. The partial derivative with respect to $x$ is $\frac{\partial}{\partial x} f(\bar{x}, \mu) = \mp 2\sqrt{\mu}$. Thus $\bar{\mu} = 0$ is the only bifurcation value, and $\bar{x} = 0$ is the bifurcation point.

Phase-line analysis for (3.30) for $\mu < 0$, $\mu = 0$, and $\mu > 0$ is shown in Figure 3.15, and reveals the fashion in which this bifurcation occurs. We observe that no steady state exists when $\mu < 0$, a unique steady state exists at $x = 0$ when $\mu = 0$, and two steady states appear when $\mu > 0$. The steady state $\bar{x}_1 = -\sqrt{\mu}$ is unstable and $\bar{x}_2 = \sqrt{\mu}$ is stable. This information can be summarized in a *bifurcation diagram*, as shown in Figure 3.16. Each curve in the bifurcation diagram represents a branch of the bifurcating steady states. This type of bifurcation is called a *saddle-node bifurcation*. Stable steady states are denoted by a solid curve; unstable steady states are denoted by a dashed curve.

### 3.7.2  Transcritical Bifurcation

Consider
$$x' = f(x, \mu) = \mu x - x^2, \qquad \mu \in \mathbb{R}, \ x \in \mathbb{R}. \tag{3.31}$$
We have $f(\bar{x}, \mu) = 0$ if $\bar{x} = 0$ or $\bar{x} = \mu$.

Differentiating (3.31) gives
$$f_x(\bar{x}, \mu) = \begin{cases} \mu, & \bar{x} = 0, \\ -\mu, & \bar{x} = \mu. \end{cases}$$

Therefore, $\bar{x} = 0$ is a bifurcation point and $\mu = 0$ is a bifurcation value. Phase-line analysis of (3.31) is shown in Figure 3.17. The corresponding bifurcation diagram is shown in

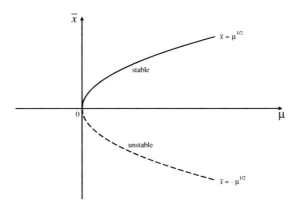

**Figure 3.16.** *Bifurcation diagram of a saddle-node bifurcation.*

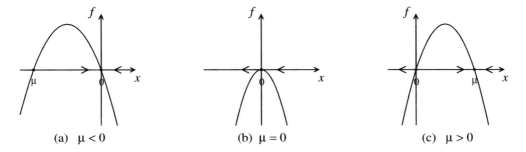

   (a)  $\mu < 0$                             (b)  $\mu = 0$                             (c)  $\mu > 0$

**Figure 3.17.** *Phase-line analysis of a transcritical bifurcation.*

Figure 3.18. In this bifurcation, two branches of equilibria exchange their stability as $\mu$ passes through the bifurcation value $\mu = 0$, and this bifurcation is called a *transcritical bifurcation.*

### 3.7.3  Pitchfork Bifurcation

Consider

$$x' = f(x, \mu) = \mu x - x^3, \qquad \mu \in \mathbb{R}, \ x \in \mathbb{R}. \tag{3.32}$$

We have $f(\bar{x}, \mu) = 0$ if $\bar{x} = 0$ or $\bar{x} = \pm\sqrt{\mu}$.
    Differentiating (3.32) gives

$$f_x(\bar{x}, \mu) = \begin{cases} -2\mu, & \bar{x} = -\sqrt{\mu}, \\ \mu, & \bar{x} = 0, \\ -2\mu, & \bar{x} = \sqrt{\mu}. \end{cases}$$

Therefore, a bifurcation occurs at the point $(\bar{x}, \bar{\mu}) = (0, 0)$. The phase-line analysis of (3.32) is shown in Figure 3.19. The corresponding bifurcation diagram is shown in Figure 3.20.

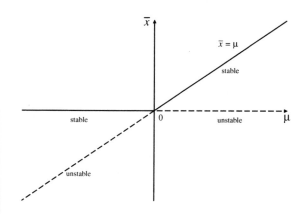

**Figure 3.18.** *Bifurcation diagram of a transcritical bifurcation.*

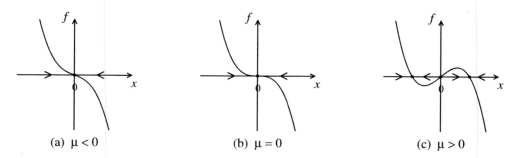

$\qquad$ (a) $\mu < 0$ $\qquad\qquad\qquad$ (b) $\mu = 0$ $\qquad\qquad\qquad$ (c) $\mu > 0$

**Figure 3.19.** *Phase-line analysis of a pitchfork bifurcation.*

As $\mu$ increases through 0 the stable steady state at the origin becomes unstable, and two new stable steady states, along the parabola $\mu = x^2$, are born. The bifurcation is called a *pitchfork bifurcation*.

### 3.7.4  Hopf Bifurcation

Consider the two-dimensional system

$$x_1' = -x_2 + x_1(\mu - x_1^2 - x_2^2),$$
$$x_2' = x_1 + x_2(\mu - x_1^2 - x_2^2),$$
(3.33)

with $(x_1, x_2) \in \mathbb{R}^2$, $\mu \in \mathbb{R}$.

$\qquad$ Using polar coordinates,

$$x_1 = r \cos\theta, \qquad x_2 = r \sin\theta,$$

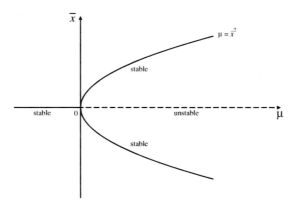

**Figure 3.20.** *Bifurcation diagram of a pitchfork bifurcation.*

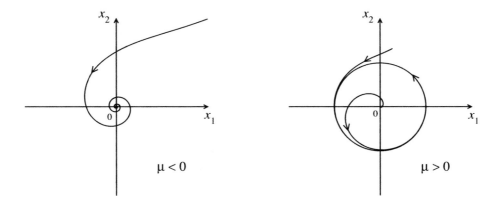

**Figure 3.21.** *Phase portraits of a Hopf bifurcation.*

we can rewrite (3.33) as

$$r' = r(\mu - r^2),$$
$$\theta' = 1. \tag{3.34}$$

Note that the equation for $r$ in (3.34) is the normal form for a pitchfork bifurcation, (3.32). Thus, as $\mu$ passes through the bifurcation value 0, (3.34) undergoes a pitchfork bifurcation (see Figures 3.19 and 3.20). The steady state $\bar{r} = 0$ of (3.34) corresponds to the steady state (0,0) of (3.33), and the other steady state, $\bar{r} = \sqrt{\mu}$, corresponds to a periodic orbit $\sqrt{x_1^2 + x_2^2} = \sqrt{\mu}$ of (3.33).

Figure 3.21 results from a translation of Figure 3.19 from the polar coordinate $r$ into rectangular coordinates $(x_1, x_2)$. The corresponding bifurcation diagram is shown in Figure 3.22. As $\mu$ increases through 0, the branch of steady states at the origin, given by $(\bar{x}_1, \bar{x}_2) = (0, 0)$, loses its stability, and a branch of stable periodic orbits emerges. This bifurcation is called a *Hopf bifurcation.*

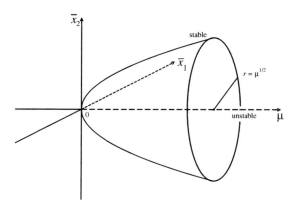

**Figure 3.22.** *Bifurcation diagram of a Hopf bifurcation.*

Note that the Jacobian matrix $Df(0,0)$ is given by

$$\begin{bmatrix} \mu & -1 \\ 1 & \mu \end{bmatrix},$$

which has a pair of complex eigenvalues, namely,

$$\lambda = \mu \pm i.$$

At the bifurcation value $\bar{\mu} = 0$, the eigenvalues are purely imaginary. The occurrence of purely imaginary eigenvalues for a set of parameter values is an important indicator for Hopf bifurcation.

### 3.7.5   The Spruce Budworm Model

In Strogatz [151], a bifurcation analysis is used to study Ludwig et al.'s [111] model for the outbreak of a spruce budworm pest. The model for the spruce budworm dynamics reads (in a non-dimensionalized form):

$$x' = rx\left(1 - \frac{x}{K}\right) - \frac{x^2}{1+x^2}\gamma = f(x), \tag{3.35}$$

where $x$ denotes the budworm population size and, as usual, $r$ is the growth rate and $K$ the carrying capacity. The right-hand side of the equation has either two, three, or four zeros. If we fix $K = 6$ and consider $r$ as a bifurcation parameter, then we can obtain a saddle-node bifurcation where a stable and an unstable steady state appears as $r$ is decreased. In Figure 3.23, we show the right-hand side of (3.35) (function $f$) for values of $r = 0.622, 0.617, 0.608$. For $r = 0.622$, we obtain two steady states: an unstable steady state at $x = 0$ and a stable steady state at about $x = 3.37$ (the outbreak steady state). As $r$ is decreased, the curve touches the $x$-axis at $x = 1.35$ and an additional steady state is born. As $r$ is reduced further, this new steady state splits into two, one a stable steady state (the refuge steady state) and the other an unstable steady state (the threshold steady state).

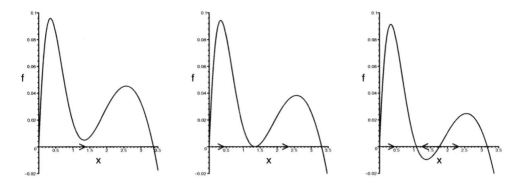

**Figure 3.23.** *Saddle-node bifurcation in a spruce budworm model. The carrying capacity is $K = 6$ and the growth rate is $r = 0.622$ on the left, $r = 0.617$ in the center, and $r = 0.608$ on the right. The right-hand panel exhibits bistable behavior. Populations starting below the threshold steady state approach the refuge steady state, whereas populations starting above approach the outbreak steady state.*

This means that the outbreak of the budworm can be reduced significantly if $r$ is reduced, providing the population is first driven below the threshold steady state. These ideas, and more complex extensions, are developed in Strogatz [151].

## 3.8    Further Reading

Most of the textbooks in mathematical biology that are reviewed in the appendix focus on modeling with differential equations. In particular, the texts of Edelstein-Keshet [51], Murray [121, 122], Britton [29], and Jones and Sleeman [95] cover the classical ODE models in mathematical biology and discuss their qualitative analysis in great detail.

The methods that have been introduced in this chapter were developed in the theory of dynamical systems. That approach follows the tradition of Coddington and Levinson [39], Hartman [79], Hirsch and Smale [86], Strogatz [152], and Perko [132]. In particular, the textbooks of Strogatz and Perko are highly recommended. While Perko focus more on the mathematical theory of dynamical systems, Strogatz discusses more applications to biology and also to other sciences.

Also, a warning is in order. Some of the standard introductory textbooks on ODEs that are commonly used in North American undergraduate education are of limited usefulness. The contents of these books focus on explicit solution methods of differential equations. Although these methods are important as well, the dynamical systems approach has been proven to be much more useful for mathematical biology.

In this chapter on ODEs we covered many aspects only on the surface. For more details on reaction kinetic models we refer to Keener and Sneyd [99]; more information on population models can be found in Kot [102]; and epidemiology is studied in detail in Brauer and Castillo-Chávez [26]. For the qualitative analysis of ODEs, the use of linear algebra, and bifurcation theory we recommend Perko [132], Hirsch and Smale [86], and Strogatz [152].

In the application of ODEs to model biological processes it is always useful to have a computer language available that can solve ODEs numerically. In this textbook we present Maple as a modeling tool. Equally useful are languages and software packages like *Mathematica*, MATLAB, or XPP-AUT.

## 3.9  Exercises for ODEs

**Exercise 3.9.1: The $C^{14}$-method.** *The $C^{14}$-method is used to estimate the age of archaeological objects. It is known that living objects accumulate the radioactive $C^{14}$-isotope during their lifetime, to a certain concentration $c_0$. If the organism dies, then the radioactive $C^{14}$ decays with a half-life of $T_{1/2} = 5760$ years.*

*Archaeologists found a piece of wood in the Nile delta which showed a concentration of 75% of $c_0$. Estimate the age of this piece of wood. Could Tutankhamen have been sitting in a boat made from the same tree as the one from which this piece of wood came?*

**Exercise 3.9.2: Learning curves.** *Psychologists interested in learning theory study learning curves. A learning curve is the graph of a function $P(t)$, the performance of someone learning a skill as a function of the training time $t$.*

(a) *What does $dP/dt$ represent?*

(b) *Discuss why the differential equation*

$$\frac{dP}{dt} = k(M - P),$$

*where $k$ and $M$ are positive constants, is a reasonable model for learning. What is the meaning of $k$ and $M$? What would be a reasonable initial condition for the model? Include a graph of $dP/dt$ versus $P$ as part of your discussion.*

(c) *Make a qualitative sketch of solutions to the differential equation.*

**Exercise 3.9.3: Harvesting.** *The Verhulst (logistic) model for population growth reads*

$$\dot{u} = au\left(1 - \frac{u}{K}\right),$$

*where $K$ denotes the carrying capacity and $a$ is a reproduction rate. We assume that the amount harvested is proportional to population size, with proportionality constant $c$. The modified model then reads*

$$\dot{u} = au\left(1 - \frac{u}{K}\right) - cu.$$

*Plot the vector field and find the steady states. For what values of $c$ does the population die out? When does the population persist? Give a biological explanation.*

**Exercise 3.9.4: Fishing.** *In this exercise, you will be considering three simple models of a fishery. Let $N(t)$ be the population of fish at time $t$. In the absence of fishing, the population* *is assumed to grow logistically, that is,*

$$\dot{N} = rN\left(1 - \frac{N}{K}\right),$$

*where $r > 0$ is the intrinsic growth rate of the population, and $K > 0$ is the carrying capacity for the fish population. The effects of fishing are modeled with an additional term in the equation for $\dot{N}$. The three models are as follows:*

**Model 1:** $\dot{N} = rN\left(1 - \frac{N}{K}\right) - H_1;$

**Model 2:** $\dot{N} = rN\left(1 - \frac{N}{K}\right) - H_2 N;$

**Model 3:** $\dot{N} = rN\left(1 - \frac{N}{K}\right) - H_3\frac{N}{A + N},$

*where $H_1$, $H_2$, $H_3$, and $A$ are positive constants.*

(a) *For each model, give a biological interpretation of the fishing term. How do they differ? What is the meaning of the constants $H_1$, $H_2$, $H_3$, and $A$?*

(b) *Critique Model 1. Why is it not biologically realistic?*

(c) *Which of Models 2 or 3 do you think is best and why?*

**Exercise 3.9.5: A metapopulation model.** *Levins [107] suggested modeling not the number of individuals but the fraction of patches that a population occupies. He suggested the following equation:*

$$P' = cP(h - P) - \mu P,$$

*where $P(t)$ denotes the fraction of occupied patches. The number $h$ denotes the fraction of patches that is actually habitable for the population and, hence, $h - P$ is the number of empty but habitable patches. Note that $0 \le P \le h \le 1$. The population colonizes empty patches with rate $c$. Occupied patches become empty with rate $\mu$.*

(a) *Find the steady states of the system.*

(b) *Assume that $h$ can be varied (e.g., construction takes up habitable patches). Draw the bifurcation diagram with $h$ as the parameter. Do all the habitable patches have to be destroyed before the population dies out?*

**Exercise 3.9.6: Gene activation.** *Consider a gene that is activated by the presence of a biochemical substance $S$. Let $g(t)$ denote the concentration of the gene product at time $t$, and assume that the concentration of $S$, denoted by $s_0$, is fixed. A model describing the dynamics of $g$ is as follows:*

$$\frac{dg}{dt} = k_1 s_0 - k_2 g + \frac{k_3 g^2}{k_4^2 + g^2}, \tag{3.36}$$

*where the $k$'s are positive constants.*

(a) *Interpret each of the three terms on the right-hand side of the equation (be sure to mention the meaning of the k's).*

(b) *Show that (3.36) can be put in the dimensionless form*

$$\frac{dx}{d\tau} = s - rx + \frac{x^2}{1 + x^2},\tag{3.37}$$

*where $r > 0$ and $s \geq 0$ are dimensionless groups. What are $r$ and $s$ in terms of the original model parameters?*

(c) *A graph of $\frac{dx}{d\tau}$ versus x is shown in Figure 3.24 for the case $s = 0$ and $r = 0.4$. On the same set of axes, sketch graphs of $\frac{dx}{d\tau}$ versus x for various values of $s > 0$. We will keep r fixed at 0.4 throughout the remainder of this question.*

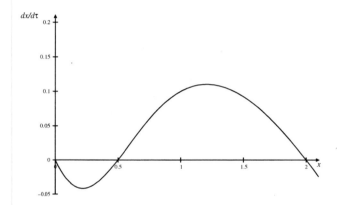

**Figure 3.24.** *Graph of $\frac{dx}{d\tau}$ versus x for (3.37).*

(d) *Make a qualitative sketch of the bifurcation diagram, showing the location and stability of the steady states of (3.37) as a function of the parameter s. Identify any bifurcation(s).*

(e) *Assume that initially there is no gene product, that is, $x(0) = 0$, and suppose that s is slowly increased from zero (i.e., the biochemical substance S is slowly introduced).*

(f) *What happens to $x(\tau)$? Why?*

(g) *What happens if s then goes back to zero? Does the gene turn off again? Why?*

**Exercise 3.9.7: Linear systems.** *We study $2 \times 2$ systems of linear ODEs:*

$$y' = Ay, \quad y = \begin{pmatrix} y_1 \\ y_2 \end{pmatrix}, \quad A = \begin{pmatrix} a & b \\ c & d \end{pmatrix}.$$

*Classify the origin $\begin{pmatrix} 0 \\ 0 \end{pmatrix}$ as a stable/unstable spiral, node, or saddle, and plot (or sketch) the phase portrait for each of the following cases:*

$$A = \begin{pmatrix} 1 & 1 \\ 3 & -1 \end{pmatrix}, \begin{pmatrix} 2 & 1 \\ 2 & 3 \end{pmatrix}, \begin{pmatrix} -1 & -2 \\ 2 & -1 \end{pmatrix}, \begin{pmatrix} 1 & 2 \\ -2 & 1 \end{pmatrix}, \begin{pmatrix} 0 & -2 \\ 2 & 0 \end{pmatrix}.$$

**Exercise 3.9.8: A linear system with complex eigenvalues.** *Show that*

$$x^{(1)}(t) = e^{\alpha t}\begin{pmatrix} \cos \beta t \\ -\sin \beta t \end{pmatrix}, \quad x^{(2)}(t) = e^{\alpha t}\begin{pmatrix} \sin \beta t \\ \cos \beta t \end{pmatrix}$$

*are two solutions of the linear differential equation* (3.16).

*The superposition principle for linear equations (see Boyce and DiPrima [25]) ensures that each solution can be written as a linear combination of $x^{(1)}$ and $x^{(2)}$:*

$$x(t) = c_1 x^{(1)}(t) + c_2 x^{(2)}(t).$$

*Write $x(t)$ in the following form:*

$$x(t) = ae^{\alpha t}\begin{pmatrix} \cos(\beta t + \phi) \\ -\sin(\beta t + \phi) \end{pmatrix},$$

*with $a = \sqrt{c_1^2 + c_2^2}$. The parameter $\phi$ in the solution is called the phase. Find an expression for the phase $\phi$ depending on $c_1$ and $c_2$.*

**Exercise 3.9.9: The trace-determinant formula.** *Prove formula* (3.22).

**Exercise 3.9.10: Using the trace-determinant formula.** *Use formula* (3.22) *to classify the stability of $\begin{pmatrix} 0 \\ 0 \end{pmatrix}$ with A given by*

$$A = \begin{pmatrix} 1 & 5 \\ 3 & 2 \end{pmatrix}, \begin{pmatrix} 0 & -2 \\ 1 & -3 \end{pmatrix}, \begin{pmatrix} -2 & 4 \\ -3 & 4 \end{pmatrix},$$

$$\begin{pmatrix} 2 & 1 \\ 1 & 3 \end{pmatrix}, \begin{pmatrix} -2 & -1 \\ 1 & 2 \end{pmatrix}, \begin{pmatrix} -1 & -2 \\ 2 & 1 \end{pmatrix}.$$

**Exercise 3.9.11: Two-population model.** *For the two-population model,* (3.8), *sketch the phase portraits for the remaining sign patterns:*

$$(+ + + -), (+ + - -), (- + - -), (+ + + +), (+ + - +), (+ - + -), (- - - -).$$

*Give a biological interpretation for each case.*

**Exercise 3.9.12: Predator–prey model.** *Suppose that an insect population, $x(t)$, is controlled by a natural predator population, $y(t)$.*

(a) *Write down a model describing the interaction of these two populations.*

(b) *Suppose an insecticide is used to reduce the population of insects, but it is also toxic to the predators; hence, the poison kills both predator and prey at rates proportional to their respective populations. Modify your model from* (a) *to take this into account.*

**Exercise 3.9.13: Inhibited enzymatic reaction.** *Write down the differential equations describing the following enzymatic reaction, where enzyme E is inhibited by inhibitor I:*

$$S + E \underset{k_{-1}}{\overset{k_1}{\rightleftharpoons}} B_1 \overset{k_2}{\rightarrow} E + Q,$$

$$E + I \underset{k_{-1}}{\overset{k_1}{\rightleftharpoons}} B_2.$$

**Exercise 3.9.14: A feedback mechanism for oscillatory reactions.** *Write down a differential equation model for the following pathway:*

$$A \underset{k_{-1}}{\overset{k_1}{\rightleftharpoons}} B \underset{k_{-2}}{\overset{k_2}{\rightleftharpoons}} C \underset{k_{-3}}{\overset{k_3}{\rightleftharpoons}} A.$$

**Exercise 3.9.15: Enzymatic reaction with two intermediate steps.** *Write down the equations describing the following reaction:*

$$S + E \underset{k_{-1}}{\overset{k_1}{\rightleftharpoons}} C_1 \underset{k_{-2}}{\overset{k_2}{\rightleftharpoons}} C_2 \underset{k_{-3}}{\overset{k_3}{\rightleftharpoons}} E + P.$$

**Exercise 3.9.16: Self-intoxicating population.** *Some populations produce waste products, which in high concentrations are toxic to the population itself. For example, algae or bacteria show the structure in Figure 3.25.*

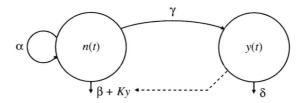

**Figure 3.25.** *Arrow diagram for a self-intoxicating population.*

*Let the population density be denoted by $n(t)$ and the toxin concentration by $y(t)$. Then*

$$\dot{n} = (\alpha - \beta - Ky)n,$$
$$\dot{y} = \gamma n - \delta y,$$

*with $\alpha, \beta, \gamma, \delta, K \geq 0$.*

(a) *Explain each term of the above system.*

(b) *Find the nullclines, the steady states, and sketch a phase portrait.*

(c) *Sketch the vector field.*

(d) *Linearize the system and characterize each of the steady states (stable/unstable, saddle, node, spiral, center, etc.). Find the regions in parameter space such that the nontrivial (coexistence) equilibrium is either a node or a spiral.*

(e) *Sketch some trajectories for the case of $\delta < 4(\alpha - \beta)$, and explain what you see in terms of the biology.*

(f) *Consider the case of higher dilution: $\delta \gg 1$, $\gamma/\delta < \infty$.*

**Exercise 3.9.17: Fish populations in a pond.** *Imagine a small pond that is mature enough to support wildlife. We desire to stock the pond with game fish, say trout and bass. Let $T(t)$ denote the population of the trout at any time t, and let $B(t)$ denote the bass population.*

(a) *Initially, assume that the pond environment can support an unlimited number of trout in isolation (i.e., growth of the trout population is exponential). Write down an equation that describes the evolution of the trout population in the absence of competition.*

(b) *Modify the equation to account for competition of the trout with the bass population for living space and a common food supply. You may assume that the growth rate of the trout population depends linearly on the bass population.*

(c) *Repeat (a) and (b) for the bass population.*

(d) *Explain the meaning of the parameters you introduced into the model.*

(e) *What are the steady states of the system? Determine the stability of the steady states using linearization.*

(f) *Perform a graphical analysis of the model. That is, find the nullclines, and sketch the phase portrait, taking into account the information obtained in (e).*

(g) *Is coexistence of the two species in the pond possible? If so, how sensitive is the final solution of the population levels to the initial stocked levels and external perturbations? Explain.*

(h) *Replace the exponential growth term in each equation with a logistic growth term. Use $r_t$ and $r_b$ to denote the intrinsic growth rate of the trout and bass, respectively, and $K_t$ and $K_b$ to denote the respective carrying capacities. Analyze the following specific case: $K_t > r_b/I_b$ and $K_b > r_t/I_t$, where $I_t$ ($I_b$) represents the strength of the effect of the bass (trout) population on the rate of change of the trout (bass) population. How does the final outcome differ from before? Explain.*

(i) *The second model is a lot more realistic than the first model, as it no longer assumes unlimited growth in the absence of competition. Think of at least one further improvement to the model. How would the equations be affected? You should write down the equations, but you do not have to analyze them.*

**Exercise 3.9.18: Exact solution for the logistic equation.**

(a) *Develop the solution of the initial value problem for the logistic equation,*

$$N' = \mu N \left(1 - \frac{N}{K}\right), \qquad N(0) = N_0.$$

*Note that the differential equation is separable, and use the method of partial fractions to find*

$$N(t, N_0) = \frac{e^{\mu t} N_0}{1 + \frac{e^{\mu t} - 1}{K} N_0}.$$

*Alternatively, you can linearize the differential equation using the transformation $u = 1/N$.*

(b) *Compare the solution obtained above to the solution of the Beverton–Holt model, developed in Exercise 2.4.8.*

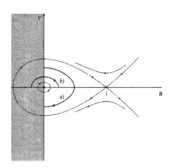

# Chapter 4

# Partial Differential Equations

Partial differential equations (PDEs) involve partial derivatives and appear when we consider quantities depending on more than one independent variable, such as space and time. In Section 4.1, we review some basic facts about partial derivatives. We then treat two of the most important examples of PDEs in mathematical biology. In Section 4.2, we discuss an age-structured model, where the independent variables are time and age. In Section 4.3, we discuss reaction-diffusion equations, where the independent variables are time and space.

## 4.1  Partial Derivatives

Let $f(x, y)$ be a function which depends on two variables, $x$ and $y$. A good way to illustrate $f$ is to draw its *graph*. Let $D \subseteq \mathbb{R}^2$ be a domain such that $f$ is defined on $D$. Then the graph of $f$ is

$$\text{graph}(f) = \{(x, y, f(x, y)) : (x, y) \in D\}.$$

**Example 4.1.1:** Consider

$$f(x, y) = \frac{1}{2}x^2 + y^2(1 - y)^2.$$

The *level sets*, or *contour lines* of $f(x, y)$, are curves in the $(x, y)$ plane such that $f(x, y) = k$, where $k$ is a constant. In Figure 4.1, we show the graph of $f$. Please note that the contour lines are plotted on the graph as $\{(x, y, k) : (x, y) \in D\}$.

Since $f$ depends on two variables, we can define two derivatives of $f$:

$$\frac{\partial}{\partial x} f(x, y) = \lim_{h \to 0} \frac{f(x + h, y) - f(x, y)}{h},$$

$$\frac{\partial}{\partial y} f(x, y) = \lim_{h \to 0} \frac{f(x, y + h) - f(x, y)}{h},$$

which we call the *partial derivatives* of $f$.

The partial derivatives as defined above are evaluated at $(x, y)$, and hence they are also functions of $(x, y)$. When we arrange these two functions in a vector, we call it the

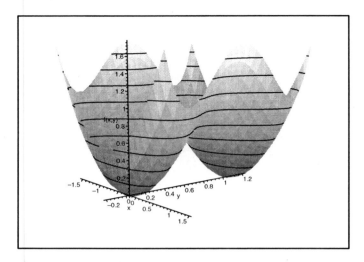

**Figure 4.1.** *Graph of the function* $f(x, y)$ *from Example* 4.1.1.

*gradient* of $f$:

$$\operatorname{grad} f(x, y) = \left( \begin{array}{c} \frac{\partial}{\partial x} f(x, y) \\[2mm] \frac{\partial}{\partial y} f(x, y) \end{array} \right).$$

The gradient has a geometric interpretation. The vector $\operatorname{grad} f(x, y)$ always points to the direction of steepest ascent of the function $f(x, y)$. Equivalently, the vector $\operatorname{grad} f(x, y)$ is always perpendicular to the level curves.

Partial derivatives can be used to find tangent planes, normal lines, etc. We recommend consulting a standard textbook on calculus for more detailed information about partial derivatives.

## 4.2  An Age-Structured Model

### 4.2.1  Derivation

In this section, we will look at an age-structured model for females in a population. The reason for being able to restrict our attention to females is that only females have the potential of giving birth. We let $u(t, a)$ denote the density of females with age $a$ at time $t$. To derive an evolution equation for $u(t, a)$, we consider the population after a small time increment $\Delta t$. The change in the number of individuals between the ages of $a$ and $a + \Delta a$ is given by

$$u(t + \Delta t, a) - u(t, a) = u(t, a - \Delta a) - u(t, a) - \mu(a)u(t, a)\Delta t. \qquad (4.1)$$

The first term on the right-hand side of the equation represents the number of females progressing *from* the previous age class, the second term represents the number of females progressing *to* the next age class, and the third term represents the number of females that die, with $\mu(a)$ being the age-dependent death rate. Dividing the equation by $\Delta t$ and applying the usual limiting process, we obtain

$$\frac{\partial u(t, a)}{\partial t} = -\frac{\partial u(t, a)}{\partial a} - \mu(a)u(t, a),$$

or

$$\frac{du(t, a)}{dt} = \frac{\partial u(t, a)}{\partial t} + \frac{\partial u(t, a)}{\partial a} = -\mu(a)u(t, a). \tag{4.2}$$

Note that since age and time progress at the same rate, we have $\Delta a = \Delta t$ and $da/dt = 1$. You may recognize the above equation as a transport or convection equation. The partial derivative with respect to $a$ is the transport term and represents the contribution to the change in $u(t, a)$ from females getting older (and the velocity with which females age is 1). We have a first-order PDE, and we need two conditions to complete the model. In particular, we need boundary conditions at $a = 0$ and $t = 0$. The distribution

$$u_0(a) = u(0, a) \tag{4.3}$$

represents the initial age distribution and can be any nonnegative function with finite number $N_0 = \int u_0(a)da$. The distribution $u(t, 0)$ represents the newborns, and it is determined by the biology as follows:

$$u(t, 0) = \int_0^\infty \beta(a)u(t, a)\, da, \tag{4.4}$$

where $\beta(a)$ is the age-dependent reproduction rate. Thus, the integral represents the total number of newborns at time $t$.

The model we have just derived is shown in the form of an arrow diagram in Figure 4.2.

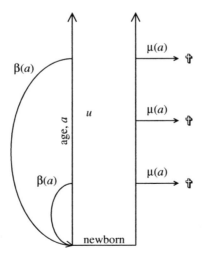

**Figure 4.2.** *Arrow diagram of an aging population. $\mu(a)$ is the age-dependent death rate, and $\beta(a)$ is the age-dependent reproduction rate.*

## 4.2.2   Solution

To solve (4.2) with boundary conditions (4.3) and (4.4), we first introduce index notation for partial derivatives

$$u_t := \frac{\partial}{\partial t} u(t, a), \qquad u_a := \frac{\partial}{\partial a} u(t, a).$$

A common and powerful method to treat PDEs is to write it so that it looks like an ODE. Here, we write (4.2) as

$$u_t = \mathcal{L}u, \quad \text{with} \quad \mathcal{L}u = -u_a - \mu u, \tag{4.5}$$

where $\mathcal{L}$ is a linear differential operator which acts on functions. $\mathcal{L}$ is linear since for two continuously differentiable functions $z_1$ and $z_2$, and for two real constants $\alpha_1$ and $\alpha_2$, we obtain

$$\mathcal{L}(\alpha_1 z_1 + \alpha_2 z_2) = \frac{\partial}{\partial a}(\alpha_1 z_1 + \alpha_2 z_2) - \mu(a)(\alpha_1 z_1 + \alpha_2 z_2)$$

$$= \alpha_1 \frac{\partial}{\partial a} z_1 + \alpha_2 \frac{\partial}{\partial a} z_2 - \alpha_1 \mu(a) z_1 - \alpha_2 \mu(a) z_2$$

$$= \alpha_1 \mathcal{L} z_1 + \alpha_2 \mathcal{L} z_2.$$

We can treat (4.5) similar to a system of linear differential equations written in matrix form. We study eigenvalues, $\lambda$, and eigenfunctions, $w(a)$, of $\mathcal{L}$, which satisfy

$$\lambda w(a) = \mathcal{L}w(a). \tag{4.6}$$

If $\lambda$ is known, then we can explicitly solve for $w(a)$ as follows. We rewrite (4.6) as

$$\lambda w(a) = -w_a - \mu(a)w(a),$$

isolate $w_a$ to get

$$w_a = -(\lambda + \mu(a))w,$$

and integrate

$$w(a) = w_0 e^{-\int_0^a (\lambda + \mu(b))db},$$

where the constant $w_0$ will be specified later.

Now that we have found the eigenfunction $w(a)$, we proceed with *separation of variables*. This approach expresses the hope that the solution, $u(t, a)$, can be written as a product $u(t, a) = g(t)w(a)$ of one function, $g(t)$, which depends only on time $t$, and another function, $w(a)$, which depends only on age, $a$. It is not clear, at the beginning, whether such a solution exists, but it is worth a try. We consider

$$u(t, a) = g(t)w(a), \tag{4.7}$$

where $w(a)$ is the eigenfunction of $\mathcal{L}$, as identified above. We substitute (4.7) into (4.2) and obtain

$$u_t = g'(t)w(a) = -u_a - \mu u = -g w_a - \mu g w$$

$$= g(-w_a - \mu w) = g(\mathcal{L}w)$$

$$= \lambda g(t)w(a).$$

Since $w(a) \neq 0$, we obtain an equation for $g(t)$:

$$g'(t) = \lambda g(t),$$

with solution

$$g(t) = g_0 e^{\lambda t},$$

where the constant $g_0$ has to be specified later.

Hence, a general solution to (4.2) is

$$u(t, a) = g(t)w(a) = g_0 w_0 e^{\lambda t} e^{-\int_0^a (\lambda + \mu(b))db}. \tag{4.8}$$

If we introduce another constant $c = g_0 w_0$, then the general solution (4.8) contains two constants which still have to be found: $c$ and the eigenvalue $\lambda$. But we have two more conditions, the initial condition (4.3) and the boundary conditions (4.4), which we can use to determine $c$ and $\lambda$.

Setting $t = 0$ in (4.8) we have

$$u(0, a) = c e^{-\int_0^a (\lambda + \mu(b))db} = u_0(a). \tag{4.9}$$

This relation cannot be valid for all arbitrarily chosen initial conditions $u_0(a)$. This simply means that the separation of variables method works only if (4.9) can be satisfied for a constant $c$. If not, this method does not work (but there might be other methods which do). Here, we assume that (4.9) is satisfied, and we continue with the boundary condition at age $a = 0$ (4.4). We substitute the general solution (4.8) into (4.4) and obtain

$$c e^{\lambda t} e^{-\int_0^0 (\lambda + \mu(b))db} = c e^{\lambda t} = \int_0^\infty \beta(a) c e^{\lambda t} e^{-\int_0^a (\lambda + \mu(b))db} da.$$

Dividing both sides by $c e^{\lambda t}$ gives a condition to find $\lambda$, namely,

$$1 = \int_0^\infty \beta(a) e^{-\int_0^a (\lambda + \mu(b))db} da. \tag{4.10}$$

If $\beta(a)$ is not zero on an interval, then there is exactly one solution to this equation.

**Lemma 4.1.** *Assume $\beta(a) \geq \beta_0 > 0$ for some interval $a \in [a_1, a_2]$. Then there is a unique $\bar{\lambda} \in \mathbb{R}$ such that* (4.10) *holds.*

*Proof.* The right-hand side (r.h.s.) of (4.10) is a strictly decreasing function in $\lambda$. We have

$$\lim_{\lambda \to -\infty} \text{r.h.s.} = +\infty, \qquad \lim_{\lambda \to +\infty} \text{r.h.s.} = 0.$$

Hence there is exactly one value $\bar{\lambda}$ where the right-hand side equals 1 (compare with Figure 4.3).                                                                                  □

We now know $\bar{\lambda}$ from (4.10) and $c$ from (4.9) and the solution $u(t, a)$ in (4.8) is specified. The time variable appears only in $e^{\lambda t}$. Hence the sign of $\lambda$ alone determines the asymptotic behavior. The following theorem has been proven in Webb [160].

**Theorem 4.2.** *If $\bar{\lambda} < 0$, then $u(t, a) \to 0$ as $t \to \infty$. If $\bar{\lambda} > 0$, then $u(t, a) \to \infty$ as $t \to \infty$. For each solution $u(t, a)$; the function $u(t, a)g_0^{-1}e^{-\lambda t}$ converges to $w(a)$ as $t \to \infty$ for each age $a$.*

This result is also known as *renewal theorem* (Thieme [156]). It shows that over time, the population age distribution approximates a profile determined by $w(a)$, and this profile either grows or decays exponentially with time, depending on the sign of the eigenvalue $\lambda$. Since $w(a)$ is an eigenfunction of the differential operator $\mathcal{L}$ in (4.5), the asymptotic profile $w(a)$ is uniquely determined by the mortality rate $\mu(a)$. In Section 8.5, the reader is asked to simulate the age-structured model numerically using Maple.

## 4.3   Reaction-Diffusion Equations

Another very important class of PDEs are *reaction-diffusion equations*, for which the independent variables are time, $t$, and space, $x$. Reaction-diffusion equations are used whenever the spatial spread of a population or chemical species is of importance. Reaction-diffusion models have their limitations and there are more advanced models (such as correlated random walks, transport equations, or integrodifference equations), but it is always a good idea to start with a reaction-diffusion model for spatial spread. This has successfully been done in epidemic models, for pattern formation, for predator–prey systems, and in signal transport, to name a few areas. A good overview is given in Murray [122] and in Britton [28].

### 4.3.1   Derivation of Reaction-Diffusion Equations

Assume a population with density $u(x, t)$ is living and moving in a container. To describe movement, we introduce another dependent quantity, the particle flux, $J(x, t) \in \mathbb{R}^n$. At each location $x$ and at each time $t$, the flux $J(x, t)$ is a vector which points in the general direction of movement at that location. Its magnitude, $|J(x, t)|$, is proportional to the

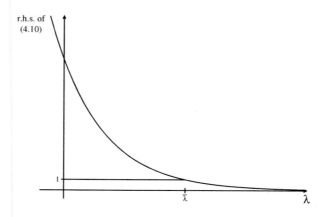

**Figure 4.3.** *There is exactly one value $\bar{\lambda}$ such that* (4.10) *is satisfied.*

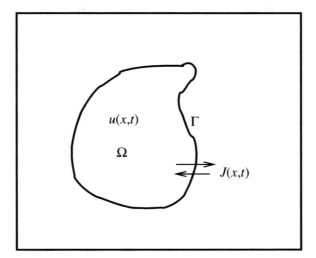

**Figure 4.4.** *Sketch of a test volume $\Omega$ with boundary $\Gamma$, population density $u(x, t)$, and flux $J(x, t)$ through the boundary.*

amount of particles which flow in that direction per unit time. Specifically, the flux $J$ plays the role of the heat flux in heat transport, or a concentration flux for a chemical reactor, and so on.

We consider a test volume $\Omega$ with boundary $\Gamma$ and we balance the fluxes inward and outward on $\Omega$ through $\Gamma$ (see Figure 4.4). In words,

*Change of u in $\Omega$ = flux through $\Gamma$ + change due to birth, death.*

Written in mathematical terms, this gives

$$\frac{d}{dt} \int_\Omega u(x, t)dV = -\int_\Gamma J(x, t)dS + \int_\Omega f(u(x, t))dV,$$

where $f(u(x, t))$ describes birth and death, $dV$ denotes integration over the space $\mathbb{R}^n$ and $dS$ denotes surface integration in dimension $\mathbb{R}^{n-1}$.

We use the divergence theorem

$$\int_\Gamma J(x, t)dS = \int_\Omega \operatorname{div} J(x, t)dV,$$

and we get

$$\int_\Omega \left( \frac{d}{dt}u - f(u) + \operatorname{div} J \right) dV = 0.$$

The above equation is satisfied in each test volume $\Omega$. Then (if the measure $dV$ is not degenerate) it follows that

$$\frac{d}{dt}u - f(u) + \operatorname{div} J = 0. \tag{4.11}$$

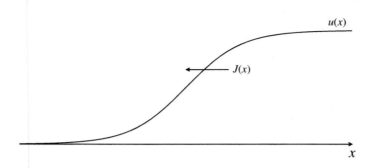

**Figure 4.5.** *Schematic of Fick's second law. A positive gradient of u gives rise to a negative flux J.*

Next, we need an expression of the flux in terms of the population distribution. As for chemical reactions, we use Fick's second law[1]

$$J = -D\nabla u. \tag{4.12}$$

We assume that the flux $J$ is proportional to the negative gradient of the particle distribution. In Figure 4.5, we show a positive gradient of $u$ ($\frac{\partial}{\partial x}u(x, t) > 0$). The flux points to the left, leading to the equilibration of $u$. If we combine the balance law (4.11) with Fick's law (4.12), we get a *reaction-diffusion equation*,

$$\frac{d}{dt}u = D\Delta u + f(u), \tag{4.13}$$

where the Laplacian $\Delta u$ is defined as

$$\Delta u(x, t) = \frac{\partial^2}{\partial x_1^2}u(x, t) + \cdots + \frac{\partial^2}{\partial x_n^2}u(x, t), \quad x = (x_1, \ldots, x_n) \in \mathbb{R}^n.$$

If $f = 0$, then (4.13) is simply the *diffusion equation* or *heat equation*.

### 4.3.2  The Fundamental Solution

The fundamental solution is a particular solution of the diffusion equation (equation (4.13) with $f = 0$) that can be used to find other solutions by convolution (see, e.g., Britton [28]). Moreover, this solution shows many of the common properties of solutions of reaction-diffusion equations in general.

The *fundamental solution* appears for a particle which starts at the origin 0. In terms of random walks on a grid (see Chapter 5), it is straightforward to start with a particle at 0. In the continuous case, however, we use a $\delta$-*distribution* $\delta_0(x)$. The $\delta$-distribution is not a function in the classical sense. It is defined by its action on smooth functions. If $f(x)$ is a

---

[1] In the interpretation of heat transport, this law is known as *Fourier's law*.

smooth function, then $\delta_0(x)$ is the one and only object which satisfies

$$\int_{\mathbb{R}} \delta_0(x) f(x) dx = f(0)$$

and

$$\int_{\mathbb{R}} \delta_0(x) dx = 1.$$

To get an idea about the shape of $\delta_0(x)$ keep in mind that

$$\delta_0(x) = \begin{cases} +\infty & \text{for } x = 0, \\ 0 & \text{for } x \neq 0, \end{cases} \tag{4.14}$$

which is, however, not a valid definition of $\delta_0(x)$.

The $\delta$-distribution is the prototype of a class of functions which are called *distributions* (we refer to Friedlander [60] for further details on distributions). For now, it is sufficient to understand the properties as described above and consider the *initial value problem* for a particle which diffuses in one dimension and starts with certainty at 0:

$$g_t = D g_{xx}, \quad g(x, 0) = \delta_0(x). \tag{4.15}$$

The *fundamental solution* (in one dimension) (see Exercise 4.5.2) is

$$g(x, t) = \frac{1}{2\sqrt{\pi D t}} e^{-\frac{x^2}{4Dt}}. \tag{4.16}$$

In Figure 4.6, we show this solution at times $t = 0$, $t = t_1 > 0$, and $t = t_2 > t_1$, for $D = 1$. Although the initial condition is not continuous, the solution (4.16) is continuous for all

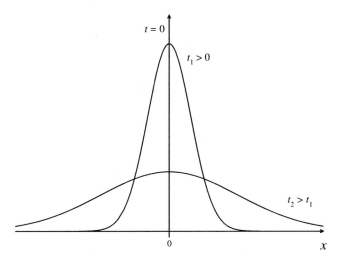

**Figure 4.6.** *Solutions of the diffusion equation (4.15) at three times, $t = 0$, $t = t_1 > 0$, and $t = t_2 > t_1$.*

$t > 0$. In fact, it is infinitely often continuously differentiable, a property which is known as the *regularizing* property of the diffusion equation.

At $t = 0$, we have $\delta_0(x) = 0$ for all $x \neq 0$. However, as soon as $t > 0$, we have $g(x, t) > 0$ for all $x \in \mathbb{R}$. There is a minimal chance to find the particle very far from its starting point. The diffusion equation allows for infinitely fast propagation.

If we study the diffusion equation with a general initial condition,

$$u_t = Du_{xx}, \quad u(x, 0) = f(x), \tag{4.17}$$

then the solution can be found by convolution with $g$ (see Evans [54]):

$$u(x, t) = (f * g(\cdot, t))(x),$$

where the convolution integral is given by

$$
\begin{aligned}
(f * g(\cdot, t))(x) &= \int_{-\infty}^{\infty} f(y)\, g(x - y, t)\, dy \\
&= \frac{1}{2\sqrt{\pi Dt}} \int_{-\infty}^{\infty} f(y) e^{-\frac{(x-y)^2}{4Dt}}\, dy.
\end{aligned}
\tag{4.18}
$$

### 4.3.3 Critical Domain Size

Reaction-diffusion equations can be used to estimate the size of a habitat that can support a population. In general, it is not possible to establish a stable surviving population on an island that is too small. For pests, like the spruce budworm (see Murray [120]), information about the critical domain size can be used to determine how to split a woodland into small enough patches so as to prevent the budworms from settling in.

To illustrate the use of reaction-diffusion equations in this context, we introduce Fisher's equation, which shows all necessary features. Fisher [56] proposed the following model for the spread of an advantageous gene in a population:

$$u_t = Du_{xx} + \mu u(1 - u), \tag{4.19}$$

where $u(x, t)$ is the density of the gene in the population at time $t$ and location $x$. The term $\mu u(1 - u)$ is already familiar to us: it is Verhulst's law of growth with saturation. Fisher's equation also applies for population growth of randomly moving individuals. We will study this equation on a one-dimensional domain of size $l$, $I = [0, l]$.

A PDE on a bounded domain needs boundary conditions. Here we are guided by the application, and we discuss the most common possibilities.

The case of an island has already been mentioned. Appropriate *island boundary conditions* are

$$u(0, t) = 0, \quad u(l, t) = 0. \tag{4.20}$$

These are also called *homogeneous Dirichlet boundary conditions* (see Figure 4.7). We can also study a valley or a box, or a patch with sealing walls. Then no individual can leave the patch. Appropriate *box boundary conditions* are

$$u_x(0, t) = 0, \quad u_x(l, t) = 0. \tag{4.21}$$

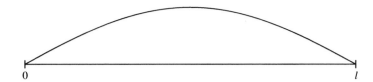

**Figure 4.7.** *A typical solution of (4.19) with homogeneous Dirichlet boundary conditions* (island conditions).

**Figure 4.8.** *A typical solution of (4.19) with homogeneous Neumann boundary conditions* (box conditions).

These are also called *homogeneous Neumann boundary conditions* (see Figure 4.8). Obviously, combinations of island and box boundary conditions can occur if, for example, the patch is bounded by a wall on the one side and by water on the other. We could also include some semipermeable walls such that only a fraction of the population can leave the domain, etc. We restrict our attention to the first two cases given above. Note that we need *one set* of boundary conditions, either (4.20) or (4.21), but not both at the same time.

The question we are investigating is,

*How large must an island or box be to support a population?*

It has been shown in research articles (see, e.g., Britton [28]) that it is equivalent to ask when the trivial solution $u(x, t) = 0$ is unstable. If $u(x, t) \equiv 0$ would be stable, then each solution (near 0) would converge to 0, and the population would die out. Hence, $u(x, t) \equiv 0$ has to be unstable to allow for a surviving population. We are not introducing the notion of *stable* or *unstable* for PDEs here, but we can use them in the same way as for ODEs (see Chapter 3).

For Fisher's equation (4.19), the following questions are equivalent (Grindrod [73]).

  (i) *How large must an island or box be to support a population?*

 (ii) *What is the critical domain size $l^*$ such that $u \equiv 0$ is stable for $l < l^*$ and unstable for $l > l^*$?*

(iii) *What is the critical domain size $l^*$ such that a nontrivial stationary solution (steady state) exists for $l > l^*$?*

We investigate (iii); that is, we will seek nontrivial steady-state solutions of Fisher's equation (4.19).

A steady-state solution satisfies $u_t = 0$, and hence

$$u_{xx} = -\frac{\mu}{D}u(1 - u). \tag{4.22}$$

We are looking for solutions $u(x) \neq 0$ which satisfy the correct boundary conditions, and we will use phase-plane analysis from Chapter 3 to study (4.22). With a new variable, $v := u_x$, we obtain the system

$$u_x = v,$$
$$v_x = -\frac{\mu}{D}u(1 - u), \tag{4.23}$$

with Dirichlet boundary conditions (4.20)

$$u(0) = 0, \quad u(l) = 0,$$

*or* with Neumann boundary conditions (4.21),

$$v(0) = 0, \quad v(l) = 0.$$

Note that (4.23) is a $2 \times 2$ system of ODEs:

$$u' = v,$$
$$v' = -\frac{\mu}{D}u(1 - u). \tag{4.24}$$

In the previous chapter, we considered ODEs with $t$ being the independent variable. Here, $x$ is the independent variable, but the same methods apply.

The equilibria of (4.24) are

$$P_1 = (0, 0), \quad P_2 = (1, 0).$$

The Jacobian of (4.24) is

$$Df(u, v) = \begin{pmatrix} 0 & 1 \\ 2\frac{\mu}{D}u - \frac{\mu}{D} & 0 \end{pmatrix}.$$

The linearization of (4.24) at $P_1$ is

$$Df(0, 0) = \begin{pmatrix} 0 & 1 \\ -\frac{\mu}{D} & 0 \end{pmatrix},$$

which has purely imaginary eigenvalues $\lambda_{1/2} = \pm i\sqrt{\frac{\mu}{D}}$. Hence, $(0, 0)$ is a center.

At $P_2$, we find

$$Df(1, 0) = \begin{pmatrix} 0 & 1 \\ \frac{\mu}{D} & 0 \end{pmatrix},$$

with eigenvalues $\lambda_{1/2} = \pm\frac{\mu}{D}$. Hence, $(1, 0)$ is a saddle.

Since $(1, 0)$ is a saddle for the linearization, it is also a saddle for the full, nonlinear system (4.24). This follows from the Hartman–Grobman theorem (see Theorem 3.6). Unfortunately, the Hartman–Grobman theorem does not apply to the center case. We cannot decide yet whether $(0, 0)$ is a stable spiral, an unstable spiral, or indeed a center for the nonlinear system (4.24).

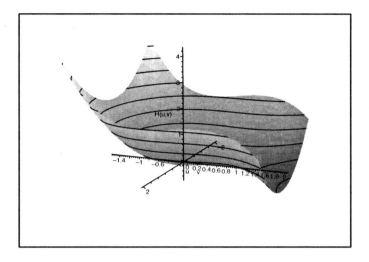

**Figure 4.9.** *Hamiltonian function $H(u, v)$ and level curves.*

We can obtain the missing information from a *Hamiltonian function*, which is a function $H(u, v)$ that satisfies

$$\frac{\partial H}{\partial v} = u' \quad \text{and} \quad \frac{\partial H}{\partial u} = -v'. \tag{4.25}$$

For solutions $(u(x), v(x))$ of (4.24), we get via the chain rule

$$\frac{d}{dx} H(u(x), v(x)) = \frac{\partial H}{\partial u} \cdot u' + \frac{\partial H}{\partial v} \cdot v' = -v'u' + u'v' = 0. \tag{4.26}$$

For (4.24), we can write down the Hamilton function explicitly:

$$H(u, v) = \frac{1}{2} v^2 + \frac{\mu}{D} \frac{u^2}{2} - \frac{\mu}{D} \frac{u^3}{3}.$$

From (4.26), it follows that the value of $H$ does not change along solution curves $(u(x), v(x))$.

In Figure 4.9, we show $H$ as a function of $(u, v)$. Since $H$ does not change along solution curves, the solution curves must follow the level curves of $H$. Since we have a Hamiltonian function, it follows that each bounded solution is either

1. an equilibrium point,

2. a connection of equilibrium points, or

3. a closed orbit.

From Figure 4.9 then, we conclude that the steady state $(0, 0)$ is a center. We now have enough information to sketch the phase portrait of (4.24) in Figure 4.10. Although the phase portrait includes regions of $u < 0$, we consider only solutions which satisfy $u \geq 0$. Since

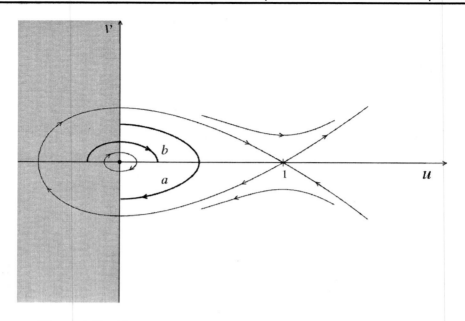

**Figure 4.10.** *Phase portrait of system* (4.24). *The curve a represents a solution which satisfies homogeneous Dirichlet (island) boundary conditions, whereas curve b represents a solution with homogeneous Neumann (box) boundary conditions. The gray area is not biologically relevant because u < 0.*

$u(x)$ is a population density, it cannot be negative. We refer to the region $u < 0$ as *not biologically relevant*.

To find relevant solutions, we have to consider the boundary conditions. A solution satisfying Dirichlet (island) boundary conditions is a solution that starts with $u(0) = 0$ and ends with $u(l) = 0$, while $u \geq 0$ for all $x$. Curve a in Figure 4.10 is one such solution. Similarly, a solution satisfying Neumann (box) boundary conditions is a solution that connects $v(0) = 0$ with $v(l) = 0$. A typical solution is indicated by curve b in Figure 4.10. Of course, this solution is not biologically relevant. The only relevant solutions for the Neumann case are $u \equiv 0$ and $u \equiv 1$.

Hence, we already can answer our original question if the domain is a box. Since $u(x, t) \equiv 1$ exists for any value of $l$, *a box of any size supports a population up to the carrying capacity.* We conclude that the critical domain size under box conditions is $l^* = 0$.

What is the critical domain size for the island or Dirichlet problem? Let's take a closer look at the Dirichlet solutions. Each solution has a unique intersection with the $u$-axis, say at $\bar{u}$ (see Figure 4.11). As $\bar{u} \to 1$, the solution approaches the saddle point. Very close to the saddle point, it takes longer and longer to move forward. Hence, $l \to \infty$ for $\bar{u} \to 1$.

One might guess that $l \to 0$ for $\bar{u} \to 0$, but this is not correct. For $\bar{u} \to 0$, we enter the range close to $(0, 0)$, where the linearization describes the behavior of the solutions. Recall that $(0, 0)$ is a center with eigenvalues $\lambda_{1/2} = \pm i \sqrt{\frac{\mu}{D}}$. As was shown in (3.17), the general solution near $(0, 0)$ is given as $(u(x), v(x))^T = c_1 (\cos(\sqrt{\frac{\mu}{D}} x + \phi), \sin(\sqrt{\frac{\mu}{D}} x + \phi))^T$.

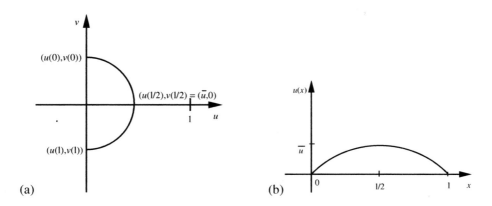

**Figure 4.11.** (a) *For each possible Dirichlet solution, there is a unique intersection with the u-axis intersection at $\bar{u}$.* (b) *The same Dirichlet solution shown as a function of $x$.*

Thus, near $(0, 0)$, a Dirichlet solution corresponds to a half-circle starting at $(u(0), v(0)) = (0, c_1)$ and ending at $(u(l), v(l)) = (0, -c_1)$. From $(u(0), v(0)) = (0, c_1)$, we obtain the phase shift $\phi = \pi/2$. Similarly, from $(u(l), v(l)) = (0, -c_1)$, we obtain the condition that $\sqrt{\frac{\mu}{D}} l = \pi$, or $l = \pi \sqrt{\frac{D}{\mu}}$. In the limit $\bar{u} \to 0$, we get a critical domain size of $l^* = \pi \sqrt{\frac{D}{\mu}}$.

If $l > l^*$, we get a population distribution of the form shown in Figure 4.7 and Figure 4.11 (b). If $l < l^*$, the domain cannot support the population. Note that the case $l = l^*$ cannot be decided by linear analysis.

Now we are able to answer our original question as well if the domain is an island. *An island can support a population if its length $l$ satisfies $l > l^* = \pi \sqrt{\frac{D}{\mu}}$. If $l < l^*$, each initial population will die out.*

### 4.3.4 Traveling Waves

Another important problem in spatial ecology is if and how species can invade new habitats. Our method for studying this is to look for traveling wave solutions of a reaction-diffusion equation. To illustrate this, we again study Fisher's equation,

$$u_t = D u_{xx} + \mu u (1 - u), \tag{4.27}$$

but now on the whole line $\mathbb{R}$. We seek solutions which describe the invasion of the population into a new habitat. In particular, we seek solutions $u(x, t)$ that have the form shown in Figure 4.12, and then move with constant speed $c$. A solution of this type can be expressed as

$$u(x, t) = \phi(x - ct).$$

For $c > 0$, the function $\phi(x - ct)$ is the function $\phi(x)$ shifted to the right by $ct$; see Figure 4.13. The parameter $c$ is the *wave speed*, the new variable $z := x - ct$ is called the *wave variable*, and the function $\phi(z)$ is called the *wave profile*.

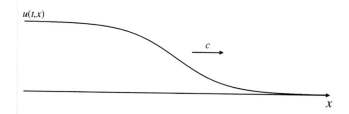

**Figure 4.12.** *A typical invasion traveling wave.*

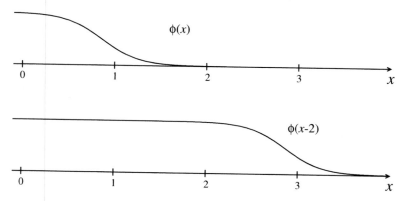

**Figure 4.13.** *The profile $\phi(x)$ from the top figure is shifted by 2 to the right (bottom).*

We make the *traveling wave ansatz*

$$u(x, t) = \phi(x - ct), \quad \phi(-\infty) = 1, \quad \phi(+\infty) = 0, \tag{4.28}$$

where instead of boundary conditions, we now have conditions at $\pm\infty$. For $x \to -\infty$, the population already has grown to its carrying capacity (1 in this case), and for $x \to +\infty$, the population has not arrived yet.

From (4.28), we obtain

$$\frac{\partial}{\partial t} u(x, t) = -c\phi', \quad \frac{\partial^2}{\partial x^2} u(x, t) = \phi'',$$

and (4.27) reduces to the following ODE for $\phi(z)$:

$$-c\phi' = D\phi'' + \mu\phi(1 - \phi). \tag{4.29}$$

As in the previous section, we introduce a new variable, $\psi := \phi'$, and write (4.29) as a $2 \times 2$ system

$$\phi' = \psi,$$

$$\psi' = -\frac{c}{D}\psi - \frac{\mu}{D}\phi(1 - \phi). \tag{4.30}$$

The equilibria of (4.30) are $P_1 = (0, 0)$ and $P_2 = (1, 0)$. Using the linearization, we find that the point $P_1 = (0, 0)$ is stable for $c > 0$. It is a stable spiral for $c < 2\sqrt{D\mu}$ and a stable node for $c > 2\sqrt{D\mu}$. The point $P_2 = (1, 0)$ is always a saddle.

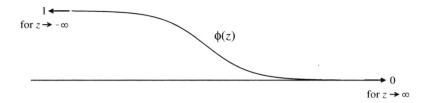

**Figure 4.14.** *The traveling wave as a function of the wave variable z.*

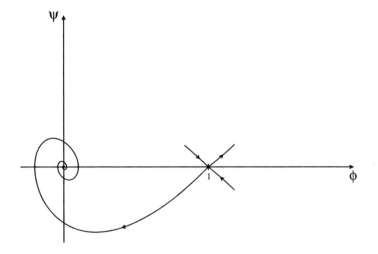

**Figure 4.15.** *Heteroclinic connection from the saddle at* $(1, 0)$ *to the stable spiral at* $(0, 0)$. *Here* $\mu = D = 1$ *and* $c < 2$. *There is no nonnegative traveling wave.*

Recall that the boundary conditions for the wave profile are $\phi(-\infty) = 1$ and $\phi(+\infty) = 0$. Moreover, from the form of $\phi$ as shown in Figure 4.14, it is clear that $\psi(-\infty) = \psi(+\infty) = 0$. Hence, in the phase portrait of system (4.30), we have to find a connection from the saddle $(1, 0)$ to the stable point $(0, 0)$. We show these connections for $c < 2\sqrt{D\mu}$ in Figure 4.15, and for $c > 2\sqrt{D\mu}$ in Figure 4.16.

The function $\phi$ is the profile of the population density; hence it has to be nonnegative. Thus solutions for $c < 2\sqrt{D\mu}$ are not biologically relevant. They correspond to an oscillating front (see Figure 4.17). We obtain that the minimal speed $c^*$ for which a wave front solution exists is given by $c^* = 2\sqrt{D\mu}$ (here we argued graphically; a proof can be found in Källén, Arcuri, and Murray [97]).

### General Fisher Equation

The above result on the minimal wave speed of traveling fronts can be generalized to general Fisher equations

$$u_t = Du_{xx} + f(u),$$

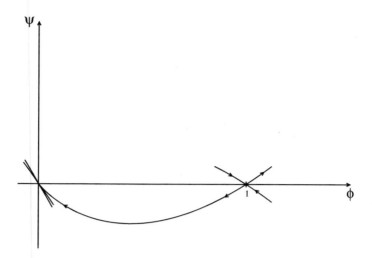

**Figure 4.16.** *Heteroclinic connection from the saddle at* $(1,0)$ *to the stable node at* $(0,0)$. *Here* $\mu = D = 1$ *and* $c > 2$. *There exists a nonnegative traveling wave.*

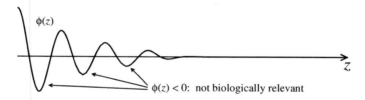

**Figure 4.17.** *Oscillations of the leading edge of the wave from Figure* 4.15.

where $f(u)$ has a shape similar to $\mu u \left(1 - \frac{u}{K}\right)$. The exact conditions on $f$ are as follows: There is a $K > 0$ such that

$$f(0) = 0, \quad f(K) = 0,$$
$$f(u) > 0 \quad \text{for all } 0 < u < K,$$
$$f'(0) > 0, \quad f'(K) < 0.$$

Moreover, if we assume that $f(u)$ satisfies the subtangential condition,

$$f'(0)u > f(u) \quad \text{for all } 0 < u < \infty,$$

then the minimal wave speed is

$$c^* = 2\sqrt{Df'(0)}.$$

### The Linear Conjecture

As we saw in the previous sections, the minimal wave speed $c^*$ is exactly that value where $(0,0)$ changes from spiral into node. If we consider the traveling wave solution close to

$(0, 0)$, then the behavior is described by the linearization around $(0, 0)$. The Jacobian of (4.30) at $(0, 0)$ is

$$Df(0, 0) = \begin{pmatrix} 0 & 1 \\ -\frac{\mu}{D} & -\frac{c}{D} \end{pmatrix},$$

which has trace $-\frac{c}{D}$ and determinant $\frac{\mu}{D}$. Hence, $(0, 0)$ is a node if and only if

$$c^2 - 4D\mu > 0,$$

or $c > 2\sqrt{D\mu}$. The eigenvalues are then given by

$$\lambda_{1/2} = -\frac{c}{2D} \pm \frac{1}{2}\sqrt{\frac{c^2}{D^2} - 4\frac{\mu}{D}},$$

and for $c^* = 2\sqrt{D\mu}$, we have an eigenvalue of multiplicity 2:

$$\lambda_1 = \lambda_2 = -\frac{c^*}{2D}.$$

The solution near $(0, 0)$ behaves like $e^{-\frac{c^*}{2D}x}$ for $x \to \infty$. Hence, $-\frac{c}{2D}$ is the decay rate at the wave front.

Indeed, in many cases, it is enough to measure the decay rate of the profile for large $x$ to get a good approximation for the minimal wave speed $c^*$. This is known as the *linear conjecture*.

## 4.4 Further Reading

There are a number of introductory textbooks on PDEs, such as the books by Haberman [75] and Keane [98]. The contents of these and similar books have been developed in the context of applications in engineering and physics. Most of the material deals with separation and series solutions (see also Exercise 4.5.6). Although these methods are very important, they do not play a major role in applications to biological systems. For PDEs in mathematical biology, a more modern approach is used, which is based on dynamical system theory and nonlinear dynamics. For example, the material in Section 4.3.3 cannot be found in any of the classical introductory textbooks, although it can be understood easily with a basic background in ODEs.

The text of Webb [160] is an introductory text and also a standard reference for age-structured population models. The material from Section 4.2 is based on Webb. For reaction-diffusion equations (including the Fisher equation), a standard reference is Murray [122]. A very good introduction to critical domain size and traveling waves can be found in Britton [28]. The traveling wave problem is also discussed in detail in Grindrod [73].

The material on critical domain size and on traveling waves is also covered in the introductory biomath textbooks which are mentioned in the appendix, "Further Reading": Britton [29], Jones and Sleeman [95], and Taubes [155].

Pattern formation, *Turing* instabilities, and *activator–inhibitor* systems have not been discussed. We refer to the aforementioned texts of Murray, Britton, or Grindrod. Okubo and Levin [127] give a detailed overview of the manifold applications of reaction-diffusion and reaction-advection-diffusion equations to biological problems (advection refers to directed movement).

Two more recent books on reaction-diffusion equations and related models applied to population dynamics are by Thieme [156] and Cantrell and Cosner [37]. Both texts give a comprehensive treatment of the underlying theory of dynamical systems, bifurcations, and functional analysis. Thieme's book deals with stage-structured population models, and Cantrell and Cosner study questions about *permanence* and *persistence* in spatially nonhomogeneous ecological systems.

To obtain a good basic knowledge of the theory of PDEs and their mathematical properties, we recommend the following textbooks: Evans [54], McOwen [116], and Renardy and Rogers [135]. These texts are pure PDE courses and they do not feature biological applications. They are appropriate for a beginning graduate student, and they are not too easy. To properly derive a solution theory for PDEs, one has to introduce appropriate function spaces and one needs some function-analytical tools.

## 4.5   Exercises for PDEs

**Exercise 4.5.1: Diffusion through a membrane.** *This question deals with diffusion through a membrane. We assume that a membrane of width $L$ separates two regions (e.g., the interior and exterior of a cell). Consider a chemical that has a concentration $c_1$ inside the cell and $c_2$ outside the cell. The transport through the membrane can be described by a one-dimensional diffusion equation $u_t = D u_{xx}$. We assume that the solution settles onto an equilibrium.*

(a) *Find the equilibrium and sketch the concentration at equilibrium as a function of position.*

(b) *Using Fick's law, the flux across the membrane is given as*

$$J(x, t) = -D u_x(x, t),$$

*where $u(x, t)$ is a solution of the diffusion equation. Find the flux at equilibrium. The quotient $D/L$ is known as the* permeability *of the membrane. Why do you think this is so?*

**Exercise 4.5.2: Fundamental solution.**

(a) *Show that the function*

$$g(x, t) = \frac{1}{2\sqrt{\pi D t}} \, e^{-\frac{x^2}{4Dt}}$$

*solves the diffusion equation $u_t = D u_{xx}$.*

(b) *Make sure that $g(x, t) \geq 0$ for all $t \geq 0$ and $x \in \mathbb{R}$ and investigate the limits of $x \to \pm\infty$ and $t \to \infty$.*

**Exercise 4.5.3:  Signaling in ant populations.** *Certain ant species (such as* Pogonomyrmex badius*) use pheromones as a signal for danger. A good model for the spread of the pheromones in the tube is the one-dimensional diffusion equation. In experiments, Bossert* *and Wilson released ants in a long tube and stimulated one ant until it released a pheromone. They measured within which distance and after which time delay the other ants would react to the signal. We assume that at time $t = 0$ a signal of strength $\alpha$ is released. The diffusion constant is $D = 1$. Other ants react to the stimulus if the concentration they perceive is 10% of $\alpha$ or higher.*

(a) *For each $t > 0$, find the region in the tube $0 \le x \le x(t)$ where the ants would react to the stimulus* (region of influence).

(b) *Sketch the time evolution of $x(t)$.*

(c) *Find the time $t^*$ such that the region of influence is empty for all $t > t^*$.*

**Exercise 4.5.4: Dingoes in Australia.** *A dingo population which lives in the eastern parts of Australia is prevented from invasion to the west by a fence which runs north–south. In this exercise, we study the case in which the fence breaks somewhere (at time $t = 0$).*

    *Two farms, A and B, are located on the west side of the fence. The distance from farm A to the fence is 100 miles, and the distance from farm A to B is another 100 miles. The farmers would like to know how long it would take for the dingoes to reach their farms. We model the spread of the dingo population with Fisher's equation*

$$u_t = Du_{xx} + ku\left(1 - \frac{u}{K}\right),$$

*with $k = 1$ (1/month), and $K = 1$ (in the units of $u$).*

(a) *The region between farm A and the fence is flat and the diffusion constant is $D_1 = 100$ (miles$^2$/month). When does the dingo population reach farm A?*

(b) *The region between farm A and B has rocks and slope; hence there the diffusion constant is $D_2 = 50$ (miles$^2$/month). When does the dingo population reach farm B?*

*Hint: For part* (a) *consider a traveling wave and calculate the wave speed corresponding to $D_1$ and $k$. Find the spatial decay rate $\lambda_1$ of this wave. For* (b), *take the exponentially decaying wave from part* (a) *and use $D_2$ to find the wave velocity which corresponds to a decay rate of $\lambda_1$.*

**Exercise 4.5.5: Signal transport in the axon.** *Fitzhugh [57] and Nagumo, Arimoto, and Yoshizawa [123] derived a model for signal transduction in the axon,*

$$u_t = u_{xx} + u(1 - u)\left(u - \frac{1}{2}\right), \tag{4.31}$$

*where $u$ represents the membrane potential. We study this model on $0 \le x \le l$ with homogeneous Neumann boundary conditions,*

$$u_x(0, t) = 0, \qquad u_x(l, t) = 0.$$

(a) *Determine the system of two ODEs which describe the steady-state solutions of (4.31).*

(b) *Find the equilibrium points of the system of* (a) *and study their stability.*

(c) *Show that*

$$H(u, u_x) = \frac{1}{2}(u_x)^2 - \frac{1}{4}u^4 + \frac{1}{2}u^3 - \frac{1}{4}u^2$$

   *is a Hamiltonian function for the system you found in (a).*

(d) *Sketch a phase portrait in the $(u, u_x)$ plane.*

(e) *Find the steady-state solutions that satisfy the Neumann boundary conditions, and sketch them as a function of $x$.*

(f) *Give a biological interpretation of these steady-state solutions.*

**Exercise 4.5.6: Separation.** *In this exercise, we flesh out the details of one of the standard solution methods for linear PDEs,* separation of variables. *Here we consider the diffusion equation,*

$$u_t = Du_{xx},$$

*on an interval* $[0, 1]$ *with homogeneous* Dirichlet boundary conditions,

$$u(0, t) = 0, \quad u(1, t) = 0.$$

*This case and many other similar cases are studied in any introductory textbook on PDEs. We recommend that you try this exercise first before you consult the literature.*
     *We study solutions of the form*

$$\phi(x, t) = e^{\lambda t} \sin(\omega x).$$

(a) *For which values of $\omega$ is $\phi(x, t)$ a solution of the diffusion equation with homogeneous Dirichlet boundary conditions? Use a parameter $k \in \mathbb{N}$ to enumerate all possible values of $\omega$.*

(b) *Find the relationship between $\lambda_k$ and $\omega_k$. This relation is called the* dispersion *relation.*

(c) *What is the qualitative behavior of $\phi(x, t)$ as $t \to \infty$?*

(d) *In a broader context, the operator $A = \frac{\partial^2}{\partial x^2}$ is a linear map between Banach spaces, much like a matrix on $\mathbb{R}^n$. The values $\lambda_k$ calculated above can be understood as eigenvalues of $A$ and $\sin(\omega_k x)$ as corresponding eigenfunctions. For $D = 1$, plot the first five eigenvalues in the complex plane, and plot the first five eigenfunctions on $[0, 1]$.*

(e) *Given an initial condition*

$$u(x, 0) = \sum_{k=1}^{N} a_k \sin(\omega_k x)$$

   *with constant coefficients $a_k$, $k = 1, \ldots, N$, guess the solution. Prove that your guess is correct.*

**Exercise 4.5.7: Linear transport.** *We investigate the following equation for* $u(x, t), t \geq 0, x \in \mathbb{R}$:

$$u_t + cu_x = 0, \quad u(0, x) = u_0(x).$$

(a) *The solution* $u(x, t)$ *can be understood as a surface over the* $(x, t)$ *plane. What is the gradient of* $u$? *Write the above equation in the following form:* $vector \cdot grad\ u = 0$. *Give an interpretation in terms of the solution* $u(x, t)$.

(b) *For each constant* $k$, *the curve* $x - ct = k$ *is called the* characteristic curve. $\Gamma_k = \{(t, x(t)) : x(t) = k + ct, t \geq 0\}$. *Show that solutions are constant on these characteristic curves.*

(c) *Solve the above initial value problem.*

**Exercise 4.5.8: Correlated random walk.** *In the next chapter, we will derive the diffusion equation from an uncorrelated random walk. This means that the movement direction of a random walker at time t is independent of time* $t - \Delta t$. *If correlation is included, we are led to the following system (see Zauderer [169] for more details on the derivation):*

$$u_t^+ + \gamma u_x^+ = \mu(u^- - u^+),$$
$$u_t^- - \gamma u_x^- = \mu(u^+ - u^-),$$

*where* $u^\pm(x, t)$ *denote densities for right/left moving particles,* $\gamma$ *is the movement speed, and* $\mu$ *is the rate of switching direction.*

(a) *Derive an equivalent system of two equations for the total population density* $u = u^+ + u^-$ *and the population flux* $v = \gamma(u^+ - u^-)$.

(b) *For the* $(u, v)$-*system from* (a), *consider the* parabolic limit:

$$\gamma \to \infty, \quad \mu \to \infty, \quad \lim \frac{\gamma^2}{2\mu} = D < \infty.$$

*Which equation for* $u$ *follows?*

(c) *From the* $(u, v)$-*system from* (a), *derive a single second-order PDE for* $u$ *alone. Again, study the parabolic limit for this second-order equation.*

(d) *Interpret your results.*

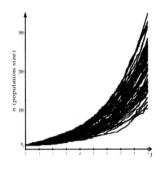

# Chapter 5

# Stochastic Models

## 5.1 Introduction

In this chapter, we consider models where outcomes are uncertain. Even though outcomes are uncertain, we can still assign probabilities to different outcomes, and then study how these probabilities change with time. An excellent reference text in this area is [3].

Many biological systems change from one state to another over time. Nerves change from quiescent to excited and back, cells change from healthy to diseased, or one plant population replaces another. While changes between states can be uncertain, probabilities of transitioning from one state to the next can be assigned nonetheless. If we know the transition probabilities between states, then we can evaluate changes in the system over time.

In preparation for the mathematical formulation of a model with uncertain outcomes, we first define the mathematical quantities that will go into the model. If $S$ is a *sample space* (collection of all possible outcomes of an "experiment") with a probability measure, and $X$ is a real-valued function defined over the elements of $S$, then $X$ is a *random variable*. For example, if $X$ were the fork length of a captured fish in cm, then $S$ would be the nonnegative real numbers.

We can follow the change in a random variable as a parameter, such as time, increases. A family of random variables $\{X(t)\}$, indexed by a parameter $t$, is called a *stochastic process*.

We start this chapter with an example of a memoryless stochastic process, or *Markov process* (Section 5.2). We model ecosystem succession dynamics via a Markov process. Here a *Markov chain* model, describing transitions from one state to the next, can be understood with matrix theory. Next we focus on random variables (Section 5.3). We introduce probability density (Section 5.3.1) and probability mass (Section 5.3.2) as measures for sample space $S$, and discuss descriptive statistics (Section 5.3.3) and probability generating functions (Section 5.3.4) as means to characterize random variables. The last part of the chapter concerns applications and extensions of tools developed earlier in the chapter. We consider random motion via diffusion processes (Section 5.4), branching processes (Section 5.5), linear birth and death processes (Section 5.6), and nonlinear birth and death processes (Section 5.7). These can be used to describe animal movement (Section 5.4.2), the extinction of family names (Section 5.3.4), the polymerase chain reaction (Section 5.5.2), population extinction (Section 5.6.2), and dynamics of the common cold (Section 5.7.1).

## 5.2   Markov Chains

The simplest stochastic processes are those which can be completely characterized by their current state, and where past states of a variable do not affect future outcomes. A stochastic process $\{X(t)\}$ is called a *Markov process* if it is history-independent. In the case with $t$ being a discrete sequence $t_1, t_2, \ldots$, it is a one-step memory process, that is,

$$
\begin{aligned}
\Pr\{X(t_i) = x_1 | X(t_{i-1}) = x_2 \cap X(t_{i-2}) = x_3 \cap \cdots\} \\
= \Pr\{X(t_i) = x_1 | X(t_{i-1}) = x_2\}.
\end{aligned} \tag{5.1}
$$

Here, Pr denotes the probability associated with an event, $\cap$ means "and," and | means "given that." Markov processes are sometimes referred to as being *memoryless*; that is, the next state for the stochastic process depends only upon the current state. A *Markov chain* is a model which tracks the progression of a Markov process from time step to time step.

   One example of a Markov chain involves succession in plant communities. As plant communities mature to a climax ecosystem, certain plant species replace others. In this section, we study succession in plant communities with Markov chains. We begin with a two-tree forest in Section 5.2.1. We generalize the example to formulate a Markov theory in Section 5.2.2. A large forest is considered in Section 5.2.3.

### 5.2.1   A Two-Tree Forest Ecosystem

By way of example, consider a population comprising red oak and hickory. At any point in space, the sample space of possible outcomes is $\mathcal{S} = \{RO, HI\}$, where RO represents red oak and HI represents hickory. We assume that the life spans of the two trees are similar. In each generation, red oak may be replaced by itself or by hickory, and hickory may be replaced by itself or red oak. This is a Markov process, with the index $t$ indicating the generation.

   For example, suppose that when a red oak tree dies, it is equally likely to be replaced by hickory or red oak, and that when a hickory tree dies, it has probability 0.74 of being replaced by red oak and 0.26 of being replaced by hickory. These transitions can be shown in either a graphical (Figure 5.1) or a tabular (Table 5.1) format. Note that the columns of the table sum to 1.

   The table format can be translated into a transition matrix. For example, the entries of Table 5.1 can be written in a transition matrix P,

$$
P = \begin{pmatrix} 0.5 & 0.74 \\ 0.5 & 0.26 \end{pmatrix}.
$$

To track the changes in the system over time, we define a vector $\mathbf{u}_t = (o_t, h_t)^T$ which describes the probability of red oak and of hickory at a given location in the forest after $t$ generations. In the case of a large homogeneous forest, the same transition model would

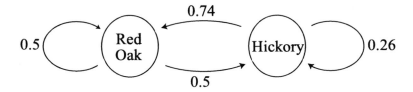

**Figure 5.1.** *Transitions between red oak and hickory vegetation shown as a graph.*

**Table 5.1.** *Transitions between red oak and hickory vegetation shown as a table.*

|            | Canopy   |         |
|------------|----------|---------|
| % Saplings | Red oak  | Hickory |
| Red oak    | 0.50     | 0.74    |
| Hickory    | 0.50     | 0.26    |

apply at every point in space. Hence, $o_t$ and $h_t$ can be interpreted as the *proportions* of red oak and hickory in a large statistically homogeneous forest ecosystem.

If we suppose that the forest is initially 50% red oak and 50% hickory, then $\mathbf{u}_0 = (0.5, 0.5)^T$. To find $\mathbf{u}_1$, we calculate as follows:

$$
\begin{array}{ccccccc}
o_1 = & \begin{array}{c}\text{proportion}\\\text{of red oak at}\\\text{time 0}\end{array} & \cdot & \begin{array}{c}\text{probability}\\\text{red oak is replaced}\\\text{by red oak}\end{array} & + & \begin{array}{c}\text{proportion}\\\text{of hickory}\\\text{at time 0}\end{array} & \cdot & \begin{array}{c}\text{probability}\\\text{hickory is}\\\text{replaced by red oak}\end{array}\\
\end{array}
$$

$$= \quad (0.5) \quad \cdot \quad (0.5) \quad + \quad (0.5) \quad \cdot \quad (0.74)$$

$$= \quad 0.62$$

$$
\begin{array}{ccccccc}
h_1 = & \begin{array}{c}\text{proportion}\\\text{of red oak at}\\\text{time 0}\end{array} & \cdot & \begin{array}{c}\text{probability}\\\text{red oak is replaced}\\\text{by hickory}\end{array} & + & \begin{array}{c}\text{proportion}\\\text{of hickory}\\\text{at time 0}\end{array} & \cdot & \begin{array}{c}\text{probability}\\\text{hickory is}\\\text{replaced by hickory}\end{array}\\
\end{array}
$$

$$= \quad (0.5) \quad \cdot \quad (0.5) \quad + \quad (0.5) \quad \cdot \quad (0.26)$$

$$= \quad 0.38.$$

In terms of the matrix formulation, we can write

$$\mathbf{u}_1 = P\mathbf{u}_0,$$
$$\mathbf{u}_2 = P\mathbf{u}_1,$$
$$\vdots$$

etc.

If we continue this process, we observe that the forest approaches an equilibrium value. For example,

$$\mathbf{u}_4 = \begin{pmatrix} 0.596 \\ 0.404 \end{pmatrix}, \qquad \mathbf{u}_5 = \begin{pmatrix} 0.597 \\ 0.403 \end{pmatrix}.$$

The forest has reached an equilibrium $\mathbf{u}^*$ when

$$P\mathbf{u}^* = \mathbf{u}^*.$$

Here, $\mathbf{u}^*$ is an *eigenvector* corresponding to eigenvalue $\lambda = 1$. To calculate the eigenvector we solve

$$(P - I)\mathbf{u}^* = 0 \Rightarrow \begin{pmatrix} -0.5 & 0.74 \\ 0.5 & -0.74 \end{pmatrix} \begin{pmatrix} o^* \\ h^* \end{pmatrix} = 0,$$

which has solution $o^* = 0.597$, $h^* = 0.403$.

A more complex model is given below in Section 5.2.3. However, before considering this complex model, we derive the general theory.

## 5.2.2  Markov Chain Theory

We consider a system with $n$ possible states for the system. Given that a transition occurs from state $j$, the *transition probability* $p_{ij}$ describes the probability of the transition taking the system to state $i$, $1 \leq i, j, \leq n$. When the transition probabilities are entered into a *transition matrix* $P = (p_{ij})$, the matrix columns sum to 1 because a transition occurring from state $j$ takes the system to some state $i$, $1 \leq i \leq n$, with probability 1. Finally, to track the probability associated with being in each state we define a *probability vector*, a vector $\mathbf{u} = (u_1, \ldots, u_n)^T$ whose nonnegative entries sum to 1.

A general Markov model for transitions then takes the form of a discrete-time dynamical system (Section 2.3),

$$\mathbf{u}_{t+1} = P\mathbf{u}_t, \qquad \mathbf{u}_0 \text{ given,} \tag{5.2}$$

where $\mathbf{u}_t$ is a probability vector and $P$ is a transition matrix. To calculate the long-term probabilities associated with each state we can use the eigenvector of $P$ corresponding to eigenvalue $\lambda = 1$. This result is made precise in the following theorem (see also [3]).

**Theorem 5.1.** *Providing some power of $P$ has all positive entries, then for any probability vector $\mathbf{u}_0$ and model $\mathbf{u}_{t+1} = P\mathbf{u}_t$, $\mathbf{u}_t \to \mathbf{u}^*$ as $t \to \infty$, where $P\mathbf{u}^* = \mathbf{u}^*$.*

The requirement that some power of $P$ has all positive entries ($P$ is primitive) ensures that, given enough time-steps, one can transition from any state to any other state and hence the result is independent of the original state $\mathbf{u}_0$.

## 5.2.3  The Princeton Forest Ecosystem

A more complex model for successional dynamics was made for the well-studied Princeton forest ecosystem [90, 91]. Here the transitional probabilities between five dominant trees were measured in terms of which species replaced resident trees, once they died. Results were as given in Figure 5.2 and Table 5.2.

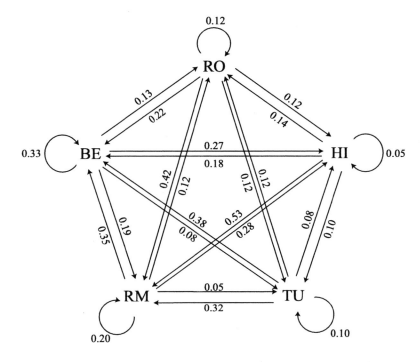

**Figure 5.2.** *Transitions between tree species in the Princeton forest shown as a graph. Based on* [91, 90].

**Table 5.2.** *Transition probabilities for the Princeton forest ecosystem shown as a table. Based on* [91, 90].

|                 | % Saplings | Canopy | | | | |
|-----------------|------------|--------|-------|-------|-------|-------|
|                 |            | RO     | HI    | TU    | RM    | BE    |
| RO = Red oak    | Red oak    | 0.12   | 0.14  | 0.12  | 0.12  | 0.13  |
| HI = Hickory    | Hickory    | 0.12   | 0.05  | 0.08  | 0.28  | 0.27  |
| TU = Tulip tree | Tulip tree | 0.12   | 0.10  | 0.10  | 0.05  | 0.08  |
| RM = Red maple  | Red maple  | 0.42   | 0.53  | 0.32  | 0.20  | 0.19  |
| BE = Beech      | Beech      | 0.22   | 0.18  | 0.38  | 0.35  | 0.33  |

For this full forest succession model, knowledge of the transitional probabilities can be translated into a prediction for the climax successional community, $u^*$, by applying Theorem 5.1 (see Exercise 5.8.2):

$$\mathbf{u}^* = \begin{pmatrix} 0.128 \\ 0.197 \\ 0.080 \\ 0.298 \\ 0.297 \end{pmatrix} \begin{matrix} \longleftarrow \\ \longleftarrow \\ \longleftarrow \\ \longleftarrow \\ \longleftarrow \end{matrix} \begin{matrix} \text{RO} \\ \text{HI} \\ \text{TU} \\ \text{RM} \\ \text{BE} \end{matrix} . \tag{5.3}$$

A comparison of the prediction and the observed proportions of trees in the climax forest areas was made in [90, 91], where it was shown that the prediction and observation were closely correlated.

## 5.3   Working with Random Variables

In the previous section, our mathematical analysis of the forest succession involved calculating changes in probabilities associated with the random variable $X(t)$ taking on different values. In that example, $X(t)$ described the event of having a particular sort of tree at a given location in the forest. In different problems, a random variable could describe any number of things. For example, the random variable $A$ could be used to describe the length of time a cell remains alive.

Random variables can be broken down into two main classes: continuous (the set $S$ of values that $X$ takes is continuous) and discrete (the set $S$ of values that $X$ takes is discrete). For each of these classes, we need a mathematical formalism which assigns a probability to the different values in $S$ that the random variable can attain.

To derive more complex models for stochastic processes, we need some basic laws of probability that can be applied to random variables. We motivate discussion of the laws by a simple example which involves cell death.

Consider the length of time $A$ for which a cell lives. $A$ is a random variable defined on the nonnegative real numbers which describes the age of the cell. We may be interested in relating events associated with random variables, for example, the event in which the cell dies by age $a_2$, given that it was alive at age $a_1 < a_2$.

The *law of conditional probability* can be used to relate conditional events to other events. Specifically, if $E_1$ and $E_2$ are events which occur with nonzero probability, then

$$\Pr\{E_2|E_1\} = \frac{\Pr\{E_2 \cap E_1\}}{\Pr\{E_1\}}. \tag{5.4}$$

In the above example, if we let $E_2$ be the event that $A \in (a_1, a_2)$ and $E_1$ be the event that $A > a_1$, then the probability that the cell dies during age interval $(a_1, a_2)$, given that it was alive at age $a_1 < a_2$, is

$$\begin{aligned}
\Pr\{A \in (a_1, a_2)|A > a_1\} &= \frac{\Pr\{A \in (a_1, a_2) \cap A > a_1\}}{\Pr\{A > a_1\}} \\
&= \frac{\Pr\{A \in (a_1, a_2)\}}{\Pr\{A > a_1\}}.
\end{aligned} \tag{5.5}$$

### 5.3.1   Probability Density

In the above instance of cell death, the age of cell death $A$ is a continuous random variable defined over the nonnegative real numbers. Here, the age of death is described by a *probability density function* $f(a)$ such that

$$\Pr\{a_1 \le A < a_2\} = \int_{a_1}^{a_2} f(\alpha)\, d\alpha. \tag{5.6}$$

The related *cumulative density function* describing the probability that the cell dies by age $a$ is

$$F(a) = \Pr\{A < a\} = \int_0^a f(\alpha)\, d\alpha. \tag{5.7}$$

The fundamental theorem of calculus relates the two functions by $f(a) = F'(a)$. The actual probability density function $f$ depends upon the model we choose for cell death.

To show how $f$ can be derived from first principles, we consider a simple model for cell death. Suppose the age-dependent death rate is given by $\mu(a)$. Our model assumes that the probability of the cell dying in time interval $(a, a + \Delta a)$, given that it was alive at age $a$, is equal to $\mu(a)\Delta a + o(\Delta a)$, where $o(x)$ is the Landau symbol, designating lower order terms ($\lim_{x \to 0} o(x)/x = 0$). In terms of the random variable $A$, this model can be written as

$$\Pr\{A \in (a, a + \Delta a) | A > a\} = \mu(a)\Delta a + o(\Delta a). \tag{5.8}$$

Now

$$\Pr\{A > a\} = \Pr\{A \in (a, a + \Delta a)\} + \Pr\{A > a + \Delta a\}. \tag{5.9}$$

Using (5.5) and (5.8), we rewrite the last statement as follows:

$$\begin{aligned} \Pr\{A > a\} &= \Pr\{A \in (a, a + \Delta a) | A > a\} \Pr\{A > a\} + \Pr\{A > a + \Delta a\} \\ &= \Pr\{A > a\}\mu(a)\Delta a + o(\Delta a) + \Pr\{A > a + \Delta a\}, \end{aligned} \tag{5.10}$$

so that

$$\Pr\{A > a + \Delta a\} - \Pr\{A > a\} = -\Pr\{A > a\}\mu(a)\Delta a + o(\Delta a). \tag{5.11}$$

Dividing both sides of (5.11) by $\Delta a$ and taking the limit as $\Delta a \to 0$ yields

$$\frac{dp}{da} = -\mu(a)p(a), \qquad p(0) = 1, \tag{5.12}$$

where $p(a) = \Pr\{A > a\}$. This differential equation for the probability of surviving to age $a$ has solution

$$p(a) = \exp\left(-\int_0^a \mu(\tau)\, d\tau\right). \tag{5.13}$$

Returning to the definition of the cumulative density function for the age of cell death (5.7), we observe that $F(a) = 1 - p(a)$; hence the probability density for cell death is

$$f(a) = -p'(a) = \mu(a)\exp\left(-\int_0^a \mu(\tau)\, d\tau\right). \tag{5.14}$$

For the case where $\mu(a)$ is a constant, we have an exponentially distributed waiting time or cell death, $f(a) = \mu \exp(-\mu a)$.

## 5.3.2  Probability Mass

When the random variable $X$ is discrete, the probabilities associated with each outcome in $S = \{x_0, x_1, x_2, \ldots\}$ are given by a *probability mass function* with values

$$p_n = \Pr\{X = x_n\}. \tag{5.15}$$

The simplest discrete random variable is a *Bernoulli* random variable. This random variable $X$ has only two possible outcomes: 0 (failure) or 1 (success). For example, it could be used to describe whether a given cell is alive or dead at some fixed age $a$. Here, we associate success with the outcome of being alive at the fixed age $a$. The sample space is $S = \{0, 1\}$, and the probabilities associated with each outcome are $p_0 = \Pr\{X = 0\} = 1 - p$ and $p_1 = \Pr\{X = 1\} = p$, where $p = \Pr\{\text{success}\}$. For the above cell death model, $p = p(a)$, given by (5.13), describes the probability of being alive (success) at age $a$. A simpler example is given by the outcome from flipping a coin, which yields $X = 0$ (tails) or $X = 1$ (heads). Here, $X$ is distributed as a Bernoulli random variable and, if the coin is fair, $p = 0.5$.

We now consider the case where there are many identical cells or, alternatively, many coin flips. The sum of $m$ independent identically distributed Bernoulli random variables is a *binomial* random variable. The probability of $n$ successes in $m$ trials is

$$p_n = \text{Bin}(n; m, p) = \binom{m}{n} p^n (1 - p)^{m-n}, \qquad n = 0, 1, 2, \ldots, m. \qquad (5.16)$$

For example, the probability of seven heads in ten coin flips is $p_7 = \text{Bin}(7; 10, 0.5) = 0.117$. Note that the binomial theorem ensures that the probabilities sum to 1:

$$\sum_{n=0}^{m} p_n = \sum_{n=0}^{m} \binom{m}{n} p^n (1 - p)^{m-n} = (p + (1 - p))^m = 1. \qquad (5.17)$$

If we consider a population of $m$ independent, identical cells, and denote

$$X_i(a) = \begin{cases} 1 & \text{if cell } i \text{ is alive at age } a, \\ 0 & \text{if cell } i \text{ is dead at age } a, \end{cases} \qquad i = 1, \ldots, m, \qquad (5.18)$$

then we can count the number of cells alive at age $a$ as

$$Y(a) = \sum_{i=1}^{m} X_i(a). \qquad (5.19)$$

At any fixed age $a$, $Y(a)$ is the sum of $m$ independent, identical cells. Hence $Y(a)$ is a stochastic process described by a binomial random variable, with $p_n = \text{Bin}(n; m, p(a))$ and $p(a)$ as given in (5.13).

If each trial is an independent identically distributed Bernoulli random variable, then the probability of the $k$th success occurring on the $n$th trial is governed by the negative binomial distribution

$$p_n = \text{NB}(n; k, p)$$

$$= \Pr\{k - 1 \text{ successes on the first } n - 1 \text{ trials}\} \cdot \Pr\{\text{success}\}$$

$$= \binom{n-1}{k-1} p^{k-1} (1 - p)^{n-k} \cdot p$$

$$= \binom{n-1}{k-1} p^k (1 - p)^{n-k}, \qquad n = k, k + 1, k + 2, \ldots. \qquad (5.20)$$

For example, the probability of the seventh head occurring on the tenth coin flip is $p_7 =$ NB$(7; 10, 0.5) = 0.082$. Returning to the example of cell death, the probability of finding the $k$th living cell on the $n$th cell checked is $p_n = $ NB$(n; k, p(a))$.

### 5.3.3 Descriptive Statistics

When we use stochastic processes to describe the uncertain behavior of biological models, it is convenient to have summary statistics to describe the qualitative features. The most commonly used measures are the mean, which describes average values, and the variance, which describes variability about the mean.

Given a random variable $X$, whose sample space is the natural numbers and whose probabilities are $p_n = \Pr\{X = n\}$, $n = 0, 1, 2, \ldots$, and any real-valued function of $X$, denoted by $\phi$, the *expected value* of $\phi(X)$ is

$$E(\phi(X)) = \sum_{n=0}^{\infty} \phi(n) p_n. \tag{5.21}$$

For some functions $\phi$, the sum will not converge. When this is true, we say that $E(\phi(X))$ does not exist.

We note that expectation is a linear operator, so that for any functions $\phi$ and $\psi$, the expectation of a linear combination of $\phi$ and $\psi$ is the linear combination of the expectations: $E(a\phi(X) + b\psi(X)) = aE(\phi(X)) + bE(\psi(X))$ for all real numbers $a$ and $b$.

When $\phi = X^m$, the *mth moment* $M_m$ is

$$M_m = E(X^m) = \sum_{n=0}^{\infty} n^m p_n. \tag{5.22}$$

The first moment $M_1$ is also referred to as the *expected value of $X$*, $E(X)$, or *mean of $X$*, $\mu$. The variance is the expected value of the squared deviations about the mean,

$$\mathrm{var}(X) = \sigma^2 = E((X - \mu)^2) = \sum_{n=0}^{\infty} (n - \mu)^2 p_n. \tag{5.23}$$

Using the linear operator property of expectation, we note that the variance can be rewritten as $\mathrm{var}(X) = \sigma^2 = E(X^2 - 2\mu X + \mu^2) = E(X^2) - 2\mu E(X) + \mu^2 = M_2 - M_1^2$. This is sometimes referred to as the computational form of the variance.

While the variance gives a measure of squared deviations about the mean, we may be interested in a measure for the typical spread about the mean. This is the standard deviation, $\sigma = \sqrt{\sigma^2}$. The *coefficient of variation*, a scaled measure of the spread, is c.v.$= \sigma/\mu$.

We now consider some examples taken from the distributions discussed in earlier sections. The first and second moments of a Bernoulli random variable are both given by $p$, and hence the variance is $M_2 - M_1^2 = p - p^2 = p(1 - p)$, which is highest for values of $p$ intermediate between 0 and 1.

To calculate the mean of the binomial distribution, we must evaluate

$$M_1 = E(X) = \sum_{n=0}^{m} n p_n = \sum_{n=0}^{m} n \binom{m}{n} p^n (1-p)^{m-n}$$

$$= \sum_{n=1}^{m} \frac{m!}{(n-1)!(m-n)!} p^n (1-p)^{m-n}$$

$$= mp \sum_{n=1}^{m} \binom{m-1}{n-1} p^{(n-1)} (1-p)^{m-n}$$

$$= mp. \tag{5.24}$$

A similar calculation gives

$$M_2 - M_1 = E(X(X-1)) = m(m-1)p^2. \tag{5.25}$$

Hence the variance is $\sigma^2 = M_2 - M_1^2 = E(X(X-1)) + M_1 - M_1^2 = mp(1-p)$.

Note that the mean and variance of the binomial distribution are simply $m$ times the mean and variance for the Bernoulli distribution. In general, if $Y = \sum X_i$, then $E(Y) = \sum E(X_i)$, and if the $X_i$'s are independent ($E(X_i X_j) = E(X_i)E(X_j)$), then var$(Y) = \sum$ var$(X_i)$. The proof of this is left as an exercise (Exercise 5.8.3).

The expected values for continuous random variables are defined in an analogous way to those for discrete random variables. Given a random variable $X$, whose sample space is the nonnegative real numbers and whose probability density function is $f(x)$, the expectation of $\phi(X)$ is

$$E(\phi(X)) = \int_0^\infty \phi(x) f(x) \, dx, \tag{5.26}$$

providing it exists. As with the previous discussion of discrete random variables, the mean and variance can be calculated from the moments of $X$.

In our cell death example, we may be interested in the mean and variance in the age of death $A$ for a cell where $A$ is exponentially distributed. Here, integration by parts yields

$$M_1 = E(A) = \int_0^\infty a\mu \exp(-\mu a) \, da = \mu^{-1},$$

$$M_2 = E(A^2) = \int_0^\infty a^2 \mu \exp(-\mu a) \, da = 2\mu^{-2}, \tag{5.27}$$

and hence $A$ has a mean of $\mu^{-1}$ and the computational formula for the variance yields var$(A) = M_2 - M_1^2 = \mu^{-2}$.

### 5.3.4  The Generating Function

One of the workhorses in stochastic processes is the generating function. Given any discrete random variable $Y$ that assumes values in the natural numbers $n$ with probability $p_n$, the *generating function* is defined as

$$g(s) = \sum_{n=0}^{\infty} s^n p_n, \qquad 0 \le s \le 1. \tag{5.28}$$

Formally, we may write

$$g(s) = E(s^Y) \tag{5.29}$$

to denote the generating function. Note that the power series in (5.28) converges for all $0 \leq s \leq 1$ and is an increasing function of $s$, with $g(0) = p_0$, $g(1) = 1$, and hence $p_0 \leq g(s) \leq 1$.

All the information about the random variable $Y$ is contained within its generating function $g(s)$. To observe this, note that Taylor's theorem permits us to expand a function in terms of its derivatives at zero. Thus, by calculating derivatives, we regain the probabilities $p_n$ associated with the random variable as follows:

$$p_n = \frac{1}{n!} \frac{d^n g}{ds^n} \bigg|_{s=0}. \tag{5.30}$$

The generating function allows us to compute the mean as

$$E(Y) = \sum_{n=0}^{\infty} n p_n = g'(1). \tag{5.31}$$

The variance takes only slightly more effort. We can write

$$g''(1) = \sum_{n=0}^{\infty} n(n-1) p_n, \tag{5.32}$$

yielding

$$\text{var}(Y) = g''(1) + g'(1) - g'^2(1). \tag{5.33}$$

In other words, the generating function allows us to compute all the probabilities and statistics we need in a straightforward way. As we will show in several cases, it may be easier to compute the generating function $g(s)$ for a discrete random variable than to compute the values of $p_n$ directly.

By way of example, we consider the Bernoulli random variable of Section 5.3.2. Recall that the random variable $Y$ has only two possible outcomes: 0 with probability $1 - p$, and 1 with probability $p$. The generating function is simply $g(s) = (1 - p) + sp$. The derivatives, evaluated at $s = 1$, are $g'(1) = p$ and $g''(1) = 0$. Hence formula (5.31) yields the mean for the Bernoulli random variable as $p$, and formula (5.33) yields the variance as $p(1 - p)$. These formulae were derived, using different methods, in Section 5.3.3.

Just as we earlier calculated the variance for a sum of independent random variables, we can also calculate the generating function for a sum of independent random variables. Whereas the variances add, the generating functions multiply. Specifically, if $X$ and $Y$ are independent random variables with probability mass functions $g_n$ and $h_n$, and generating functions $g(s)$ and $h(s)$, respectively, then $X + Y$ has generating function $g(s)h(s)$. To show this, we first observe that

$$\Pr\{X + Y = n\} = \sum_{k=1}^{n} g_k h_{n-k}. \tag{5.34}$$

Hence the generating function for $X + Y$ is

$$\sum_{n=0}^{\infty} \left( \sum_{k=0}^{n} g_k h_{n-k} \right) s^n = \sum_{k=0}^{\infty} \sum_{n=k}^{\infty} g_k h_{n-k} s^n = \sum_{k=0}^{\infty} g_k s^k \sum_{n=k}^{\infty} h_{n-k} s^{n-k} = g(s)h(s). \quad (5.35)$$

Note that the calculation is made possible by the careful switching of the limits for the double sum after the first "equals" sign of the above line. You may want to check this to make sure you agree with it. This result can be extended in a straightforward way to the sum of $m$ random variables. The generating function for the sum is the $m$-fold product of the individual generating functions. We use this result when analyzing branching processes in Section 5.5.

We now extend the previous example of using generating functions to calculate the mean and variance of a Bernoulli random variable. Recall that the sum of $m$ independent Bernoulli random variables is the binomial random variable $\text{Bin}(n; m, p)$ (equation (5.16)). Hence the generating function for $\text{Bin}(n; m, p)$ is simply the $m$-fold product of the Bernoulli random variable generating function, $(1 - p + sp)$, with itself, $g(s) = (1 - p + sp)^m$. This generating function $g(s)$ has derivatives $g'(1) = mp$ and $g''(1) = m(m - 1)p^2$, and so formulae (5.31) and (5.33) yield $mp$ and $mp(1 - p)$ for the mean and variance, respectively. These formulae were derived, using different methods, in Section 5.3.3.

In the remainder of the chapter, we apply the methods from this section to develop and analyze a series of stochastic models. We motivate each model with a biological problem, but the mathematical tools that are brought to bear on the problem have general application to the analysis of stochastic processes.

## 5.4   Diffusion Processes

Most living organisms move in space. Given that we have some information about how an organism moves over short time scales, can we determine where it is likely to be over long time scales? If movement rules are simple, mathematical models can be used to translate the movement rules into equations. As we will show in this section, analysis of the resulting equations yields a probability density function that can be used to track the changing location of the animal over time.

We consider an individual executing a random walk in one-dimensional space. At each time step, the individual jumps to either the right or the left, and its new position is determined by its current position plus a random increment to the left or right. This is another example of a Markov process, because the current location plus the random increment is sufficient to determine the next position. The precise path taken to get to the current location plays no role in determining future positions.

In the next section, we calculate the probability mass function for the location of the individual after a given number of time steps. We show that, after a sufficiently large number of time steps, the probability mass function can be approximated by a Gaussian probability density function.

| time step | 1 | 2 | 3 | 4 | 5 | 6 | 7 | 8 | 9 | 10 |
|-----------|---|---|---|---|---|---|---|---|---|----|
|           | L | R | L | R | L | L | L | L | R | L |

| 11 | 12 | 13 | 14 | 15 | 16 | 17 | 18 | 19 | 20 |
|----|----|----|----|----|----|----|----|----|----|
| R | L | R | L | L | L | R | L | L | L |

**Figure 5.3.** *Jumps to the left and right, for 20 time steps, are generated by flipping a fair coin 20 times.*

## 5.4.1 Random Motion in One Dimension

Suppose an individual, released at $x = 0$, moves back and forth randomly along a line in fixed steps $\lambda = \Delta x$ at fixed time intervals $\tau = \Delta t$, and that this movement is unbiased (equal probability of moving right and left). After $k$ time steps (time $= k\tau$), the individual is anywhere from $x = -k\lambda$ to $x = k\lambda$.

We describe the stochastic process with $p_n(k)$, the probability the individual reaches $n$ space steps to the right ($x = n\lambda$) after $k$ time steps ($t = k\tau$). Suppose that to reach $n\lambda$, the individual has moved $a$ steps to the right and $b$ steps to the left.

$$n = a - b, \quad k = a + b,$$

implying that $k + n = 2a$ is even. Thus $k$ odd implies $n$ odd, and $k$ even implies $n$ even. After $n$ time steps, only every other point ("evens" or "odds") can be occupied.

For example, consider the case where, after $k = 20$ time steps, the jumps to the left $L$ and right $R$ are given by the sequence shown in Figure 5.3. These were generated by flipping a fair coin 20 times, and writing $R$ for heads and $L$ for tails. Here, $a = 6$, $b = 14$, $k = 20$, $n = -8$.

Other orderings of $a = 6$ jumps to the right and $b = 14$ jumps to the left in the $k = 20$ time steps would describe other paths that would still lead to location $n = -8$ after $k = 20$ time steps. The number of possible ways to insert $a = 6$ $R$'s into a sequence of length 20 is given by the following combinatorial expression:

$$\frac{20 \cdot 19 \cdot 18 \cdot 17 \cdot 16 \cdot 15}{6 \cdot 5 \cdot 4 \cdot 3 \cdot 2 \cdot 1} = \frac{20!}{14! \, 6!} = \binom{20}{6}.$$

In general, the number of possible paths that an individual can take to reach $x = n\lambda$ in $k$ time steps (time $= k\tau$) is given by

$$\frac{k!}{a! \, b!} = \frac{k!}{a! \, (k - a)!} = \binom{k}{a},$$

with $\binom{k}{a}$ being the number of possible combinations of $a$ moves to the right in $k$ time steps (and $a = \frac{k+n}{2}$). The expression $\binom{k}{a}$ is referred to as the *binomial coefficient*. It features in both the binomial distribution, (5.16), and the *binomial expansion*

$$(x + y)^k = \sum_{a=0}^{k} \binom{k}{a} x^{k-a} y^a.$$

The total number of possible $k$-step paths is $2^k$, so

$$p_n(k) = \frac{\text{no. of possible paths to reach } x = n\lambda \text{ in } k \text{ time steps}}{\text{no. of possible paths in } k \text{ time steps}} = \frac{1}{2^k}\binom{k}{a}.$$

This is a binomial distribution, (5.16), where the probability of success is $\frac{1}{2}$. Note that for $n + k$ even,

$$\sum_{n=-k}^{k} p_n(k) = \sum_{a=0}^{k}\binom{k}{a}\left(\frac{1}{2}\right)^{k-a}\left(\frac{1}{2}\right)^{a} = \left(\frac{1}{2} + \frac{1}{2}\right)^{k} = 1,$$

where we used the facts that $n = -k$ implies $a = 0$ and $n = k$ implies $a = k$.

Providing $k$ is sufficiently large (i.e., a sufficiently large number of time steps have taken place), the binomial distribution can be approximated with a normal (Gaussian) distribution with variance $k$ [3],

$$p_n(k) \approx \left[\frac{2}{\pi k}\right]^{\frac{1}{2}}\exp\left[\frac{-n^2}{2k}\right].$$

We can translate this result in terms of continuous time and space by recalling that $n = \frac{x}{\lambda}$ and $k = \frac{t}{\tau}$. What happens when $\lambda, \tau \to 0$, but $x = n\lambda$ and $t = k\tau$ are finite? The relevant quantity is

$$p(x,t) = \lim_{\substack{\lambda \to 0 \\ \tau \to 0}} \frac{p_{(x/\lambda)}(t/\tau)}{2\lambda} \approx \left[\frac{\tau}{2(\lambda)^2}\frac{1}{\pi t}\right]^{\frac{1}{2}}\exp\left[\frac{-\tau}{2(\lambda)^2}\frac{x^2}{t}\right]$$

$$= \left[\frac{1}{4\pi Dt}\right]^{\frac{1}{2}}\exp\left[\frac{-x^2}{4Dt}\right] \tag{5.36}$$

if we assume

$$D = \lim_{\substack{\lambda \to 0 \\ \tau \to 0}} \frac{(\lambda)^2}{2\tau} \neq 0. \tag{5.37}$$

This assumption implies that individuals move very quickly ($\lambda/\tau \to \infty$), but switch direction very frequently ($\tau \to 0$). This limit is often referred to as the *parabolic limit* because, as we will see below, it can be connected to the parabolic diffusion PDE. Since the speeds and times can be scaled by changing the units used to measure them, we can say, equivalently, that the parabolic limit is valid when we are describing movement over small spatial scales ("small" units for space) and large time scales ("large" units for time), relative to the characteristic space and time steps taken by the individual.

Note that (5.36) is the fundamental solution of the diffusion equation as shown in Section 4.3.2. The coefficient $D$ is called the *diffusion coefficient*. It has dimensions length$^2$/time and measures how efficiently individuals disperse from high to low density. For example, hemoglobin in blood has diffusion coefficient $D \approx 10^{-7}$ cm$^2$/sec, whereas oxygen in blood has diffusion coefficient $D \approx 10^{-5}$ cm$^2$/sec.

In the next section, we consider the limiting case of small time and space steps and large velocity. In this case, the probability density function, describing the location of the individual, satisfies a parabolic PDE called the diffusion equation.

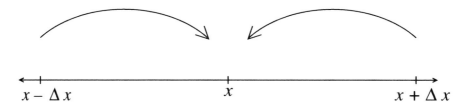

**Figure 5.4.** *Movement on the lattice giving rise to the diffusion equation. Here an individual arriving at x can come from locations $x - \Delta x$ and $x + \Delta x$, where $\Delta x = \lambda$ in equation (5.38).*

## 5.4.2  Diffusion Equation

An alternate approach to the random walk derives the diffusion PDE for a probability density function describing the location of the individual directly from a *master equation*. Let $X(t)$ be a stochastic process describing the location of an individual at time $t$, which is released at location $x = 0$ at time $= 0$ ($X(0) = 0$). We define a time-dependent probability density function $p(x, t)$ and small space interval $\lambda$, such that $p(x, t)\lambda = $ probability an individual released at $x = 0$ and time $= 0$ is between $x$ and $x + \lambda$ at time $t$.

Assume the random walk is unbiased. The *master equation* which describes movement on the lattice of points situated a distance $\lambda$ apart is

$$p(x, t + \tau) = \frac{1}{2} \, p(x - \lambda, t) + \frac{1}{2} \, p(x + \lambda, t) \tag{5.38}$$

(see Figure 5.4).

Expanding in Taylor series gives

$$p(x, t) + \tau \frac{\partial p}{\partial t}(x, t) + \frac{(\tau)^2}{2} \frac{\partial^2 p}{\partial t^2}(x, t) + \text{h.o.t.}$$

$$= \frac{1}{2} \left\{ p(x, t) - \lambda \frac{\partial p}{\partial x}(x, t) + \frac{(\lambda)^2}{2} \frac{\partial^2 p}{\partial x^2}(x, t) + \text{h.o.t.} \right.$$

$$+ \left. p(x, t) + \lambda \frac{\partial p}{\partial x}(x, t) + \frac{(\lambda)^2}{2} \frac{\partial^2 p}{\partial x^2}(x, t) + \text{h.o.t.} \right\},$$

where h.o.t. indicates higher-order terms in the Taylor series. The above equation can be simplified to yield

$$\frac{\partial p}{\partial t} + \frac{\tau}{2} \frac{\partial^2 p}{\partial t^2} = \frac{(\lambda)^2}{2\tau} \frac{\partial^2 p}{\partial x^2}.$$

Taking the limit as $\lambda, \tau \to 0$ so that $\frac{(\lambda)^2}{2\tau} \to D$ yields the diffusion equation

$$\frac{\partial p}{\partial t} = D \frac{\partial^2 p}{\partial x^2}. \tag{5.39}$$

The solution to (5.39), corresponding to the point release of an individual, is a Gaussian centered about zero, with variance $2Dt$:

$$p(x, t) = \left[\frac{1}{4\pi Dt}\right]^{\frac{1}{2}} \exp\left[\frac{-x^2}{4Dt}\right], \tag{5.40}$$

as we found in (5.36), and also earlier in (4.16). This solution can be verified by substitution and by noting that as $t \to 0$, $p(x, t) \to \delta(x)$ (see Exercise 4.5.2). Because (5.40) is an even function, the first moment $M_1 = E(X)$ is equal to zero (see (5.26)), and hence the variance is the second moment. We can derive a differential equation for the second moment,

$$M_2(t) = \int_{-\infty}^{\infty} x^2 p(x, t) \, dx,$$

by using (5.39) and integrating by parts,

$$\dot{M}_2 = \int_{-\infty}^{\infty} x^2 p_t \, dx = \int_{-\infty}^{\infty} x^2 D p_{xx} \, dx = x^2 D p_x \Big|_{-\infty}^{\infty} - \int_{-\infty}^{\infty} 2x D p_x \, dx$$

$$= -2xp \Big|_{-\infty}^{\infty} + \int_{-\infty}^{\infty} 2Dp \, dx = 2D \int_{-\infty}^{\infty} p(x, t) \, dx = 2D.$$

Solving this differential equation subject to the initial condition $M_2(0) = 0$ gives

$$M_2 = 2Dt, \tag{5.41}$$

which implies that the second moment $M_2$ grows linearly with time at rate $2D$. This linear growth in the second moment describes increasing uncertainty as to the location of the individual as time progresses (see Figure 4.6).

## 5.5   Branching Processes

A branching process is a stochastic process that describes a reproducing population. The random variable is the number of individuals in each generation, where it is assumed that the behavior of the younger generation is (stochastically) independent from the older generation. The simplest and most important example of a branching process is the Galton–Watson process, which is discussed in Section 5.5.1; an application to a polymerase chain reaction is given in Section 5.5.2.

### 5.5.1   Galton–Watson Process

Reverend H.W. Watson and Francis Galton [159] were interested in the extinction of family names. In an age where family name inheritance was restricted to males, it was possible that a run of "bad luck" would result in a family name going extinct even if, on average, a man were to have more than one son. Watson and Galton supposed that $p_0, p_1, p_2, \ldots$ were the probabilities that a man has 0, 1, 2, $\ldots$ sons, and determined the probability that the direct male line (i.e., family name arising from that original man) was extinct after $r$ generations.

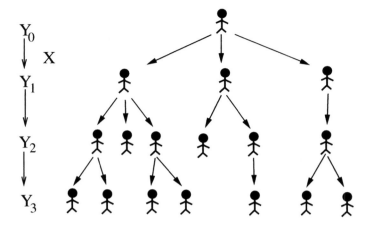

**Figure 5.5.** *The Galton–Watson process.*

More generally, they determined the probability associated with any specified number of direct male descendants in a given generation. Whereas rules for inheritance of family names have changed since the times of Galton and Watson, the mathematics behind the calculations remains relevant today and can be applied to a variety of interesting stochastic problems.

The Galton–Watson process is a Markov process in discrete time. Here, $Y_t$ denotes the number of parents at time step $t$, $t = 0, 1, 2, \dots$. Initially, there is a single parent ($Y_0 = 1$), who has a random number of offspring $X$ with $\Pr(X = n) = p_n$. After a single generation, $Y_1 = X$. In the next generation, each of the $Y_1$ offspring becomes a parent and has a random number of offspring, and so on (Figure 5.5). Here the conditional distribution of $Y_{t+1}$, given that $Y_t = m$, is the sum of $m$ independent variables, each with the same distribution as $X$. This is referred to as a branching process. An excellent discussion of branching processes is given in [102], and we follow the approach given there.

Our investigation into Watson and Galton's question starts by calculating the change in the generating function for the number of parents, $Y_t$, from one time step to the next. By knowing the original number of parents, $Y_0$, and by tracking the change in generation from time step to time step, we can calculate the generating function for the number of parents at every time step. As discussed in Section 5.3.4 and shown in (5.30), the generating function will then give us the probability mass function for the number of parents in each time step. If we are interested in the likelihood of extinction of the family name by generation $t$, then this can be calculated as the probability mass associated with $Y_t = 0$.

We suppose that the generating function for the number of offspring, $X$, from a parent is given by $g(s)$, and that there are $Y_t$ identical parents at time step $t$. For example, if each parent had 0, 1, or 2 offspring with probability 1/3, then $p_0 = p_1 = p_2 = 1/3$, $p_n = 0$ for $n > 2$, and $g(s) = 1/3 + s/3 + s^2/3$.

The parents in generation $t + 1$ are the offspring from generation $t$. The number of offspring arising from each of the $Y_t$ parents in generation $t$ is assumed to be independent from the number of offspring arising from the other $Y_t - 1$ parents in generation $t$. Hence the number of parents in generation $t + 1$ is given by the sum

$$\sum_{i=1}^{Y_t} X_i, \tag{5.42}$$

where the $X_i$'s are independent, identically distributed random variables, each with generating function $g(s)$.

We know that the generating function for the sum of $m$ independent, identically distributed random variables, each with generating function $g(s)$, is simply $(g(s))^m$ (see Section 5.3.4). Therefore, if we knew the number of parents in generation $t$ to be $Y_t = m$, then the generating function for the number of parents in the next generation would be $(g(s))^m$.

However, our calculation is complicated by the fact that the number of parents in generation $t$, $Y_t$, is a random variable. So as to keep track of this random variable, we denote the probability mass for the random variable $Y_t$ as $q_{tm} = \Pr\{Y_t = m\}$ and the generating function for $Y_t$ as $h_t(s)$. Thus the generating function for the number of parents in generation $t + 1$ is $(g(s))^m$, conditioned upon the different values that $m$ can attain in generation $t$,

$$h_{t+1}(s) = \sum_{m=0}^{\infty} \Pr\{Y_t = m\}(g(s))^m = \sum_{m=0}^{\infty} q_{tm}(g(s))^m = h_t(g(s)) = h_t \circ g(s), \tag{5.43}$$

where $\circ$ indicates functional composition.

At $t = 0$, there is a single parent ($Y_0 = 1$) so that $h_0(s) = s$. Applying (5.43), we observe that $h_1(s) = g(s)$. Continuing to the next generation, we have $h_2(s) = g \circ g(s)$. Returning to our above example where each parent has 0, 1, or 2 offspring with probability 1/3, we calculate $h_0(s) = s$, $h_1(s) = g(s) = 1/3 + s/3 + s^2/3$, $h_2(s) = g \circ g(s) = 1/3 + (1/3+s/3+s^2/3)/3+(1/3+s/3+s^2/3)^2/3 = 13/27+5s/27+6s^2/27+2s^3/27+s^4/27$, and so forth. Using either the definition of the generating function (5.28) or equation (5.30), we can deduce that, after two generations, the probabilities of having 0, 1, 2, 3, or 4 parents are 13/27, 5/27, 6/27, 2/27, and 1/27, respectively. Hence the probability of extinction after two generations is 13/27.

With each new generation, we iterate with the generating function to obtain

$$h_t(s) = h_{t-1} \circ g(s) = g^t(s), \tag{5.44}$$

a $t$-fold composition of the generating function $g(s)$. While the generating function $h_t(s)$ may not have such a simple form as given in the example above, (5.44) gives us a straightforward method for calculating $h_t(s)$. From this generating function, we can use the methods of Section 5.3.4 to calculate the probability mass function for the number of parents at each time step.

The expected number of parents in the $t$th generation is

$$E(Y_t) = h'_t(1) = g'(h_{t-1}(1))h'_{t-1}(1) = g'(1)h'_{t-1}(1) = R_0 E(Y_{t-1}), \tag{5.45}$$

where $R_0 = g'(1) = E(Y_1)$, and hence $E(Y_t) = R_0^t$, so the expectation grows geometrically with *reproduction ratio* $R_0$.

Calculation of the variance from (5.33) by similar methods is a little more involved and yields

$$
\operatorname{var}(Y_t) = \begin{cases} \frac{(R_0^t-1)R_0^t}{R_0(R_0-1)}\sigma^2, & R_0 \neq 1, \\ \sigma^2 t, & R_0 = 1, \end{cases} \tag{5.46}
$$

where $\sigma^2$ is the variance of the generating function $g(s)$ (see Exercise 5.8.7). Higher moments also can be found in a similar manner.

To determine the chance of a lineage going extinct, we can use the cobwebbing method of Section 2.2. The probability of being extinct in generation $t$, $x_t$, is $x_t = h_t(0)$. Equation (5.44) tells us that $x_t$ satisfies

$$
x_{t+1} = g(x_t), \tag{5.47}
$$

with initial condition $x_0 = 0$. Recall that $R_0 = g'(1)$, and hence there are two generic behaviors, depending upon whether $R_0 < 1$ (subcritical case; see Figure 5.6, top panel) or $R_0 > 1$ (supercritical case; see Figure 5.6, bottom panel). In the subcritical case, eventual extinction is inevitable. In the supercritical case, $x_t \rightarrow x^*$, the unique root to $g(x) = x$, and eventual extinction is possible, but not inevitable.

## 5.5.2 Polymerase Chain Reaction

Polymerase chain reaction (PCR) is a standard technique of molecular biology in which a small amount of nucleic acid (DNA or RNA) taken from a probe is multiplied so that it can be detected. This method is the first step in DNA fingerprinting, preceding the sequencing of the amplified nucleic acid. However, most investigations are interested not only in determining the sequencing of the nucleic acid, but also in calculating the amount of the sequenced strands present in the original probe. Quantitative PCR allows the user to calculate the starting amounts of the nucleic acid template by analyzing the amount of DNA produced during each cycle of PCR. The technique relies on the fluorescence of a reporter molecule that increases as product accumulates with each cycle of amplification.

The strings of nucleic acid encoding the DNA or RNA are incubated with a mixture of primers and nucleotides. This mixture allows the strings to replicate. String replication is stochastic, with the amplification factor (probability of doubling, or $p_d$) ranging between 0.6 and 0.8. Each string is assumed to be independent of the others. Hence this can be described as a Galton–Watson process with $p_0 = 1 - p_d$, $p_2 = p_d$ (given), and $p_n = 0$, $n \neq 0, 2$. The generating function associated with the PCR process is $g(s) = (1 - p_d) + p_d s^2$, and therefore the reproductive ratio is $R_0 = g'(1) = 2p_d$.

Using $Y_t$ to denote the number of copies of a given strand at time step $t$, we now consider the question of estimating the number of strands in the original probe, $Y_0$, given estimates for the number of strands in two successive probes $\tilde{Y}_t$, and $\tilde{Y}_{t+1}$ for $t$ large (typically $10 \leq t \leq 20$). Here, the tilde indicates that we do not know the precise number of strands $Y_t$ and $Y_{t+1}$, only estimates based on fluorescence levels.

In the Galton–Watson process (above), we assumed a single parent at time $t = 0$. If we modify the analysis to allow for $Y_0$ parents at time $t = 0$, the equation for the expected

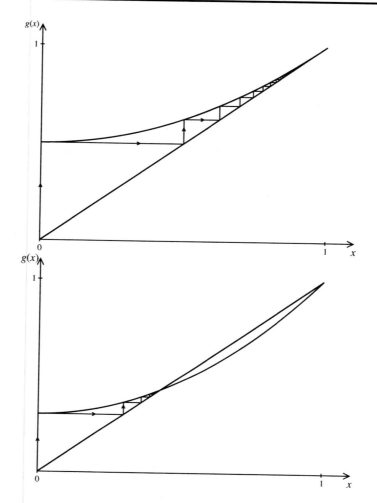

**Figure 5.6.** *Sketch of the extinction probability for the subcritical case (top) and the supercritical case (bottom). Cobwebbing tells us how the probability that the lineage goes extinct changes with time (equation (5.47)). $x_t$ indicates the probability of being extinct after $t$ generations. The initial condition is $x_0 = 0$. Top: In the subcritical case ($R_0 < 1$), extinction is certain. Here, cobwebbing shows $x_t \to 1$. Bottom: In the supercritical case ($R_0 > 1$), extinction is possible, but not inevitable. Here, cobwebbing shows $x_t \to x^* \neq 1$.*

number of strands after $t$ generations becomes $E(Y_t) = R_0^t Y_0$ and so $R_0 = E(Y_{t+1})/E(Y_t)$. We define the estimator for the reproductive ratio, $\hat{R}_0$, as follows:

$$\hat{R}_0 = \frac{\tilde{Y}_{t+1}}{\tilde{Y}_t}. \tag{5.48}$$

Knowing $R_0$ and $t$, we find $Y_0 = E(Y_t)/R_0^t$. This yields a simple estimator for the original number of strands, namely,

$$\hat{Y}_0 = \frac{\tilde{Y}_t}{\hat{R}_0^{\,t}} = \tilde{Y}_t \left( \frac{\tilde{Y}_{t+1}}{\tilde{Y}_t} \right)^{-t} = \frac{\tilde{Y}_t^{\,t+1}}{\tilde{Y}_{t+1}^{\,t}}. \qquad (5.49)$$

Of course, each time the PCR experiment is repeated, this estimator will give slightly different values for $\hat{Y}_0$. This is because the $\tilde{Y}_t$ values will vary between replicates. It is possible to analyze the variance of the estimator (5.49) by using simulations and other methods and to calculate confidence intervals, although we do not pursue this here.

## 5.6  Linear Birth and Death Process

Populations are subject to two primary types of stochasticity. *Environmental stochasticity* refers to variation and uncertainty in the environmental conditions in which a population finds itself. These conditions include effects of temperature, rainfall, competition from other species, and so forth. *Demographic stochasticity* refers to variation and uncertainty arising from the unpredictable behavior of the individuals that make up a population. It is relevant when population sizes are small (e.g., fewer than 25). Here, populations with a positive net growth rate can still go extinct due to a "run of bad luck," where insufficient individuals reproduce before they die. In this section, we consider how to model demographic stochasticity in continuous time using a linear birth and death model. Here it is assumed that individuals act independently from one another, so there are no nonlinear interaction terms in the equations. We derive formulae for the mean and variance of a population undergoing stochastic birth/death, and calculate the probability of extinction. We start by considering a pure birth process in Section 5.6.1 and include death in Section 5.6.2.

### 5.6.1  Pure Birth Process

To start, we ignore death and consider a *pure birth* process where individuals give birth at rate $b$. In the absence of demographic stochasticity, the underlying model equation would be

$$\frac{dn}{dt} = bn, \qquad n(0) = n_0, \qquad (5.50)$$

where $n(t)$ is the number of individuals in the population at time $t$, and $n_0$ is the number of individuals at time $t = 0$. This equation has the solution $n(t) = n_0 \exp(bt)$.

We define the underlying stochastic process by

$$N(t) = \text{number of individuals at time } t \text{ (random variable)},$$

$$p_n(t) = \Pr\{N(t) = n\}, \ n = 0, 1, 2, \ldots.$$

We assume that the birth event is a *Poisson process*, namely, that the probability of the event occurring in a short period of time $\tau$ is proportional to $\tau$, and the probability of two events occurring during the short period of time is $o(\tau)$.

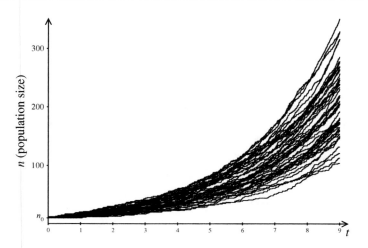

**Figure 5.7.** *Stochastic simulation of the pure birth process given in* (5.51). *Parameters are* $b = 1/3$ *and* $n_0 = 10$. *Fifty different trajectories are given. They differ only in the seed for the random number generator used.*

For one individual, we have

$$\Pr\{1 \text{ birth in } [t, t + \tau)\} = b\tau + o(\tau),$$
$$\Pr\{> 1 \text{ birth in } [t, t + \tau)\} = o(\tau),$$
$$\Pr\{0 \text{ births in } [t, t + \tau)\} = 1 - b\tau + o(\tau).$$

For $n$ individuals, we have

$$\Pr\{1 \text{ birth in } [t, t + \tau)\} = nb\tau(1 - b\tau)^{n-1} = nb\tau + o(\tau),$$
$$\Pr\{m \text{ births in } [t, t + \tau)\} = \binom{n}{m}(b\tau)^m(1 - b\tau)^{n-m} = o(\tau), \ 1 < m \leq n, \quad (5.51)$$
$$\Pr\{0 \text{ births in } [t, t + \tau)\} = 1 - nb\tau + o(\tau).$$

Figure 5.7 shows a stochastic simulation of this process.

To translate this stochastic process into a differential equation, we require a *master equation* that relates probabilities at different time steps,

$$p_n(t + \tau) = p_{n-1}(t) \cdot \Pr\{1 \text{ birth in } [t, t + \tau)\} + p_n(t) \cdot \Pr\{0 \text{ births in } [t, t + \tau)\}$$
$$= p_{n-1}(t)(n - 1)b\tau + p_n(t)(1 - bn\tau) + o(\tau).$$

After rearranging, we can write

$$\frac{p_n(t + \tau) - p_n(t)}{\tau} = b\{(n - 1)p_{n-1}(t) - np_n(t)\}.$$

As $\tau \to 0$, we obtain an *infinite* system of ODEs,

$$\frac{d}{dt}p_n(t) = b\{(n-1)p_{n-1}(t) - np_n(t)\}, \quad n = n_0, n_0+1, n_0+2, n_0+3, \ldots, \qquad p_{n_0-1} = 0,$$
$$(5.52)$$

with initial data, describing $n_0$ individuals present at time $t = 0$, as follows:

$$p_n(0) = \begin{cases} 1 & \text{if } n = n_0, \\ 0 & \text{otherwise.} \end{cases} \tag{5.53}$$

It is possible to solve this system exactly (see [102] or Exercise 5.8.9). However, we will focus on calculating the mean $M_1$ and variance $\sigma^2 = M_2 - M_1^2$ of $N(t)$ (see (5.21) and (5.23)), by first deriving differential equations for $M_1$ and $M_2 - M_1$:

$$\begin{aligned} \frac{dM_1}{dt} &= \sum_{n=1}^{\infty} n \dot{p}_n = \sum_{n=1}^{\infty} bn\{(n-1)p_{n-1} - np_n\} \\ &= b \sum_{n=0}^{\infty} \{(n+1)np_n - n^2 p_n\} \\ &= b \sum_{n=1}^{\infty} np_n \\ &= bM_1. \end{aligned} \tag{5.54}$$

Together with the initial condition $M_1(0) = n_0$, equation (5.54) has the solution

$$M_1(t) = n_0 e^{bt},$$

with the same solution as given by the linear deterministic model (5.50).

To find a differential equation for the variance $\sigma^2$ we consider first $M_2 - M_1 = E(X(X-1))$ (5.25) and obtain

$$\begin{aligned} \frac{d(M_2 - M_1)}{dt} &= \frac{d}{dt} \sum_{n=1}^{\infty} n(n-1)p_n = b \sum_{n=1}^{\infty} n(n-1)\{(n-1)p_{n-1} - np_n\} \\ &= b \sum_{n=1}^{\infty} \{(n+1)n^2 p_n - (n-1)n^2 p_n\} \\ &= 2b \sum_{n=1}^{\infty} n^2 p_n. \end{aligned}$$

Hence $\frac{d}{dt}\{M_2 - M_1\} = 2bM_2$. Using (5.54), this can be rewritten as

$$\frac{dM_2}{dt} = 2bM_2 + bM_1. \tag{5.55}$$

Thus, the differential equation for the variance is

$$\begin{aligned} \frac{d\sigma^2}{dt} &= \frac{d}{dt}(M_2 - M_1^2) = \frac{dM_2}{dt} - 2M_1 \frac{dM_1}{dt} \\ &= \frac{dM_2}{dt} - 2bM_1^2 \\ &= 2bM_2 + bM_1 - 2bM_1^2 \\ &= 2b\sigma^2 + bM_1. \end{aligned}$$

The initial condition (5.53) yields an initial variance of zero, $\sigma^2(0) = 0$. The above equation, with $\sigma^2(0) = 0$ and $M_1(0) = n_0$, can be solved with an integrating factor to yield

$$\sigma^2(t) = n_0 e^{bt} \left( e^{bt} - 1 \right), \tag{5.56}$$

which implies that the variance increases exponentially for large time $t$. For example, for the simulation shown in Figure 5.7, the values $n_0 = 10$ and $b = 1/3$ yield a final variance of $\sigma^2(9) = 3833$ and standard deviation of $\sigma(9) = 61.9$ by the final time $t = 9$.

In the pure birth process, the probability of the population going extinct is zero, because there is no death included in the model. To include death, we consider a simple birth and death process in the next section.

## 5.6.2   Birth and Death Process

In the previous section, we neglected death of individuals, but could derive a simple system that can be solved explicitly (see Exercise 5.8.9) and whose mean and variance can be calculated in a straightforward way. Death cannot be ignored in realistic biological models. When it is included, the model becomes only slightly more complex, but the analysis of the model becomes considerably more challenging.

When we extend the analysis from the previous section to a population of individuals that give birth at a rate $b$ and die at a rate $d$, the transitions for $n$ individuals in a time step of length $\tau$ become

$$\Pr\{1 \text{ birth in } [t, t + \tau)\} = nb\, \tau + o(\tau),$$
$$\Pr\{1 \text{ death in } [t, t + \tau)\} = nd\, \tau + o(\tau), \tag{5.57}$$
$$\Pr\{\text{no change in } [t, t + \tau)\} = 1 - n(b + d)\tau + o(\tau),$$

and the probability of having more than one birth or death in the time step is $o(\tau)$. Figure 5.8 shows these transitions diagrammatically, and Figure 5.9 shows a stochastic simulation of the process.

The master equation for this process is

$$p_n(t + \tau) = (n - 1)\, b\tau\, p_{n-1}(t)$$
$$+ (n + 1)\, d\tau\, p_{n+1}(t) + (1 - n\tau\, (b + d))\, p_n(t) + o(\tau),$$

and the corresponding differential equation is

$$\frac{dp_n}{dt} = (n - 1)\, b\, p_{n-1} + (n + 1)\, d\, p_{n+1} - (b + d)\, n\, p_n, \quad n = 0, 1, 2, \ldots, \quad p_{-1} = 0, \tag{5.58}$$

with initial condition

$$p_n(0) = \begin{cases} 1 & \text{if } n = n_0, \\ 0 & \text{otherwise.} \end{cases}$$

For the simple birth process, $p_n(t)$ depends only upon $p_n(t)$ and the preceding $p_{n-1}(t)$. In the birth-death process, $p_n(t)$ depends not only on $p_n(t)$ and $p_{n-1}(t)$, but also on the as yet unknown $p_{n+1}(t)$.

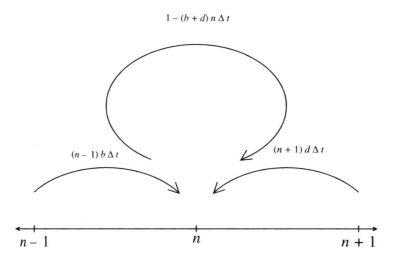

**Figure 5.8.** *The linear birth-death process can be described by transitions between the natural numbers. Here, in a small time step $\tau$, a population of size n can arise because of single birth in a population of size $n-1$, because of single death in a population of size $n+1$, or because of no change in the population size. These transitions are shown in (5.57).*

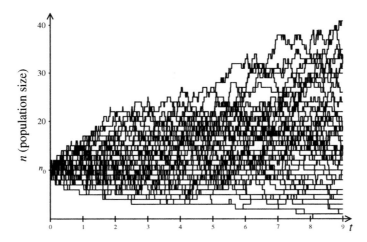

**Figure 5.9.** *Stochastic simulation of the birth and death process given in (5.57). Parameters are $b = 1/3$, $d = 0.3$, and $n_0 = 10$. Fifty different trajectories are given. They differ only in the seed for the random number generator used.*

To get around this problem, we solve for all $p_n$'s at once using the probability generating function,

$$g(s, t) = \sum_{n=0}^{\infty} s^n p_n(t), \qquad\qquad g(s, 0) = s^{n_0}.$$

Here, the original definition of the generating function (5.28) has been extended to allow for time-dependency so that the function $g$ depends upon $t$ as well as $s$. The probability of extinction at time $t$ can be calculated directly from $g$:

$$p_0(t) = \Pr\{\text{extinction by time } t\} = g(0, t). \tag{5.59}$$

The formulae for probabilities, mean, and variance follow from (5.30), (5.31), and (5.33):

$$p_n(t) = \frac{1}{n!} \left. \frac{\partial^n g}{\partial s^n} \right|_{s=0}, \tag{5.60}$$

$$g_s(1, t) = \sum_{n=0}^{\infty} n \, p_n(t) = E(N(t)) = M_1(t), \tag{5.61}$$

$$g_{ss}(1, t) = \sum_{n=0}^{\infty} n(n - 1) \, p_n(t) = M_2(t) - M_1(t), \tag{5.62}$$

$$\operatorname{var}(N(t)) = M_2(t) - M_1(t)^2$$
$$= \left[ \, g_{ss} + g_s - g_s^2 \, \right]_{s=1}. \tag{5.63}$$

To calculate $g(s, t)$, we first derive a PDE satisfied by $g$, as follows:

$$\frac{\partial g}{\partial t} = \sum_{n=0}^{\infty} s^n \, \dot{p}_n(t)$$

$$= b \sum_{n=0}^{\infty} (n - 1) \, s^n \, p_{n-1}(t) + d \sum_{n=0}^{\infty} (n + 1) \, s^n \, p_{n+1}(t) - (b + d) \sum_{n=0}^{\infty} n \, s^n \, p_n(t)$$

$$= b \sum_{k=1}^{\infty} k \, s^{k+1} \, p_k(t) + d \sum_{k=1}^{\infty} k \, s^{k-1} \, p_k(t) - (b + d) \sum_{k=1}^{\infty} k \, s^k \, p_k(t)$$

(where $k = n - 1$, $k = n + 1$, and $k = n$, respectively, for the three sums)

$$= (bs - d)(s - 1) \sum_{k=1}^{\infty} k \, s^{k-1} \, p_k(t) = (bs - d)(s - 1) \frac{\partial g}{\partial s}.$$

Thus, the generating function satisfies a first-order PDE,

$$\frac{\partial g}{\partial t} - (bs - d)(s - 1) \frac{\partial g}{\partial s} = 0, \qquad g(s, 0) = s^{n_0}. \tag{5.64}$$

The solution to this equation can be found using the *method of characteristics*. While this is a very useful method for solving first-order PDEs, it is beyond the scope of this book. Details of the method applied to this equation are given in [102]. The interested reader is encouraged to look up the solution method from [102]. Alternatively, the solution,

$$g = \left[ \frac{d - c(s) \, e^{-t(b-d)}}{b - c(s) \, e^{-t(b-d)}} \right]^{n_0}, \qquad c(s) = \frac{bs - d}{s - 1}, \qquad b \neq d,$$

can be verified by substituting directly into (5.64). The case $b = d$ is covered in detail in [102].

Substitution of the solution $g(s, t)$ into the formula for the mean yields

$$E(N(t)) = g_s(1) = n_0 \, e^{(b-d)t} = n_0 \, e^{rt}, \qquad r = b - d,$$

as predicted by the deterministic model. The variance is

$$\mathrm{var}(N(t)) = \left[\, g_{ss} + g_s - g_s^2 \,\right]_{s=1} = \frac{n_0 \, (b+d)}{r} \, e^{rt} \, (e^{rt} - 1).$$

Growth or decay of the variance over time depends upon the sign of $r = b - d$: the variance grows if $b > d$ and decays if $b < d$. The variance also scales with the birth and death rates. It increases with $b$ and $d$, even if $r$ is held constant.

The probability of extinction is

$$p_0(t) = g(0, t) = \left[ \frac{d(1 - e^{-rt})}{b - de^{-rt}} \right]^{n_0}.$$

For example, for the simulation shown in Figure 5.9, values of $b = 1/3$, $d = 3/10$, and $n_0 = 10$ give $p_0(9) = 0.028$. Thus, approximately 1 out of the 50 simulations should have gone extinct by time 9. If we are interested in whether the population eventually goes extinct, we must consider two cases, namely, $b > d$ ($r > 0$) and $b < d$ ($r < 0$). If $b > d$ ($r > 0$), then

$$\lim_{t \to \infty} p_0(t) = \left[ \frac{d}{b} \right]^{n_0},$$

which implies that the probability of extinction is greater than zero, even though the birth rate is greater than the death rate. For example, for the simulation shown in Figure 5.9, the probability of eventual extinction is $(d/b)^{n_0} = 0.34$, so approximately 17 out of the 50 simulations in Figure 5.9 should eventually go extinct. On the other hand, if $b < d$ ($r < 0$), then

$$\lim_{t \to \infty} p_0(t) = 1,$$

which implies certain extinction.

## 5.7  Nonlinear Birth-Death Process

In this section, we develop techniques for dealing with a nonlinear birth-death process. We already know *linear* birth-death processes, where individuals act independently of each other. For many systems in biology, this hypothesis is not completely appropriate: a growing population eventually reaches the limits of the carrying capacity of the ecosystem (individuals start to compete for resources); an infection cannot grow exponentially (eventually all susceptibles are infected). If the correlations between individuals are strong, then a linear model is no longer appropriate and the dependence between the individuals must be taken into account.

In the following two sections, we model the common cold in households as a nonlinear birth-death process. Development of the model is given in Section 5.7.1, and analysis of the model as an embedded discrete-time Markov process is given in Section 5.7.2.

**Table 5.3.** *Data for the final size distribution (taken from* [15]*).* $F_i$ *denotes the number of households of size* $N = 5$ *with* $i$ *infected members at the end of the epidemic.*

| Total number of infected members | Number of households |
|---|---|
| 1 | $F_1 = 112$ |
| 2 | $F_2 = 35$ |
| 3 | $F_3 = 17$ |
| 4 | $F_4 = 11$ |
| 5 | $F_5 = 6$ |

### 5.7.1   A Model for the Common Cold in Households

As an example, we consider the common cold in households (see [10, 15]). Consider households with $N$ members and assume that one member catches a cold. At least in principle, the disease can spread to other members of the household. Data collected from 181 families were reported in [83] (see Table 5.3). Our aim is to use a model to describe the number of infected individuals in households, given in Table 5.3. Since our "population dynamics" involves nonlinear interactions between infectives and susceptibles, a linear birth-death process ("birth" means infection and "death" means recovery) will not suffice. We have to look into the mechanisms of the spread in more detail.

**State of a household:** At each point of time, we find a certain number of susceptibles, infected, and recovered (see Section 3.3.3). We may characterize the state of a household by $(i, r)$, where $i$ and $r$ denote the number of infected and recovered members, respectively (see Figure 5.10). The number of susceptibles is $N - i - r$.

**Dynamics:** Next, we specify the rates by which the system changes state. As in the classical SIR model, we assume that the recovery rate per individual is a constant, $\alpha$. At a transition from state $(i, r)$ to state $(i - 1, r + 1)$, one of the $i$ infected individuals recovers. Hence, the rate of recovery in a household is $i\alpha$. Applying the Law of Mass Action, the rate at which one susceptible individual will become infected is proportional to the number of infected individuals. Accordingly, the rate for the transition from $(i, r)$ to $(i + 1, r)$ is $\beta i(N - i - r)$.

**Model equations:** Let $p_{i,r}(t)$ be the probability of finding the system in state $(i, r)$ at time $t$. The master equation for $p_{i,r}(t)$ is then

$$\frac{d}{dt}p_{i,r}(t) = -(i\alpha + \beta i(N - i - r))p_{i,r}(t)$$
$$+(i + 1)\alpha p_{i+1,r-1}(t) + (i - 1)(N - i - r)\beta p_{i-1,r}(t), \tag{5.65}$$

with $i = 0, \ldots, N$ and $r = 0, \ldots, N - i$. By convention, if an index exceeds $N$ or is below 0, formally the corresponding probability is taken to be zero. We start with one infected and $N - 1$ susceptible persons, i.e.,

$$p_{1,0}(0) = 1, \qquad p_{i,r}(0) = 0 \quad \text{for} \quad (i, r) \neq (1, 0).$$

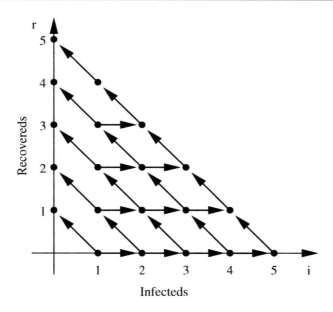

**Figure 5.10.** *Possible states and transitions for a household of size $N = 5$. We start with one infected person, that is, in state $(1, 0)$.*

The data in Table 5.3 show the total number of infected individuals during the epidemic which we refer to as the *final size of the epidemic*. A good test of our model is to see how well it reproduces the data shown in Table 5.3. In the next section, we show how to calculate the distribution of final sizes from (5.65).

### 5.7.2 Embedded Time-Discrete Markov Process and Final Size Distribution

In this section, we consider probabilities associated with the system being state $(i, r)$ under the condition that the epidemic has come to an end. In other words, $\lim_{t\to\infty} p_{i,r}(t)$. As time tends to infinity, all infectives will have moved into the recovered class, so no infected individuals are present anymore and the total mass of the probability $p_{i,r}$ is contained in the states $(0, r), r = 1, \ldots, 5$. This means $p_{i,r} = 0$ if $i \neq 0$.

One way to obtain the distribution for $t \to \infty$ is to simulate the system of ODEs (5.65) with a computer program over a long time interval. Faster, and more elegant, is not to use the differential equations directly. Instead of using time, we count the *events*. One event is either "infection" or "recovery" of an individual. An event is a transition from one state $(i, r)$ to a different state $(i', r')$. We can use the number of events as a new "time" variable. Then we obtain a discrete-time dynamical system. This system is called the *embedded discrete-time Markov process* or the *embedded Markov chain*. After a finite number of iterations, we will end up with a probability distribution that has no mass at all in states with infected individuals.

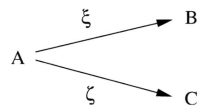

**Figure 5.11.** *We go from state A to state B with rate $\xi$, and from state A to state C with rate $\zeta$. In which state will we end up?*

To develop this *embedded discrete-time Markov* model, we need the transition *probabilities* from one state $(i, r)$ into another $(i', r')$ (note that these are *probabilities*, not *rates*). Assume that we are in state $(i, r)$. Which event will take place first? Recovery or infection?

We formulate this problem in a slightly more abstract way (see Figure 5.11). Assume we are in state $A$, going to state $B$ with rate $\xi$ and to state $C$ with rate $\zeta$. What is the probability of moving from state $A$ to state $B$ ($\Pr(A \to B)$) rather than moving from state $A$ to state $C$ ($\Pr(A \to C)$)? Let $p_A(t)$, $p_B(t)$, and $p_C(t)$ be the probabilities of being in state $A$, $B$, and $C$ at time $t$, given that we are in state $A$ at time $t = 0$. Then,

$$\frac{d}{dt} p_A(t) = -(\xi + \zeta) p_A(t), \qquad \frac{d}{dt} p_B(t) = \xi \, p_A(t), \qquad \frac{d}{dt} p_A(t) = \zeta \, p_C(t),$$
$$p_A(0) = 1, \quad p_B(0) = 0, \quad p_C(0) = 0.$$

The solution to this system of linear equations is

$$p_A(t) = e^{-(\xi + \zeta)t}, \quad p_B(t) = \frac{\xi}{\xi + \zeta} \left( 1 - e^{-(\xi + \zeta)t} \right), \quad p_C(t) = \frac{\zeta}{\xi + \zeta} \left( 1 - e^{-(\xi + \zeta)t} \right).$$

Hence,

$$\Pr(A \to B) = \lim_{t \to \infty} p_B(t) = \frac{\xi}{\xi + \zeta}, \qquad \Pr(A \to C) = \lim_{t \to \infty} p_C(t) = \frac{\zeta}{\xi + \zeta}.$$

We observe that the transition probability from one state to another is the rate of this transition to occur ($\xi$ or $\zeta$) divided by the sum of all rates of all possible transitions leaving the first state ($\xi + \zeta$).

With this rule in mind, we come back to our problem and define a discrete Markov chain. Let $P_{(i,r),(i',r')} = \Pr((i, r) \to (i', r'))$ be the transition probability from state $(i, r)$ to state $(i', r')$. As in our example with states $A$, $B$, and $C$, here we also have two possible paths by which to leave a state $(i, r)$: recovery and infection. Then,

$$
P_{(i,r),(i+1,r)} = \frac{\beta \, i (N - i - r)}{\beta \, i (N - i - r) + i\alpha} = \frac{(\beta/\alpha) \, i (N - i - r)}{(\beta/\alpha) \, i (N - i - r) + i}
$$
$$
= \frac{R_0 \, (N - i - r)}{R_0 \, (N - i - r) + 1} \tag{5.66}
$$

for $i = 1, \ldots, N$ and $r = 0, \ldots, N - i - 1$, and where $R_0 = \beta/\alpha$. Similarly,

$$P_{(i,r),(i-1,r+1)} = \frac{\alpha i}{\beta i (N - i - r) + i\alpha} = \frac{i}{(\beta/\alpha) i (N - i - r) + i}$$

$$= \frac{1}{R_0 (N - i - r) + 1} \tag{5.67}$$

for $i = 1, \ldots, N$ and $r = 0, \ldots, N - i$. Note that if we are in one of the states $(0, r)$, then we will not leave them. These states are called *absorbing states*. Hence,

$$P_{(0,r),(0,r)} = 1 \qquad\qquad \text{for } r = 0, \ldots, N, \tag{5.68}$$

and $P_{(i,r),(i',r')} = 0$ for all index combinations not mentioned so far. There is only one parameter, $R_0$, in these equations, although our original model includes two parameters, $\alpha$ and $\beta$. The reduction of the number of parameters is possible since the final size of the epidemic (the total number of individuals infected during the epidemic) is the only information we need, and hence we can ignore the time course or the disease.

Finally we use the transition probabilities $P_{(i,r),(i',r')}$ to find the probability $q_{i,r}(n)$ that the system is in state $(i, r)$ after $n$ events,

$$q_{i,r}(n) = \sum_{(k,l)} P_{(k,l),(i,r)} q_{k,l}(n - 1) \tag{5.69}$$

for $i, r = 1, \ldots, N$. In (5.69), we sum over all possible states $(k, l)$. The system of equations (5.66)–(5.68) forms a discrete-time model for our household. In Section 8.6, we will implement this model and compare the model output with the data.

## 5.8 Exercises for Stochastic Models

**Exercise 5.8.1: Forest ecosystem succession.** *Assume that one starts with a forest comprised of beech. Work out $\mathbf{u}_1$, $\mathbf{u}_2$, and $\mathbf{u}_3$ for (5.2), with the entries of P as described in Table 5.2. How many generations does it take before the maximum difference in any component of the forest is within 5% of (5.3)?*

**Exercise 5.8.2: Princeton forest.** *Use Table 5.2 and show that the eigenvector $u^*$ of the transition matrix P with eigenvalue 1 is given by formula (5.3).*

**Exercise 5.8.3: Mean and variance for a sum of random variables.** *Show that if $X_i$ are random variables and $Y = \sum X_i$, then $E(Y) = \sum E(X_i)$. Show that if the $X_i$'s are independent $(E(X_i X_j) = E(X_i)E(X_j))$, then $\mathrm{var}(Y) = \sum \mathrm{var}(X_i)$.*

**Exercise 5.8.4: Mean and variance for a negative binomial distribution.** *Derive the following formulae for the mean and variance of the negative binomial distribution by first calculating $E(X)$ and $E(X(X + 1))$: $M_1 = k/p$ and $\sigma^2 = k(1/p - 1)/p$.*

**Exercise 5.8.5: Random walk derivation of a diffusion-advection equation.** *Describe, by means of a master equation, a random walk with movement to nearest neighbors on a two-dimensional lattice with spacing $\lambda$ and time steps of size $\tau$. Here there is the probability of* *moving to the left L, right R, up U or down V. Assume that $L = 0.25 - \lambda\gamma_1$, $R = 0.25 + \lambda\gamma_1$, $U = 0.25 + \lambda\gamma_2$, and $V = 0.25 - \lambda\gamma_2$ where $\lambda$ denotes the spatial grid size. Derive a diffusion-advection equation*

$$\frac{\partial p}{\partial t} + \mathbf{c} \cdot \nabla p = D\Delta p$$

*by taking the appropriate diffusion limit of the random walk. Here $\mathbf{c}$ is an advection vector which you should describe in terms of $\gamma_1$ and $\gamma_2$.*

**Exercise 5.8.6: Spatially varying diffusion model.** *We study two master equations as models for diffusion in a spatially varying environment:*

$$p(x, t + \tau) = \alpha(x - \lambda)p(x - \lambda, t) + N(x)p(x, t) + \alpha(x + \lambda)p(x + \lambda, t), \qquad (5.70)$$

*with $N(x) + 2\alpha(x) = 1$, and*

$$p(x, t + \tau) = \alpha(x - \lambda/2)p(x - \lambda, t) + N(x)p(x, t) + \alpha(x + \lambda/2)p(x + \lambda, t), \quad (5.71)$$

*with $N(x) + \alpha(x - \lambda/2) + \alpha(x + \lambda/2) = 1$.*

(a) *For each of the models (5.70) and (5.71), give an interpretation of all terms. In particular, what is the meaning of $\alpha$ and $N$?*

(b) *For each of the above models, use Taylor series expansion and the diffusion limit to derive an equation for $p(x, t)$. For (5.70) the diffusion limit has the form*

$$\frac{\partial p}{\partial t} = \frac{\partial^2}{\partial x^2}\{A(x)p\},$$

*where $A(x)$ is a function you should determine. The second problem (5.71) leads to*

$$\frac{\partial p}{\partial t} = \frac{\partial}{\partial x}\left\{B(x)\frac{\partial p}{\partial x}\right\},$$

*where $B(x)$ is a function you should determine.*

(c) *Compare and contrast the assumptions for the two models and the resulting equations for $p(x, t)$. Show that the difference between the two models can be expressed as an advection term that appears in one model but not the other.*

*For further reading on nonhomogeneous spatial models see [9].*

**Exercise 5.8.7: Variance for a branching process.** *Show that the variance for the branching process is as given in (5.46). Hint: First derive the recursion relation*

$$h''_{t+1}(1) = R_0 h''_t(1) + R_0^{2t} g''(1),$$

*and then use the substitution $h''_t(1) = R_0 u_t$ to calculate $h''_t(1)$ explicitly. Finally, use the fact that $h''(1) = \sigma^2 + R_0(R_0 - 1)$ to derive the required result.*

**Exercise 5.8.8: The survival of right whales.** *A female right whale may produce* 0, 1, *or* 2 *females the following year. A female at time t produces* 0 *offspring if it dies before* $t + 1$, 1 *offspring (itself) if it survives without reproducing, and* 2 *offspring (itself and its calf) if it survives and reproduces. Let p be the survival probability and let m be the probability of producing a female calf. Thus* $p_0 = 1 - p$, $p_1 = p(1-m)$, *and* $p_2 = pm$. *In* 1980, *survival was estimated at* $p = 0.99$ *and the probability of producing a female calf at* $m = 0.063$. *By* 1999, *these parameters had dropped to* $p = 0.94$ *and* $m = 0.038$. *Determine the population growth rate and the extinction probability for* 1980 *and* 1999. *(Exercise taken from* [102] *and based on the work of* [38].)

**Exercise 5.8.9: An explicit solution for the pure birth process.** *The infinite system of differential equations* (5.52) *can be solved, starting with* $p_{n_0}$ *and proceeding to* $p_{n_0+1}$, *and so on:*

$$\frac{dp_{n_0}}{dt} = -bn_0 p_{n_0}, \qquad p_{n_0}(0) = 1,$$

$$\frac{dp_{n_0+1}}{dt} = -b(n_0 + 1)p_{n_0+1} + bn_0 p_{n_0}$$

$$= -b(n_0 + 1)p_{n_0+1} + bn_0 e^{-bn_0 t}, \qquad p_{n_0+1}(0) = 0.$$

*Use this approach to show that* $p_n(t)$ *is given by the negative binomial distribution, in which the chance of success in a single trial,* $\exp(-bt)$, *decreases exponentially in time,*

$$p_n(t) = \binom{n-1}{n_0 - 1} e^{-bn_0 t} \left(1 - e^{-bn_0 t}\right)^{n-n_0} \tag{5.72}$$

*for each* $n \geq n_0$. *Calculate the mean and variance of the pure birth process from the mean and variance of the negative binomial distribution (see Exercise 5.8.3).*

   *Hint: Use the integrating factor method (or variation of constants method) to show the base case:*

$$p_{n_0+1}(t) = e^{-bn_0 t} \left(1 - e^{-bt}\right).$$

*Then use induction. Suppose* $p_n$ *is given by* (5.72) *and show that*

$$p_{n+1}(t) = \binom{n}{n_0 - 1} e^{-bn_0 t} \left(1 - e^{-bt}\right)^{n+1-n_0}$$

*satisfies*

$$\frac{d}{dt} p_{n+1} = b\left(np_n - (n+1)p_{n+1}\right).$$

# Chapter 6

# Cellular Automata and Related Models

In Chapters 2 and 3, we saw models that describe the state of a system in time (e.g., the number of individuals as a function of time). However, in some applications, space is important as well. In Chapter 4, we saw how PDEs can be used to describe systems that depend on both time and space. In this chapter, we consider another way to include spatial information in a mathematical model, namely, with cellular automata.

## 6.1 Introduction to Cellular Automata

Cellular automata are models where all variables, both independent and dependent, take on discrete values. Their main characteristic is *locality*; that is, individuals or particles are only affected by their nearest neighbors. Cellular automata are therefore natural models for infectious diseases, forest fires, or excitable media where the contact between two individuals is typically of local nature. The dynamics of a cellular automaton is defined by *local* rules. In the case of infectious diseases, such a rule could be, "If at least one of my neighbors is infected, I will become infected myself." In a cellular automaton model, space is represented by discrete points in space, called "cells." The cells can be seen either as the locations where individuals live or, sometimes, as the individuals themselves. They can be interpreted as biological cells, as impoundments on a river with separated fish populations, as territories of birds, and so on.

Cellular automata and related models are interesting to biologists for several reasons. First, many structures in biology are discrete, and a natural model to describe such structures would be a discrete one (for example, DNA sequence data are discrete). Second, rules for birth, death, or migration can be specified in a straightforward manner. Third, the time courses and patterns of cellular automata can be interpreted directly in biological terms.

One of the first cellular automaton models was developed around 1952, by John von Neumann [32], in the context of self-replication. At that time, the question of how a complex system can create another system of the same complexity (e.g., a copy of itself) was discussed. The question seems trivial today, since we're familiar with the concept of DNA replication. But at that time, the idea was revolutionary. Von Neumann investigated his model by pencil and paper only. This is quite remarkable, since his model had 29 possible

155

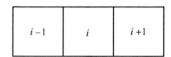

**Figure 6.1.** *One-dimensional neighborhood.*

states of the cells and, in a proof of self-replication, he outlined an initial configuration of about 200,000 cells.

The first tool to investigate cellular automata generally is to create a simulation. Until reasonably powerful computers were available, only a few results on cellular automata were published [166]. Broad exploration of cellular automata started with the availability of computers and the advent of user-friendly simulation tools.

The study of deterministic discrete systems is now a very active field. The theory is still developing and less established than the theories for continuous or stochastic systems. The behavior of discrete spatial systems is very rich and cannot be summarized with general results. From a theoretical point of view, cellular automata are very general structures. For example, there are cellular automata capable of universal computation, analogous to Turing machines [20, 62].

Before we give some specific examples of cellular automata, it helps to properly define a cellular automaton.

**Definition 6.1.** *A cellular automaton is a tuple $A = (G, E, U, f)$ of a grid $G$ of cells, a set of elementary states $E$, a set defining the neighborhood $U$, and a local rule $f$.*

In classical cellular automata, we have $G = \mathbb{Z}^d$, the $d$-dimensional square grid. Instead of grid points, we draw cells which can be colored to indicate the state of the cells. We will extend this definition later to more general graphs. In an infectious disease model, "white" could indicate a susceptible and "black" an infectious individual. The possible states of each cell is an element of the set of *elementary states* $E = \{\text{white, black}\}$. Likewise, we could use $E = \{0, 1\}$. In one dimension ($d = 1$), a cell is often influenced by adjacent cells only. If we denote the cell at position $i$ with $x_i$, then an example of a simple neighborhood would be $U(x_i) = \{x_{i-1}, x_i, x_{i+1}\}$ as shown in Figure 6.1. In two dimensions ($d = 2$), the most common neighborhoods are the *von Neumann* and the *Moore neighborhoods*. Let $x_{i,j}$ be the position of a cell. The von Neumann neighborhood (Figure 6.2, left) is described by $U(x_i) = \{x \mid \|x_i - x\|_1 \leq 1\}$. The Moore neighborhood (Figure 6.2, right) is described by $U(x_i) = \{x \mid \|x_i - x\|_\infty \leq 1\}$. Note that the use of a von Neumann or Moore neighborhood is not consistent throughout the literature. Sometimes the center cell $x_{i,j}$ is included in the neighborhood; sometimes only the surrounding cells are taken.

So far, we have a grid of cells with several possible elementary states and a neighborhood. The collection of all states of the cells of the grid is called the *state of the grid*. Mathematically, we define the *state of the grid* as a map $z : G \mapsto E$, which gives each grid cell an elementary state.

The *local rule*, $f$, describes what happens from one time step to the next. Let $z$ be the state of the grid. Then $z(x)$ is the state of cell $x$ and $z|_{U(x)}$ is the occupation of the

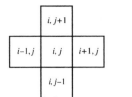

**Figure 6.2.** *Illustration of the von Neumann neighborhood (left) and the Moore neighborhood (right).*

neighborhood of cell $x$. The new state of cell $x$ depends only on the states of its neighbor cells,

$$z(x)^{t+1} = f(z^t|_{U(x)}),$$

where the superscripts $t$ and $t + 1$ denote the time steps. The local rule is applied to all cells in parallel, that is, *synchronously*, and is usually implemented in mapping two arrays alternating on each other. One of the arrays holds the current state at time $t$, while the other array holds the updated state at time $t + 1$.

### 6.1.1  Wolfram's Classification

Let us look at a class of cellular automata with $A = (\mathbb{Z}, \{0, 1\}, \{x_{i-1}, x_i, x_{i+1}\}, f)$, that is, the one-dimensional nearest-neighbor automata with states zero and one. There are $2^3 = 8$ possible occupations of the neighborhood. The local function can map to each possible neighborhood configuration to zero or one, which gives us $2^8 = 256$ different local functions, that is, 256 different automata. If we look at their dynamics, we will find quite different patterns. Wolfram [164] heuristically proposed four qualitative classes to characterize cellular automata:

**Wolfram class I:** From any initial configuration we get a fixed point; that is, the pattern becomes constant.

**Wolfram class II:** Simple stationary or periodic structures evolve. Small changes in the initial configuration affect only a finite number of cells.

**Wolfram class III:** "Chaotic" patterns emerge. Changes in the initial configuration affect a number of cells linearly growing in time.

**Wolfram class IV:** Complex localized patterns evolve with long-distance correlations. The effect of changes in the initial configuration cannot be predicted.

To identify each of the 256 local functions, Wolfram has constructed a simple system, known as Wolfram's enumeration: First, we order the eight possible neighborhood occupations $U_0 = 000, U_1 = 001, U_2 = 010, \ldots, U_7 = 111$, enumerated like three-digit binary numbers from 0 to 7. For a given local rule $f$, let $c_i = f(U_i)$ be the next state of cell $x$

**Table 6.1.** *Examples of Wolfram's enumeration.*

| $i$ | 0 | 1 | 2 | 3 | 4 | 5 | 6 | 7 | |
|---|---|---|---|---|---|---|---|---|---|
| $U_i$ | 000 | 001 | 010 | 011 | 100 | 101 | 110 | 111 | |
| $2^i$ | 1 | 2 | 4 | 8 | 16 | 32 | 64 | 128 | |
| Rule 254, $c_i$ | 0 | 1 | 1 | 1 | 1 | 1 | 1 | 1 | $\sum c_i 2^i = 254$ |
| Rule 50, $c_i$ | 0 | 1 | 0 | 0 | 1 | 1 | 0 | 0 | $\sum c_i 2^i = 50$ |
| Rule 90, $c_i$ | 0 | 1 | 0 | 1 | 1 | 0 | 1 | 0 | $\sum c_i 2^i = 90$ |

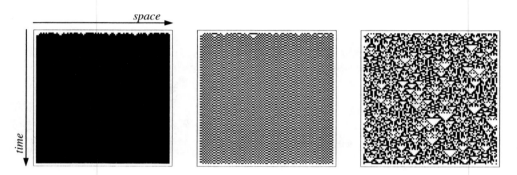

**Figure 6.3.** *Space-time patterns of the one-dimensional nearest-neighbor automata that correspond to the rules 254, 50, and 90 (from left to right). The first row in each square shows the initial configuration, which is the same for all three simulations. The following rows show successive iterates. White indicates $z(x) = 1$ and black indicates $z(x) = 0$. The boundary cells are kept constant 0 (white).*

with neighborhood $U_i$, for $i = 1, \ldots, 7$. Then the number of this local rule is $\sum_{i=0}^{7}(c_i 2^i)$, as shown in Table 6.1. Let us discuss the rules with the numbers 254, 50, and 90.

**Rule 254:** Consider a simple infection automaton where a cell becomes infected if one of its neighbors is infected. The local function $f$ for this situation is

$$f(z|_{U(x)}) = \begin{cases} 1 & \text{if } s \geq 1, \\ 0 & \text{otherwise,} \end{cases}$$

where $s = x_{i-1} + x_i + x_{i+1}$ is the number of infectious neighbors of cell $x = x_i$. As shown in Table 6.1, the number of this rule is $2 + 4 + 8 + 16 + 32 + 64 + 128 = 254$. The dynamical behavior of rule 254 is quite easily understood. If we start with a random configuration of ones and zeros, we reach a state where all cells are in state one. This means that all cells become infected, as shown in Figure 6.3 (left). If we start with all cells zero, then all cells will stay zero forever. In both cases, a stationary pattern is reached. Therefore, rule 254 is an example of an automaton from Wolfram class I.

**Rule 50:** Another simple infection model is rule 50. A cell becomes infected if one of its neighbors is infected, but infectious cells recover after one time step. Thus,

$$f(z|_{U(x)}) = \begin{cases} 1 & \text{if } z(x) = 0 \text{ and } s \geq 1, \\ 0 & \text{otherwise.} \end{cases}$$

Starting from a random configuration, we get a pattern where every cell periodically changes from zero to one and back (Figure 6.3, middle). Therefore, rule 50 is an example which belongs to Wolfram class II.

**Rule 90:** For an example of Wolfram class III, consider rule 90. It maps 000, 010, 101, and 111 to zero, and 001, 011, 100, and 110 to one. The patterns evolving have been described as "fractal-like" or chaotic (Figure 6.3, right).

There are no one-dimensional nearest-neighbor automata with class IV behavior, but we will come back to this later.

## 6.1.2 The Game of Life

In this section, we discuss one of the most popular cellular automata: the *Game of Life*, proposed by Conway (see [133]). It is a two-dimensional cellular automaton, in which each cell is either dead or alive. A living cell stays alive if it has two or three living neighbors; otherwise it dies. A dead cell becomes alive if exactly three neighbors are living. Biologically, these simple rules mimic a birth process where cells may die by isolation or overcrowding. More formally, we have $A = (\mathbb{Z}^2, \{0, 1\},$ Moore neighborhood without the cell itself, $f)$ with the local function

$$f(z|_{U(x)}) = \begin{cases} 1 & \text{for } s(x) = 3, \\ 1 & \text{for } s(x) = 2 \text{ and } z(x) = 1, \\ 0 & \text{otherwise,} \end{cases}$$

where $s$ is the number of living neighbors

$$s(x) = \sum_{y \in U(x)} z(y).$$

From these simple rules, an astonishing behavior emerges. If we start with a random configuration of dead and living cells in a rectangular area of the grid, the pattern often becomes 2-periodic (but it may take some time). Groups of three living cells alternating in shape or stationary 4-blocks often appear (see Figure 6.4). However, configurations exist with a much more interesting behavior (these often have visual names like "ship" or "beehive"). Simple configurations can expand and resemble fireworks before they collapse again. The rich patterns can hardly be described here but should be viewed like a movie. Numerous simulation tools for the Game of Life are available, and readers are invited to experiment on their own. We conclude with a summary of observations that are interesting from a more theoretical point of view:

- The evolution of most initial configurations cannot be predicted. Small changes in the initial configuration may lead to totally different patterns.

- There are patterns with different, sometimes large, periods. For example, a configuration known as "queen bee" has period 30.

 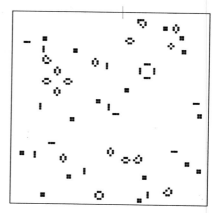

**Figure 6.4.** *Two-dimensional spatial patterns of Conway's Game of Life. Initial configuration (on the left) and after* 1300 *iterations (on the right). The boundary cells are kept constant* 0 *(white).*

- A configuration named "glider" travels diagonally across the grid. There is also the "glider gun" which generates a stream of gliders. It is an example of an unbounded pattern which expands forever, developing from an initial configuration with finite support.

- It has been shown that the Game of Life is computationally universal; that is, everything which is computable can be computed with the Game of Life [133], which indicates that this cellular automaton is very complex.

- Given the rich behavior of the Game of Life it is plausible that it is an example of Wolfram class IV.

### 6.1.3 Some Theoretical Results on Cellular Automata

After these two examples of cellular automata, the question of analytic results for cellular automata arises. Wolfram's classification is mainly phenomenological. An example of a theoretical result is the algorithm of how to construct, for every (finite) automaton $A$ with arbitrary neighborhood, another automaton $\bar{A}$ simulating it with a von Neumann neighborhood. Usually, if we simplify the neighborhood, we have to "pay" with a larger number of elementary states; that is, we usually will have $|E| \ll |\bar{E}|$. Similarly, if we construct an automaton $\bar{A}$ with only the elementary states $\{0, 1\}$, we usually have to use a much larger neighborhood $\bar{U}$ with $|\bar{U}| \gg |U|$. This sounds useful in deriving the dynamics of a new automaton to find an analogous known automaton. However, an equivalent automaton with a basic neighborhood may have a very complicated, nonintuitive local function that is not well investigated.

As we mentioned, cellular automata are a developing field. Massive simulations are used to characterize automata models by tools from statistical physics, for example. Theoretical results for numerous classes of automata have also been obtained by using a wide variety of methods. In the following sections we discuss Greenberg–Hastings automata.

## 6.2 Greenberg–Hastings Automata

We follow [71] and investigate the Greenberg–Hastings automata, which have been formulated as models for excitable media such as the cardiac muscle, cultures of *Dictyostelium discoideum*, forest fires, or infectious diseases. For an infectious disease, each cell of the automaton represents an individual or the territory an individual lives in. Individuals are classified by their epidemiological state. They are either susceptible, infectious, or immune. An important assumption is that infection can only happen by direct contact between individuals, and only neighboring cells have contact. An infectious disease can then be modeled by the following rules:

- If a susceptible cell has at least one infectious neighbor, it becomes infectious itself; otherwise it stays susceptible.

- An infected cell stays infectious for $a > 0$ time steps; then it becomes immune.

- An immune cell stays immune for $g > 0$ time steps and then becomes susceptible again.

These rules can be used to model infection with different infectious or recovery periods. Let the set of elementary states be $E = \{0, 1, \ldots, a, a+1, \ldots, a+g\} = \{0, \ldots, e\}$. Then an infection is described by the transitions

$$\underbrace{0}_{susceptible} \longrightarrow \underbrace{1 \longrightarrow \cdots \longrightarrow a}_{infectious} \longrightarrow \underbrace{a+1 \longrightarrow \cdots \longrightarrow a+g}_{immune} \longrightarrow \underbrace{0}_{susceptible} \ .$$

More formally the local function of the Greenberg–Hastings automata reads

$$f(z|_{U(x)}) = \begin{cases} 1 & \text{if } z(x) = 0 \text{ and } s \geq 1, \\ z(x) + 1 & \text{if } 0 < z(x) < e, \\ 0 & \text{otherwise,} \end{cases} \tag{6.1}$$

where $s$ is the number of infectious cells in the neighborhood of $x$. Finally, we choose $G = \mathbb{Z}^2$ and the von Neumann neighborhood.

The above rule (6.1) implies that either a cell stays susceptible or it moves through the infection–recovery cycle to finally become susceptible again.

As in [71], let us concentrate on the case $a, g \geq 1$, with $a + g \geq 3$ and $1 \leq a \leq e/2$; that is, we have at least the same number of immune as infectious stages. Possible choices for $e \leq 6$ are shown in Table 6.2. Starting with a configuration with a finite number of

**Table 6.2.** *Greenberg–Hastings automata for $e \leq 6$.*

| $e$ | $a$ = infectious stages | $g$ = immune stages |
|---|---|---|
| 3 | 1 | 2 |
| 4 | 1 or 2 | 3 or 2 |
| 5 | 1 or 2 | 4 or 3 |
| 6 | 1, 2, or 3 | 5, 4, or 3 |

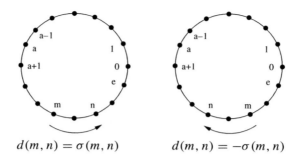

$$d(m, n) = \sigma(m, n) \qquad\qquad d(m, n) = -\sigma(m, n)$$

**Figure 6.5.** *Distance and signed distance between states.*

infected cells (finite support), there are two possibilities. Either the epidemic "dies out" (i.e., in any finite region of the grid all cells will be in state zero in finite time), or the pattern persists (i.e., there is at least one cell which becomes infected again and again). Greenberg, Greene, and Hastings [71] found a condition which allowed them to predict the fate of initial configurations. To investigate this prediction in some detail, we define the distance between states in $E$.

**Definition 6.2.**

(a) *The* distance *between two states* $m, n \in E$ *of cells is defined as*

$$d(m, n) = \min\{|m - n|, e + 1 - |m - n|\}.$$

(b) *We identify every state* $k \in E$ *with the point* $\exp(2\pi k/(e + 1))$ *on the unit circle in the complex plane. The* signed distance *between two states* $m, n \in E$ *is*

$$\sigma(m, n) = \begin{cases} d(m, n) & \text{if the arc } \overline{mn} \text{ is oriented counter-clockwise,} \\ -d(m, n) & \text{otherwise.} \end{cases}$$

In Figure 6.5, we illustrate how states can be located on the unit circle (mathematically speaking, how we calculate distances between states modulo $e + 1$).

We see that the distance is the shorter arc between two states, therefore $d(m, n) = d(n, m)$. The above rule (6.1) implies that a cell can only advance by 1 state per unit of time, which implies $0 \le d(z(x)^t, z(x)^{t+1}) \le 1$. The signed distance includes information on the orientation of the arc, so we have $\sigma(m, n) = -\sigma(n, m)$.

The following definition focuses on the topological structure of cells.

**Definition 6.3.** *A* cycle *is an ordered* $n$*-tuple* $(x_1, \ldots, x_n, x_{n+1})$ *such that* $x_1, \ldots, x_n$ *are distinguished,* $x_{n+1} = x_1$*, and* $x_{i+1}$ *is a neighbor of* $x_i$ *for* $i = 1, \ldots, n$.

Thus, a cycle is an ordered set of cells such that two successive cells are neighbors and the last and the first cells of the cycle are the same ones. To combine information on arrangement and states of the cells, we introduce the following definition.

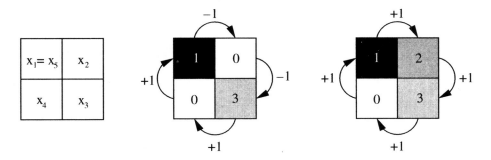

**Figure 6.6.** *Cycle with four different cells ($n = 4$). Let $a = 1, g = 2$, so that $e = 3$. Then both configurations show continuous cycles. The cycle in the middle has winding number zero; the one on the right has winding number one.*

**Definition 6.4.**

(a) *A cycle $C$ is called* continuous *at time $t$ if $d(z(x_i)^t, z(x_{i+1})^t) \leq a$ for $1 \leq i \leq n$.*

(b) *The* winding number $\mathcal{W}_t(C)$ *of a continuous cycle at time $t$ is defined as*

$$\mathcal{W}_t(C) = \frac{1}{e+1} \sum_{i=1}^{n} \sigma(z(x_i)^t, z(x_{i+1})^t).$$

Figure 6.6 gives an example of a small cycle with four different cells ($n = 4, a = 1$, $g = 2, e = 3$). In both configurations, the four cells constitute a continuous cycle, one with winding number zero and the other with winding number one. We see that from these two initial configurations, different patterns develop (Figure 6.7). To achieve persistence, we basically have to ensure that susceptible cells of a cycle will be infected again. Only a cycle with nonzero winding number ensures persistence.

Now we can formulate the theorem for persistence.

**Theorem 6.5 (Greenberg, Greene, and Hastings [71]).** *Given a Greenberg–Hastings automaton with $a + g \geq 3$ and $1 \leq a \leq e/2$ and $a, g \geq 1$, a configuration with finite support is persistent if and only if there is a time $t' \geq 0$ where we find a continuous cycle $C$ such that $\mathcal{W}_{t'}(C) \neq 0$.*

In the theorem, the "if and only if" phrase suggests that we can determine the fate of any configuration, whether it will persist or it will die out. Unfortunately, this is not true. If we find a continuous cycle with winding number unequal to zero, then the configuration will persist. But even if there is no such cycle, as in the third example in Figure 6.7, it can happen that such a cycle evolves later. Therefore, we have only a sufficient, but not a necessary, condition for persistence if only the initial configuration is checked ($t' = 0$). A proof of Theorem 6.5 can be found in [71].

To obtain persistence for finitely supported initial conditions (only a finite number of cells is nonzero), it is sufficient to check a finite number of iterations for continuous cycles

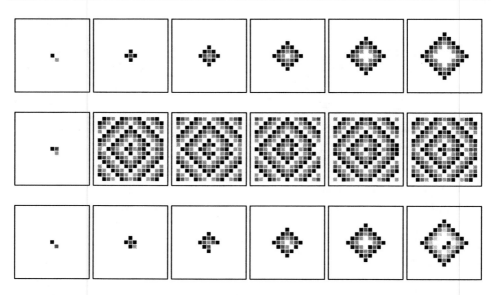

**Figure 6.7.** *Two-dimensional spatial patterns evolving from simple initial config-urations. Each row represents one simulation and shows the patterns at* 0, 13, 14, 15, 16, *and* 17 *iterations. The cycle with winding number zero from Figure 6.6 (center) is used as the initial condition for the first row. An epidemic wave travels once across the grid and then dies out. The cycle with winding number nonzero from Figure 6.6 (right) is used as initial condition of the center row. It leads to persistence of the epidemic and a periodic pattern evolves. The last row shows an example where a continuous cycle with nonzero winding number evolves and leads to persistence. For these simulations, the boundary cells were fixed at* 0.

with nonzero winding number. An upper bound for the number of iterates that need to be checked can be computed explicitly [71].

## 6.2.1   Relation to an SIR Model

It is interesting to consider the connection between Greenberg–Hastings automata and the ODE model of Kermack–McKendrick [100], namely,

$$\dot{S} = -\phi(S, I, R) + \gamma R,$$
$$\dot{I} = \phi(S, I, R) - \alpha I,$$
$$\dot{R} = \alpha I - \gamma R,$$

where we use a general *incidence function* $\phi(S, I, R)$. In the classical Kermack–McKendrick model, a mass action term $\phi(S, I, R) = \beta SI$ is chosen. Further, $\alpha$ is the rate of immuniza-tion, so $1/\alpha$ is the mean time an individual stays in the infectious compartment. Therefore, $1/\alpha$ corresponds to the number of infectious states $a$ of the Greenberg–Hastings automaton. Likewise, $1/\gamma$ corresponds to the number of recovery states $g$.

How is the incidence function related to the infection rule in the automaton? Here is the fundamental difference between the Kermack–McKenrick model and the Greenberg–Hastings automaton. The classical incidence function used in the Kermack–McKendrick model describes mass action. The individuals are well mixed, and every susceptible has the same chance to be infected by an infectious individual. Obviously this is not the case in the Greenberg–Hastings automaton. In this automaton, infectious cells can only infect neighboring susceptibles. This locality is the generator of the spatial patterns. The difference in the infection process leads to different time courses (or trajectories) of the two models. We can, however, approximate each model with the other. In particular, we can introduce a spatial mixing rule in the Greenberg–Hastings automaton, where the states of cells are exchanged at random. With increasing mixing rate, the automaton approximates the differential equation model. On the other hand, various incidence functions have been proposed for the ODE model to include the effect of local infection [84].

# 6.3 Generalized Cellular Automata

Classical cellular automata, as discussed so far, provide basic and very interesting models. However, sometimes they have to be adapted to fit a given biological process. The Greenberg–Hastings automata, for example, have been modified in the context of excitable media in many ways to get more realistic patterns. All four components of cellular automata—the grid, the elementary states, the neighborhood, and the local function—can be modified or generalized. Also, the synchronous updating method can be changed. There is an overlap between modified cellular automata and models that have been introduced in biology, physics, and chemistry under various names, such as *individual-based models* [72], *lattice gas models* [61], or *interacting particle systems* [109]. It is beyond the scope of this text to discuss which of these models may or may not be called cellular automata, but instead we introduce the most common modifications with examples.

## 6.3.1 Automata with Stochastic Rules

Returning to our rule in the Greenberg–Hastings model, namely, "a susceptible cell is infected if at least one of its neighbors is infectious," we note that infection is rare for a real infectious disease. Most diseases have a weaker level of infectivity, and the chance of becoming infected grows with the number of infected neighbors. Thus, a more realistic local rule is "the more infectious neighbors I have, the higher the probability I become infected." Since probabilities are involved, we obtain local functions with stochastic (random) components. Consider, for example,

$$f(z|_{U(x)}) = \begin{cases} 1 & \text{if } z(x) = 0 \quad \text{with probability } w(s), \\ 0 & \text{otherwise}, \end{cases}$$

where $s$ is the number of infectious neighbors. Let $p$ be the probability that an individual is infected by one of its neighbors. Then $(1 - p)^s$ is the probability that the individual is *not* infected in spite of $s$ infectious neighbors. A good choice for the probability of becoming infected by $s$ infected neighbors is $w(s) = 1 - (1 - p)^s$.

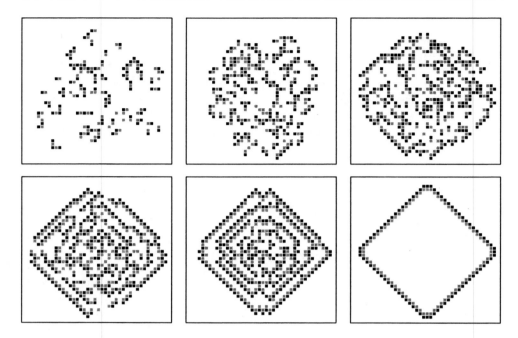

**Figure 6.8.** *Two-dimensional spatial patterns of a stochastic automaton evolving from an initial configuration of a square with nine infectious cells in the center (initial condition not shown). The infectivity parameter varies from $p = 0.5$ to $p = 1$ (in steps of 0.1 from top left to right bottom). Shown are states after 30, 23, 22, 20, 18, or 18 iteration steps, respectively.*

Let us consider a cellular automaton ($\mathbb{Z}^2$, {0, 1, 2}, von Neumann neighborhood, $f$) with 0 = susceptible, 1 = infectious, 2 = immune stage, and

$$f(z|_{U(x)}) = \begin{cases} 1 & \text{if } z(x) = 0 \quad \text{with probability } 1 - (1 - p)^s, \\ 0 & \text{otherwise.} \end{cases} \tag{6.2}$$

Figure 6.8 shows patterns evolving for different values of the parameter $p$. With $p = 1$, that is, immediate infection, one of the deterministic Greenberg–Hastings automata is recovered. With this particular initial configuration, the epidemic travels only once across the grid. The behavior for $p < 1$ is different. The epidemic is persistent, even for $p = 0.9$. It is plausible that the automata with $p \neq 1$ are more stable models; that is, the patterns produced are at least qualitatively the same when adding some stochasticity like a stochastic component in the local function, for example. This implies that models with stochastic components are the more realistic models. We see in Figure 6.8 also that the epidemic spreads slower with decreasing $p$ and the epidemic will die out fast for very small $p$. To investigate the probability that the epidemic persists up to time $t$ for a given value of $p$, we would need tools from stochastic theory, which exceed the scope of this book.

## 6.3.2   Grid Modifications

There are various possible modifications to the grid of a cellular automaton. First, when implementing automata models, often a finite grid $G \subset \mathbb{Z}^d$ is used. This is a serious change from the theoretical point of view, but in many simulations this has no significant effect. However, the "border" cells of finite grids have to be considered separately. The border cells may be set to a fixed state (resembling Dirichlet conditions in PDE models (4.20)). In this case the resting state 0 is often used. Also common is the use of a torus (periodic boundary conditions in two dimensions) and periodic boundary condition in one dimension, where the most left and right cells are identified. Reflection or flow conditions are possible but are rarely used [161].

Other variants of the grid often used are motivated by pictures such as Figure 6.7. The square structure of the grid is reflected in the patterns and results in diamond-shaped spread. In a realistic model, we expect isotropic spread giving more or less round patterns. This may be regarded more as a problem of visualization, but isotropic spread has attracted quite some attention in the literature [141]. One common choice is the use of a hexagonal grid which has more symmetries. Various other methods have been proposed, such as introducing much additional structure into the automaton. The most satisfying method to obtain symmetry-independent spread is the use of random grids. The cells are randomly placed in the plane for $d = 2$, for example. In some applications, it may be reasonable to enforce a certain minimal distance between cells. Of course, the definition of neighborhood has to be modified for random grids. Reasonable methods are to define all cells within a certain distance as neighboring cells [141].

Figure 6.9 shows examples of a cellular automaton on two different random grids. We modified the Greenberg–Hastings automaton with one infectious and one immune stage ($a = 1$, $g = 1$). Neighbors of a cell are defined to be all cells within a certain radius. We start with nine infectious cells in the center and use the deterministic local function (6.1).

 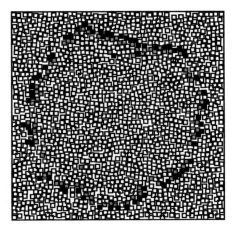

**Figure 6.9.** *Two-dimensional random grids. On the left, cells are placed purely at random; on the right, cells are placed with minimal distance between the cells. Shown are states of an epidemic automaton after 20 iterations.*

We have the same number of cells as for the simulations in Figure 6.7. On a grid where the cells are placed purely at random (Figure 6.9, left), the epidemic dies out in this particular example (the radius was chosen such that every cell has, on average, four other neighbors). But since the variance is large, the probability of reaching a state where no infected cell has a susceptible neighbor is high. On the grid with minimal distance (Figure 6.9, right), the epidemic persists. Here, the neighborhood radius around the cells is larger to ensure four additional neighbors, on average. The smaller variance of the number of neighbors lowers the probability of the epidemic dying out. We observe a more-or-less circular spread of the epidemic.

### 6.3.3   Asynchronous Dynamics

Some classical cellular automata show very interesting patterns like the Wolfram automaton with rule 90 (Figure 6.3). These are quite decorative but not necessarily realistic models. Also, in the Greenberg–Hastings example, we have seen that some patterns are sensitive to small perturbations (Figure 6.8). These issues might be the result of synchronous update. Synchronous updating is appropriate if we have discrete time steps, for example, when modeling genetics in nonoverlapping generations. But in many applications, events are not separated by generations, or by fixed time periods. Think of the spread of an epidemic. If we consider a very small time unit, then in every time step, at most one event will happen; that is, only one cell will change its state. This is modeled by *asynchronous* dynamics, that is, sequential update of the automaton. It approximates continuous time and is implemented by working with one array only, where the local rule is applied to one cell after the other. If we use asynchronous update, however, we have to select an algorithm that determines which cell is to be updated next. Various methods have been proposed, but many introduce undesired structure into the automaton.

An easy method is to choose the next cell randomly with uniform probability. More convenient from the theoretical point of view is the use of exponentially distributed waiting times, where every cell is assigned its own Poisson process (see Chapter 5). If there is an event in a cell's process, we evaluate the local function at this cell. The waiting times between two events are exponentially distributed. This method is implemented by choosing a waiting time for the next evaluation for every cell according to the exponential distribution. Once a cell is evaluated, we choose a new waiting time for it. It may happen that a cell is updated twice before another cell is updated once. This method is computationally costly. Either we have to search all waiting times for the next cell to be evaluated in every single step, or we have to maintain the order and merge the new waiting time chosen in every single step.

For a first example, we again consider the Greenberg–Hastings automaton with $A = (\mathbb{Z}^2, E = \{0, 1, 2\}$, von Neumann neighborhood, $f)$. As before, we apply the deterministic rule that "a susceptible cell is infected if at least one of its neighbors is infectious." We use asynchronous dynamics with exponentially distributed waiting times. Figure 6.10 shows the states of three runs which differ only in the seed of the random number generator for drawing the waiting times. In contrast to synchronous dynamics, different runs give different patterns. From a given state of the automaton, we observe different states depending on which cell is evaluated first. Note that this automaton with asynchronous update is related to

**Figure 6.10.** *Two-dimensional spatial patterns of three different runs of an epidemic automaton with asynchronous update. We started with nine infectious cells in the center and show states after* 4 × *grid-size asynchronous evaluations.*

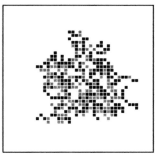
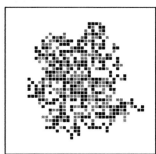

space →

time ↓

**Figure 6.11.** *Space-time patterns of one-dimensional nearest-neighbor automata with asynchronous dynamics, using Wolfram rules* 254, 50, *and* 90, *respectively. Compare to Figure* 6.3 *with synchronous dynamics. The first row of each square shows the initial configuration, which is the same for all three simulations. The following rows show successive iterates after* 1 × *grid-size single evaluations. The boundary cells are constant* 0.

interacting particle systems for which a well-known theory exists [109]. A powerful tool for analyzing asynchronous cellular automaton are *Monte-Carlo methods*. Several simulations of a specific automaton are compared. In the epidemic automaton, for example, the number of infectious, immune, and susceptible cells would be of interest and an average time course could be investigated. In our example, the states with asynchronous updating (Figure 6.10) all look quite similar, at least qualitatively.

For a second example, consider the Wolfram automata from above with asynchronous dynamics (Figure 6.11). Rule 245 leads to extinction in the synchronous case (Figure 6.3, left) and also in the asynchronous case (Figure 6.11, left); that is, after a few iterations no infectious cells are left. Rule 50 showed a 2-periodic pattern under synchronous update (Figure 6.3, middle). With asynchronous dynamics, the epidemic dies out fast (Figure 6.11, middle). With rule 90, the density of cells in state 1 (black) is similar for synchronous update (Figure 6.3, right) and asynchronous update (Figure 6.11, right), but with asynchronous dynamics the pattern no longer appears fractal-like as before.  This example shows that asynchronous dynamics can give the same patterns, qualitatively similar patterns, or totally different patterns.  One argument in favor of asynchronous dynamics is that the patterns are generally less sensitive to small random perturbations.  If one obtains highly symmetrical patterns with a strong structure when using synchronous update, one should check whether the model is realistic.  For example, it should then be tested whether the patterns persist if a little bit of random perturbation is introduced in the local function.

## 6.4   Related Models

The modeling of movement is difficult in classical cellular automata.  A possibility is to implement a left-shift rule, for example, where every cell takes the state of its left neighbor in the next time step.  However, this is not what we aim for when we are modeling biological processes such as the movement of individuals.  We will go one step further now, and discuss *Dimer automata* [142],  which are discrete spatial models which form a different class of models.  Dimer automata are appropriate when we wish to model the movement of individuals.

Consider a one-dimensional square grid, elementary states zero and one, and a nearest neighborhood.  Our rules are as follows:

1. Choose a cell $x$ at random (with uniform choice).

2. Choose a cell $y$ at random out of the neighborhood of $x$.

3. Assign new states to *both* cells $x$ and $y$.

The above rule acts on a pair of neighboring cells and as such it can be used in any dimension and with any neighborhood, even on stochastic grids.

To model particle movement, the states of *two* adjacent cells are switched in one time step, which can be seen as if we are looking at bonds instead of sites.  Since we switch the states of two cells in a Dimer automaton, we can ensure mass conservation.  A possible rule for modeling movement is $00 \mapsto 00, 01 \mapsto 10, 10 \mapsto 01, 11 \mapsto 11$.  Please note that 01 means "first cell zero," "second cell one" (the first cell does not necessarily have to be to the left).  Figure 6.12 shows patterns of this automaton.  A single cell performs a random walk on the grid.  If we start with more cells in state one, we see independent walks.  Usually not every single step is shown in these plots and the patterns may look quite different with different internal steps (Figure 6.12, middle and right).

In our final example, we use the Dimer algorithm in a simple model of alignment. Consider a two-dimensional square grid $G \in \mathbb{Z}^2$ with Moore neighborhood.  We use $E = \{0, 1, 2, 3, 4\}$, where 0 represents an empty cell and 1, 2, 3, and 4 represent particles with orientation up, left, down, and right, respectively.  Our rules are as follows:

1. Choose a cell $x$.

2. Choose a neighbor $y$ of cell $x$.

3. If $x$ is occupied and oriented towards $y$ and $y$ is empty, the particle moves from $x$ to $y$.

4. If $x$ and $y$ are occupied, then $x$ takes the orientation of $y$.

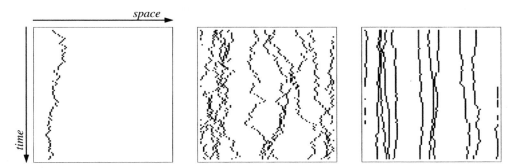

**Figure 6.12.** *Space-time patterns of Dimer automata. On the left: Starting with one cell in state one. In the middle and on the right: Starting with ten cells in state one. The first row shows the initial configuration; following rows successive iterates after* 1 × *grid-size single evaluations (left and middle), or* 0.1 × *grid-size (right). Note that the grid is closed to form a ring.*

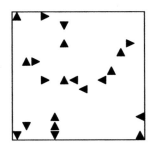 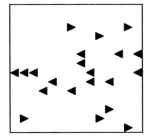

**Figure 6.13.** *Simple alignment model: Two-dimensional spatial patterns of a random initial configuration (left) and after* 400 × *grid-size single steps (right).*

Cells are chosen randomly with uniform choice on the grid, and the neighbor is also chosen with uniform choice within the neighboring cells. A simulation is shown in Figure 6.13. We observe that after some time, all particles are oriented left–right (or up–down). Particles are traveling around the torus on "lanes" separated by empty rows (or columns). More realistic models for alignment or even schooling may also be developed, but this is beyond the scope of this introduction to cellular automata.

## 6.5  Further Reading

A common general reference to cellular automata is the classical book of Wolfram [165], which is a collection of early articles. The recent book on cellular automata by Wolfram [166] has initiated interesting discussions and provides a fine section on the development of the theory of cellular automata. Toffoli and Margolus [157] and Weimar [161] both provide a very good introduction to cellular automata with an emphasis on modeling of physical phenomena. The application of cellular automata to biological pattern formation is presented in Deutsch and Dormann [49].

Due to the popularity of the Game of Life, much information can be found on the internet, including nice applets and simulation programs. Information including the construction of the proof on computational universality can be found in Poundstone's book [133].

For an example of a class of cellular automata which is well studied, consider the book by Goles and Martínez [70]. They introduce models with $E = \{0, 1\}$ where a cell becomes one if the number of neighbors is above a threshold, but stays zero otherwise. Both synchronous and asynchronous iterations are investigated. In these models, for example, in neural networks, there is a weight $a_{ij}$ for each pair of cells that describes the influence of cell $j$ on cell $i$. Cellular automata are then a special case with appropriate weights.

A summary of the classical work of people like von Neumann and Moore can be found in the book edited by Burks [32]. The Garden-of-Eden theorem is presented along with other results (published before massive simulations were possible).

For a review of cellular automata in biology, consider the article by Ermentrout and Edelstein-Keshet [53]. Modeling a broad selection of cellular automata with *Mathematica* is shown in [64].

In the field of excitable media, nice examples can be found on how cellular automata have been modified to obtain more realistic patterns. Two references to begin with are [66] and [162].

Several issues of *Physica D* are devoted to applications on cellular automata. Finally, we refer to the literature on interacting particle systems [109] and individual-based modeling [72] for related models.

## 6.6   Exercises for Cellular Automata

**Exercise 6.6.1: Wolfram rule 108.** *Consider the cellular automaton* $A = (\mathbb{Z}, \{0, 1\},$ $(x_{i_1}, x_i, x_{i+1}), f)$, *where $f$ is given by Wolfram rule* 108.

(a) *Iterate the following initial condition by hand (synchronous update):*

*To which group of the Wolfram classification does this automaton belong?*

(b) *Rule* 108 *can be interpreted as a very simple population dynamics model. Explain.*

(c) *If starting with an initial condition of randomly placed ones, what will happen? Describe the state-time patterns.*

(d) *Let us start with an initial configuration of randomly placed ones with density p. Is it possible to describe the average density $\bar{p}$ of ones finally reached? Hint: Estimate $\bar{p}$ for $p = 0$, $p = 1$, and $p = 0.5$.*

**Exercise 6.6.2: Monotonic automaton.** *We give an example of another automaton which is quite well characterized. Investigate the cellular automaton* $A = (\mathbb{Z}^2, E = \{0, 1\},$ *von Neumann neighborhood, $f$), where $f$ is given by*

$$f(z|_{U(x)}) = \begin{cases} 1 & \text{if } \sum_{y \in U(x)} z(y) \geq 3, \\ 0 & \text{otherwise.} \end{cases}$$

*If the local function is such that for any two occupations of neighborhoods $\phi$ and $\bar{\phi}$ it follows that*

$$\phi < \bar{\phi} \Rightarrow f(\phi) < f(\bar{\phi}),$$

*then the automaton is called* monotonic.

(a) *Randomly place 50 cells on a $10 \times 10$ grid and iterate synchronously (by hand). Describe what happens.*

(b) *If we start with an initial configuration with finite support, then there is a finite set of cells $M \subset G$ such that $\sup(f^t(z)) \subset M$ for all $t = 0, 1, \ldots .$*

*Construct a rectangle of ones and show that this cannot expand.*

(c) *Iterate the shown configuration first with synchronous update, then collect all states that are reachable with asynchronous iteration. What does this example show in terms of dynamical behavior of synchronous/asynchronous automata?*

**Exercise 6.6.3: Greenberg–Hastings automata.** *The Greenberg–Hastings Theorem 6.5 states that persistence is only possible if and only if a continuous cycle with winding number $\neq 0$ evolves. Consider the Greenberg–Hastings automaton $A = (\mathbb{Z}^2, E = \{0, 1, 2, 3\},$ von Neumann neighborhood, $f)$, where $f$ is given by (6.1). Stage 1 is considered infectious and stages 2 and 3 are immune stages.*

(a) *Suppose we have an initial configuration where we can select the states of two neighboring cells, all other cells are susceptible. There are $4^2 = 16$ possible combinations to choose the states for these two cells. Determine which of these initial configurations will persist. In some cases, you may look for a continuous cycle with nonzero winding number.*

(b) *Show that there is no continuous cycle in the following configuration:*

*As shown in Figure 6.7 (bottom row), a persisting pattern evolves from this configuration. Show when a continuous cycle with nonzero winding number emerges.*

(c) *Try to find an algorithm to construct a continuous cycle with nonzero winding number for arbitrary numbers of infectious stages a and immune stages g (respecting $g \geq a$).*

**Exercise 6.6.4: Game of Life.** *The fascination with the Game of Life begins when viewing it animated on a computer simulation.*

(a) *To get a feeling about what happens iterate the following configuration, called "glider," by hand or on a computer:*

(b) *If $c = 1$ cell/iteration is the "speed of light" in the Game of Life universe, at which speed does the glider travel?*

(c) *Let two gliders collide. What happens (there are several possibilities)?*

**Exercise 6.6.5: Boundaries of finite grids.** *Let us compare different boundary conditions of finite grids.*

(a) *How would you realize Dirichlet boundary conditions in a cellular automaton?*

(b) *How could a torus be programmed?*

(c) *Which of the cellular automata you know behave differently with infinite and finite grids?*

(d) *Find an automaton that behaves differently with Dirichlet conditions and on a torus.*

**Exercise 6.6.6: Stochastic epidemic automaton.** *Consider the simple epidemic automaton $(\mathbb{Z}^2, E = \{0, 1, 2\},$ von Neumann neighborhood, $f)$ with stochastic local function*

$$f(z|_{U(x)}) = \begin{cases} 1 & \text{if } z(x) = 0 \quad \text{with probability } 1 - (1 - p)^s, \\ 0 & \text{otherwise.} \end{cases}$$

*A necessary condition for persistence is that each infectious cell must be able to infect at least one other cell, on average, before it becomes immune again. Think about the critical probability $p^*$ such that the epidemic dies out for $p < p^*$ and persists for $p > p^*$. Why is $p = 1/4$ certainly too small? Look also for a proper definition of "persists" for stochastic automata.*

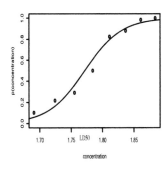

# Chapter 7

# Estimating Parameters

## 7.1 Introduction

After class, you and your friend are sitting leisurely in a cafe. Lost in thought, your friend tosses a coin. Heads, tails, tails, heads, heads, .... Since you have nothing else to do, you think about how to measure the probability of tails. *Stop!* You are much too fast. Even if you ask for something like a probability, it is necessary to set up a model; it will be a purely statistical model in this case. Implicitly, we assume that every toss takes place in an identical setting, such that the tosses are considered identical experiments. The second point—which is rather philosophical and thus should be considered in a cafe—is the impossibility of talking about things like probabilities, parameters, and so on without a model. Whenever a scientist describes an experiment, measures parameters, and identifies indices, some sort of model is used, even if, as in many cases, the model is not explicitly mentioned.

We now formulate the model for our tosses. We assume that there is for every toss the same probability $p$ for head:

$$P(\text{head}) = p, \qquad P(\text{tail}) = 1 - p.$$

The model is described by a Bernoulli random variable depicting the outcome of each toss (Section 5.3.2).

In the next step, we determine the parameter $p$ of the model. The only thing we know is $0 \le p \le 1$. We expect $p = 0.5$ (which is not true for some Euro coins; some coins are so asymmetric that they are not used at the coin toss of a soccer game). Assume that we tossed the coin $n$ times, and found heads $k$ times. Of course, the first idea is to estimate $p$ by the observation

$$\hat{p} = \frac{k}{n}.$$

We call $\hat{p}$ an *estimator* of the parameter $p$, and it is given as the relative number of heads among all tosses. How do we justify this estimator? If we choose some number $p \in [0, 1]$, we want to know how suitable this number is as a parameter for our model, given the data. Hence, we would like to have a quality measure,

$$f(p|\text{data}),$$

175

which is a measure of how well the model with parameter $p$ fits the observed data. The larger $f$, the better the model with $p$ fits the data. Given the parameter $p$, we easily derive the probability of observing the data from our model,

$$P(\text{data}|p) = P(k \text{ times heads in } n \text{ trials}|p) = \binom{k}{n} p^k (1-p)^{n-k}.$$

If the data are typical, the "true" parameter should maximize the probability for these typical data. Hence we *define*

$$f(p|\text{data}) = P(\text{data}|p).$$

From "quality of a parameter value under given data" we move silently to "probability of the data under given parameter," which is called the likelihood of the parameter. This is the standard approach of parameter estimation. There is an alternative one, the Bayesian approach (see section 7.6, "Further Reading").

Now we have our quality measure, and we want to find the best parameter, that is, the parameter that maximizes this quality measure. You can easily verify that the function $f(p|\text{data}) = P(k \text{ times head in } n \text{ trials}|p)$ assumes the maximum at $\hat{p} = k/n$, and thus this is the best estimator (see the exercises in Section 7.7).

We will exploit this idea of maximum likelihood estimators and their application to models in the following sections. To fit a model, there are basically two tasks: (1) derive a likelihood function (which includes finding an appropriate model) and (2) find the parameters that maximize the likelihood (mostly numerically). These tasks are covered in Section 7.2. Section 7.3 uses the likelihood function to compare alternative models. The task of maximizing a likelihood function can be challenging. Section 7.4 discusses optimization algroithms that can be used to find the parameters that maximize the likelihood.

## 7.2   The Likelihood Function

We need to distinguish between deterministic and stochastic models. For stochastic models, there may be cases where we do not have to incorporate measurement error (e.g., in the case of the coin); in deterministic models, we almost always expect a measurement error to occur. Hence, we find three situations: stochastic models without measurement error (discussed in Section 7.2.1), stochastic models with measurement error (not covered here), and deterministic models with measurement error (discussed in Section 7.2.2).

### 7.2.1   Stochastic Models without Measurement Error

**Example 7.2.1:** *Time to Cell Death.* Consider the cell death example from Section 5.3.1. There the length of time $a$ for a cell to die is exponentially distributed with probability mass function $f(a) = \mu \exp(-\mu a)$, where $\mu > 0$ is the mortality rate (in units 1/time). Suppose we start our observation with $n$ independent cells, and we observe ages at death to be $a_1, \ldots, a_n$. How can we use these data to estimate $\mu$? One possibility might be to estimate $\mu$ by the average of the rates $a_1^{-1}, \ldots, a_n^{-1}$, namely,

$$\hat{\mu}_1 := \frac{1}{n} \sum_{k=1}^{n} \frac{1}{a_k}. \tag{7.1}$$

Another possibility would be to first find the average rate $\bar{a}$ and then use the inverse, that is,

$$\bar{a} = \frac{1}{n} \sum_{k=1}^{n} a_k, \qquad \hat{\mu}_2 = \frac{1}{\bar{a}}. \tag{7.2}$$

Which is the better estimator, $\hat{\mu}_1$ or $\hat{\mu}_2$? As we will show, one of these estimators maximizes the *likelihood function*.

For a general model that depends on a parameter vector $p \in \mathbb{R}^n$, we define a likelihood function as follows.

**Definition 7.1.** *Let $p \in \mathbb{R}^n$ be the parameter vector of a given model. The function*

$$\mathcal{L} : \mathbb{R}^n \to \mathbb{R}_+$$

*that maps a parameter set $\pi \in \mathbb{R}^n$ to the probability (or, in the case of continuous data, the probability density) to find given data is called the likelihood. The log-likelihood is the logarithm of this function,*

$$\mathcal{LL} : \mathbb{R}^n \to \mathbb{R}, \qquad \mathcal{LL}(\pi) = \ln(\mathcal{L}(\pi)).$$

In order to estimate parameters, we have to assume that the data are typical. Without this assumption, nothing could be concluded from the data. However, with this assumption, the data are likely; that is, the probability that the data determined by the model have the "true" parameter value is large. Hence, a good estimation of the parameter vector $p$ is the maximum of the function $\mathcal{L}(\pi)$ or, equivalently, the maximum of the function $\mathcal{LL}(\pi)$ (because the logarithm ln is a monotone increasing function). As a rule, it is much simpler to maximize $\mathcal{LL}(\pi)$ than $\mathcal{L}(x)$. For example, one may use the hill-climbing algorithm for maximization (see Section 7.4 below).

**Definition 7.2.** *A parameter vector $\hat{\pi}$ that maximizes $\mathcal{L}$ is called a maximum likelihood estimation of $p$. The function that maps the data to $\hat{\pi}$ is called a maximum likelihood estimator.*

Using Definition 7.1, the likelihood function associated with the cell death problem from Example 7.2.1 is given by

$$\mathcal{L} = f(a_1) \cdots f(a_n) = \mu^n \exp\left(-\mu \sum_{k=1}^{n} a_k\right) \tag{7.3}$$

and the log-likelihood is

$$\mathcal{LL} = n \ln(\mu) - \mu \sum_{k=1}^{n} a_k. \tag{7.4}$$

Setting the first derivative of $\mathcal{LL}$ equal to zero, we have

$$\frac{n}{\mu} - \sum_{k=1}^{n} a_k = 0,$$

**Table 7.1.** *Outcome of a typical LD50 test (surviving beetles after 5h of carbon disulphide exposure). Data from* [23, 148].

| Concentration | Number of beetles | Killed beetles |
|:---:|:---:|:---:|
| $c$ | $n_i$ | $k_i$ |
| 1.6907 | 59 | 6 |
| 1.7242 | 60 | 13 |
| 1.7552 | 62 | 18 |
| 1.7842 | 56 | 28 |
| 1.8113 | 63 | 52 |
| 1.8369 | 59 | 52 |
| 1.8610 | 62 | 61 |
| 1.8839 | 60 | 60 |

or equivalently

$$\mu = \frac{1}{\bar{a}}. \tag{7.5}$$

The second derivative of $\mathcal{LL}$ at this point, $-n/\mu^2$ is negative, hence $\hat{\mu}_2 = \bar{a}^{-1}$ is the maximum likelihood estimator for the mortality rate $\mu$.

**Example 7.2.2:** *LD50 Dose for Beetles.* We want to describe toxicological properties of a certain chemical substance. One of the most important measures is the LD50 dose (LD: Lethal Dose), that is, the amount of substance where on average 50% of beetles die. The standard experimental setup is to take different dosages of the chemical, apply each dose to $n$ beetles, and count the number of dead beetles. In Table 7.1, we show an example of the outcome of an LD50 test from [23, 148].

Let $p(c)$ be the probability of one beetle being killed, given the dose $c$. We assume a logistic equation for $p(c)$,

$$\frac{d}{dc}p(c) = \alpha p(c)\,(1 - p(c)), \qquad p(\beta/\alpha) = \frac{1}{2}.$$

We will see that fixing $p = 1/2$ at the concentration $\beta/\alpha$ has the advantage that our solution looks simple. This so-called logit model is not simple to justify but suitable for the given situation [92]. The model is based on the similarity of individuals. If only a few individuals respond (i.e., $p$ is small), we expect only a few more individuals to respond if we increase the dose slightly. If we are in the range of the critical concentration ($p(c) \approx 0.5$), a small increase in $c$ should have a larger effect (more dead beetles). A few beetles may be resistant to the substance, such that we have to go to high concentrations in order to kill all beetles. Accordingly, the right-hand side of the differential equation is small if $p$ is small or close to 1, and large if $p$ is 0.5. The solution of the equation reads

$$p(c) = p(c; \alpha, \beta) = \frac{1}{1 + e^{\beta - \alpha c}}.$$

If we do experiments with a very large number of beetles, we expect a fraction $p(c)$ to survive for a dose $c$. The function $p(c)$ is a deterministic model for the fraction of survivors. This fraction depends on the independent variable $c$ and the two-dimensional parameter vector $\pi = (\alpha, \beta)^T$. We interpret $\alpha$ as a measure of the steepness of the curve $p(c)$ and $\beta$ as a measure for the death rate at zero dose, $p(0) = (1 + e^\beta)^{-1} \approx -\beta$.

However, since our sample is finite, we expect major stochastic effects. We have to include a source of variance. Let $K(c)$ be a random variable, representing the number of killed beetles $K(c)$ in a sample of size $n$. We assume that $K(c)$ follows a binomial distribution with parameter $p(c) = p(c; \alpha, \beta)$,

$$K(c) \sim \text{Bin}(n, p(c)),$$

which is another model.

Given the parameter vector $\pi$ and the concentration $c$, the likelihood of observing $k$ dead beetles in a population of $n$ healthy beetles is

$$P(K = k | \pi = (\alpha, \beta)) = \binom{n}{k} p(c; \alpha, \beta)^k (1 - p(c; \alpha, \beta))^{n-k}.$$

If we do experiments with $l$ dosages $c_1, \ldots, c_l$, number of animals $n_1, \ldots, n_l$, and number of survivors $k_1, \ldots, k_l$, we find

$$\mathcal{L}(\alpha, \beta) = P(K_i = k_i \text{ for } i = 1, \ldots, l \mid \pi = (\alpha, \beta))$$
$$= \Pi_{i=1}^{l} \binom{n_i}{k_i} p(c_i; \alpha, \beta)^{k_i} (1 - p(c_i; \alpha, \beta))^{n_i - k_i}.$$

As before, the parameter that maximizes the likelihood also maximizes the logarithm of the likelihood,

$$\mathcal{LL}(\alpha, \beta) = \sum_{i=1}^{l} \left( \ln \left( \binom{n_i}{k_i} \right) + k_i \ln(p(c; \alpha, \beta)) + (n_i - k_i) \ln(1 - p(c; \alpha, \beta)) \right).$$

For the data from Table 7.1, we find that the maximum is assumed for

$$\alpha = 33.78, \qquad \beta = 59.9.$$

The corresponding fit is shown in Figure 7.1. The LD50 dose is the concentration at which $p(c) = 0.5$, that is, $\exp(\beta - \alpha c) = 1$. Thus,

$$\text{LD50} = \beta/\alpha = 1.77.$$

If we have many data, there is a rule of thumb that yields confidence intervals for the parameters. Confidence intervals are intervals for the location of the "true" parameter value with a prescribed probability (usually 0.95 or 0.99).

**Definition 7.3.** *An interval $[\pi_{lo}, \pi_{up}]$ that covers the true parameter $\pi$ with probability $\gamma$, that is,*

$$P(\pi \in [\pi_{lo}, \pi_{up}]) = \gamma,$$

*is called a confidence interval with probability $\gamma$ (Figure 7.2).*

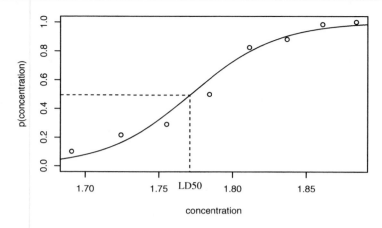

**Figure 7.1.** *Optimal fit for the LD50 experiment.  The data points are $k_i/n_i$ at concentration $c_i$, the continuous line the optimal fit for the logit model.*

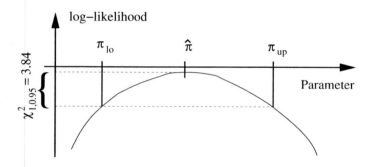

**Figure 7.2.** *Asymptotic confidence interval $[\pi_{lo}, \pi_{up}]$ for $\gamma = 0.95$.*

Such confidence intervals suggest the precision of the estimate. The point estimation, that is, the value that maximizes the likelihood, contains no such information. The confidence interval is not unique, even if the probability $\gamma$ is fixed. Any interval with the property that it covers the true parameter with given probability is a confidence interval. Usually, researchers use the confidence interval that is centered around the maximum likelihood estimator.

For simple situations, it is possible to obtain an explicit formula for these confidence intervals. In general, the scientist must use asymptotic results, that is, results that are only true if the number of data tends to infinity.

The method to obtain a confidence interval with asymptotic methods is explained in Figure 7.2. First, the maximum likelihood estimation of the parameter is determined, that is, the parameter set that maximizes the log-likelihood function. This maximum is assumed at the parameter value $\hat{\pi}$, say. Next, the $\chi^2$ value for the given confidence level $\gamma$ and one degree of freedom is determined; for example, for $\gamma = 0.95$ and one degree of freedom, we find $\chi^2_{1,0.95} = 3.84$ (Figure 7.3).

Why did we use the $\chi^2$ distribution? The reason is the universality of the normal distribution and the fact that, at a local extremum, a function can be approximated by a

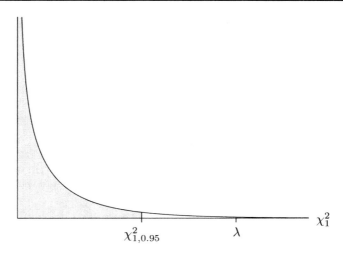

**Figure 7.3.** *When a random variable with a standard normal distribution (zero mean and variance one) is squared, it has a chi-square distribution with one degree of freedom. The $\chi_1^2$ probability density function is illustrated in this figure. Given a significance level $\alpha = 0.05$, the confidence interval $\gamma = 1 - \alpha = 0.95$ is shown as the area of the gray region. The rightmost extent of this region is $\chi_{1,0.95}^2 = 3.84$.*

quadratic—and this is precisely the definition of the $\chi^2$ distribution, which describes the distribution of the sum of squared, normally distributed random variables (Figure 7.3). However, the proof comes from asymptotic theory.

Next we determine the values of the log-likelihood function such that

$$\mathcal{LL}(\pi_{lo}) = \mathcal{LL}(\pi_{up}) = \mathcal{LL}(\hat{\pi}) - \chi_{1,\gamma}^2, \qquad \pi_{lo} < \pi_{up}. \tag{7.6}$$

The values $\pi_{lo}$ and $\pi_{up}$ are uniquely determined if the number of data is high enough, and then $[\pi_{lo}, \pi_{up}]$ is the desired confidence interval for confidence level $\gamma$.

## 7.2.2 Deterministic Models

We now consider deterministic models. There is no intrinsic source of stochasticity. Without measurement error, we might expect to be able to find model parameters so that our model perfectly matches the data. However, this is not always the case. We begin this section by discussing the identifiability of model parameters, and show that parameters cannot always be identified, even without measurement error. We then discuss how to deal with measurement errors and how to estimate model parameters. In particular, we explain the least-squares method, as well as more general methods to obtain the parameters of ODEs.

### (a) Identifiability of Parameter

In many cases, parameters cannot be identified, even though we do not have measurement errors. This may seem surprising, but in practice it is quite common.

A simple example is given by a population with per capita birth rate $b$ and per capita death rate $d$, whose dynamics are given by

$$\frac{dN}{dt} = bN - dN,$$

where $N(t)$ denotes the number of individuals in the population as a function of time. The solution, $N(t) = N(0)e^{(b-d)t}$, can be used to determine the intrinsic growth rate $r = b - d$, but not both the per capita birth and death rates, $b$ and $d$.

A more involved example arises in PET (positron emission tomography), taken from [140]. PET is an imaging technique used for measuring blood flow and metabolism within body tissues.

A person is injected with a substance (tracer) that is weakly radioactive. This injection leads to a certain concentration of tracer in the blood that changes over time, that is, can be described by a function $B(t)$. Since we have control over the injection process, we can safely assume that the function $B(t)$ is known.

Blood delivers radioactivity to tissues. In particular, radioactivity is transferred between the blood and tissue via diffusion. That is, some time after tracer injection, tissues are radioactive. It is this radioactivity that is measured with PET. The PET signal can then be processed to infer blood flow in the tissue.

The details of measuring blood flow in homogeneous tissues have been worked out and are now considered routine. Below, we will show that parameter identifiability becomes an issue when PET measures a signal from a tissue that is not homogeneous, for example, the brain, which consists of both gray and white matter.

Consider the radioactive signal from tissue in a so-called voxel (a voxel is the smallest region of the body that can be resolved [163]), and assume that two different types of tissue are located in this voxel, as shown in Figure 7.4, with relative tissue volumes $\tau$ and $1 - \tau$, respectively. The value of $\tau$ is not known.

If we let the radioactivity in the two tissues be given by $C(t)$ and $D(t)$, respectively, then we can write down a simple compartmental model for the dynamics of $C(t)$ and $D(t)$,

$$\frac{d}{dt}C(t) = -k_{C1}C(t) + k_{C2}B(t),$$

$$\frac{d}{dt}D(t) = -k_{D1}D(t) + k_{D2}B(t),$$

where $k_{C1}$ and $k_{D1}$ denote the transition rates of the tracer from the blood into the tissues, and $k_{C2}$ and $k_{D2}$ denote the transition rates of the tracer from the tissues into the blood. The time scales of diffusion between blood and tissue may be different for the two tissue types.

The signal we measure is a combination of $C(t)$ and $D(t)$, namely,

$$S(t) = \tau C(t) + (1 - \tau)D(t).$$

Given that we know the signal $S(t)$, our task is to determine $\tau$, $k_{C1}$, $k_{C2}$, $k_{D1}$, and $k_{D2}$, that is, five parameters.

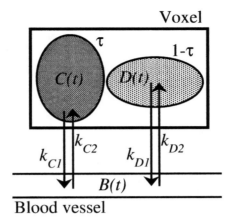

**Figure 7.4.** *Model assumption for inhomogeneous tissue. Within one voxel, we find two tissue types, with relative volumes $\tau$ and $1 - \tau$, respectively.*

We now derive a differential equation (of higher order) for $S(t)$ alone. Taking the first and the second derivatives of $S(t)$ with respect to $t$, we find

$$\frac{dS}{dt} = \tau \frac{dC}{dt} + (1 - \tau)\frac{dD}{dt}$$

$$= \tau \left[-k_{C1}C + k_{C2}B\right] + (1 - \tau)\left[-k_{D1}D + k_{D2}B\right]$$

$$= -\tau k_{C1}C - (1 - \tau)k_{D1}D + \left[\tau k_{C2} + (1 - \tau)k_{D2}\right]B,$$

$$\frac{d^2 S}{dt^2} = -\tau k_{C1}\frac{dC}{dt} - (1 - \tau)k_{D1}\frac{dD}{dt} + \left[\tau k_{C2} + (1 - \tau)k_{D2}\right]\frac{dB}{dt}$$

$$= \tau k_{C1}^2 C + (1 - \tau)k_{D1}^2 D$$

$$+ \left[\tau k_{C2} + (1 - \tau)k_{D2}\right]\frac{dB}{dt} - \left[\tau k_{C1}k_{C2} + (1 - \tau)k_{D1}k_{D2}\right]B.$$

It is possible to write a second-order ODE for $S(t)$ that does not include $C(t)$ and $D(t)$, namely,

$$\frac{d^2 S}{dt^2} + \theta_a \frac{dS}{dt} + \theta_b S = \theta_c \frac{dB}{dt} + \theta_d B,$$

where

$$\theta_a = (k_{C1} + k_{D1}),$$
$$\theta_b = (k_{C1}k_{D1}),$$
$$\theta_c = (\tau k_{C2} + (1 - \tau)k_{D2}),$$
$$\theta_d = (\tau k_{C2}k_{D1} + (1 - \tau)k_{C1}k_{D2}).$$

Even if we have perfect knowledge of both $S(t)$ and $B(t)$, the differential equation for $S(t)$ tells us that we are able to determine only the parameters $\theta_a, \dots, \theta_d$ from the data. However, from these four parameters we cannot infer the five parameters $\tau$, $k_{C1}$, $k_{C2}$, $k_{D1}$, and $k_{D2}$ unless additional information is known (see [140] for physiological constraints that can be specified to render the parameters identifiable from PET signals measured in the brain). Of course, the reason for this is the fact that we cannot observe the complete state of the system, that is, $C(t)$ and $D(t)$, but only the combined signal $S(t)$. Too much information gets lost by combining data.

The field of system theory is concerned with questions about identifiability of parameters and observability of states. In simple cases, like the case of linear models above, this task can basically be reduced to problems of linear algebra and is quite well understood. For nonlinear models, the problems become more involved [69].

### (b) Least-Squares Method

We now consider a time series measurement which constitutes time points $t_1, \dots, t_n$ and data points $x_1, \dots, x_n$. We assume a model predicts these data via a function $f$ that depends on time and on a parameter vector $p$, that is, that $x_i$ should resemble $f(t_i; p)$. We further assume that there are no problems with the identifiability of the parameter. For example, if we have an exponential decay process, we expect $x_i \approx c \exp(-\gamma t_i)$; hence, the parameter vector is $p = (c, \gamma)$ and $f(t; p) = c \exp(-\gamma t)$.

In general we will not observe the exact values, but we expect a measurement error to occur. Due to the universality of the normal distribution, the canonical assumption is that the data are normally distributed with expectation $\mu = f(t_i; p)$ and some (unknown) variance $\sigma^2$. Then the data $x_i$ are realizations of random variables $X_i$ that are distributed according to

$$X_i \sim N(f(t_i; p), \sigma^2). \tag{7.7}$$

Furthermore, we assume that the $X_i$ are independent. We often deal with a time series from a single experiment and the assumption of uncorrelated data points is questionable. In those cases, we would have to use the tools of time series analysis, a quite advanced theory [30]. However, here we assume that the data are uncorrelated. Then the likelihood for the data reads

$$\mathcal{L}(p) = P(x_i = X_i, \quad i = 1, \dots, n \mid p = (\mu, \sigma)) = \Pi_{i=1}^n \frac{1}{\sqrt{2\pi}\,\sigma} e^{-\frac{(x_i - f(t_i, p))^2}{2\sigma^2}} \tag{7.8}$$

and the log-likelihood becomes

$$\mathcal{LL}(p) = -\sum_{i=1}^n \ln(\sqrt{2\pi}\sigma) - \frac{1}{2\sigma^2} \sum_{i=1}^n (x_i - f(t_i; p))^2. \tag{7.9}$$

For any given $\sigma$, maximizing $\mathcal{LL}$ is equivalent to minimizing

$$\text{least-squares error}(p) := \sum_{i=1}^n (x_i - f(t_i; p))^2. \tag{7.10}$$

Many software packages have least-squares estimators. For an example of how the least-squares method can be used in Maple, see Section 8.1.3; another example is given for the cell competition model in Section 10.1. See also Section 7.4. The least-squares error (7.10) remaining when $p$ is chosen to minimize (7.10) is called the residual sum of squares (RSS).

The background of the usual least-squares error minimization is the assumption of normally distributed errors with a given, constant variance. If the variance is not constant but known, that is, $X_i \sim N(f(t_i; p), \sigma_i^2)$, we find with a similar argument that we should minimize the

$$\text{weighted least-squares error}(p) := \sum_{i=1}^{n} \frac{(x_i - f(t_i; p))^2}{\sigma_i^2}. \tag{7.11}$$

In this case, we have to weight every term with the inverse of the variance. Data with a small variance (that are thus more precise) gain a larger influence on the parameter estimation than terms with a large variance, where the measured data may be far away from the deterministic value. One way to choose the variance structure $\sigma_i$ is to assume that the $\sigma_i$ may depend on the expectation such that the variance is higher if the expected value is high, and vice versa. In this case,

$$\sigma_i = a \, f(t_i; p),$$

where $a$ is a positive proportionality constant.

**(c) ODEs**

Many biological models are described by ODEs. These models incorporate parameters. We expect the data to be close to a trajectory and we expect that stochastic effects lead to deviations from a single trajectory.

Suppose that the biological system is described by the ODE

$$\dot{x} = f(x; p), \qquad x(0) = x_0, \tag{7.12}$$

where $p$ denotes the parameter vector. The solution is denoted by

$$x = x(t_i; p, x_0). \tag{7.13}$$

We have data $y_i$ at time points $t_i, i = 1, \ldots, n$. Since there are stochastic effects, we usually assume that $y_i$ are realizations of random variables $Y_i$,

$$Y_i \sim x(t_i; p, x_0) + e_i, \tag{7.14}$$

where $e_i \sim N(0, \sigma^2)$ denote the error variables, which are independent normally distributed random variables with constant variance. The model we use is a combination of differential equation and error structure. We are now in the situation described above. The maximum likelihood estimator corresponds to a minimization of the least-squares error; that is, we should minimize

$$\text{Error}(p; x_0) = \sum_{i=1}^{n} (y_i - x(t_i; p, x_0))^2. \tag{7.15}$$

**Table 7.2.** *Data of the experiment of Pearson, Van Delden, and Iglewski* [131]; *see text for explanation.*

| Time [min] | 0.5 | 1 | 5 | 30 |
|---|---|---|---|---|
| Ratio | 0.5 | 1.2 | 2.5 | 2.7 |

**Example 7.2.3:** *AHL-Diffusion.* Consider the bacterium *Pseudonomas aeruginosa*, which is a pathogen in human lungs. The bacteria communicate by excreting certain substances, so-called AHLs (N-Acyl-L-Homoserinlacton). Each cell produces AHL at a low rate. If a cell recognizes that the AHL concentration in its environment exceeds a certain threshold, it changes its behavior. This communication process is called "quorum sensing." A goal of modeling is to understand the quorum sensing process. An introduction to modeling of quorum sensing can be found in the article of Dockery and Keener [47]. One aspect of the model is concerned with the question of how the substance diffuses through the cell walls. Are there active pumps and, if so, how strong are these pumps? In an experiment, Pearson, Van Delden, and Iglewski [131] added a certain dose of AHL to a culture of *Pseudonomas aeruginosa* and measured the ratio of AHL concentration in cells and in the surrounding medium over time in order to investigate this question (data shown in Table 7.2).

We use a linear model to describe this experiment. Let $u_c$ denote the concentration of AHL within the cells and $u_e$ the concentration in the medium. The dynamics is described by

$$\dot{u}_c = d_1 u_e - d_2 u_c, \qquad (7.16)$$

where $d_1$ is the rate of influx into the cells and $d_2$ the rate of efflux. We assume that AHL is added to the medium at time zero and that $u_e$ is approximately constant (e.g., the amount of AHL degraded during the experiment is negligible and the volume of the medium is much larger than the volume of all cells). We recognize (7.16) as a first-order ODE, with solution

$$u_c(t) = u_c(0)e^{-d_2 t} + \int_0^t e^{-d_2(t-\tau)} d_1 u_e \, d\tau = u_c(0)e^{-d_2 t} + \frac{d_1}{d_2}(1 - e^{-d_2 t})u_e.$$

We have data about the ratio of interior and exterior AHL concentrations; that is, we consider

$$r(t) = \frac{u_c(t)}{u_e} = u_c(0)e^{-d_2 t}/u_e + \frac{d_1}{d_2}(1 - e^{-d_2 t}). \qquad (7.17)$$

Since the dose added was quite high, we may assume that $u_c(0)/u_e \approx 0$ and end up with

$$r(t) \approx \frac{d_1}{d_2}(1 - e^{-d_2 t}). \qquad (7.18)$$

Using least squares to minimize the error $\sum(r_i - r(t_i, d_1, d_2))^2$ (see (7.15)), we fit $r(t)$ from (7.18) to the data (see Figure 7.5) and obtain

$$d_1 = 1.34/\text{minute}, \qquad d_2 = 0.5/\text{minute}.$$

Thus, the influx rate is about 2.5 times higher than the efflux rate, which allows the cells to accumulate the signaling substance. This is one of the mechanisms that allow the cells to respond to relatively low levels.

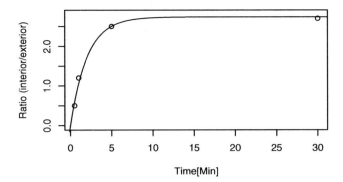

**Figure 7.5.** *Fit of model and data for active pumps (data taken from* [131]*).*

## 7.3   Model Comparison

In real world applications, there is not just one model that effectively describes a process, but many possible models. Some are simple and only account for the most basic effects; other models are quite complex, addressing many subprocesses. Which model should we choose in a certain situation? At this point, we need to compare the performance of models. Section 7.3.1 provides a common measure of model performance based on information theory. Section 7.3.2 shows how hypothesis testing methods can be used to choose between a complex model or a simpler, nested, submodel. Section 7.3.3 discusses methods of cross validation.

### 7.3.1   Akaike Information Criterion

Assume that we derived two alternative models to explain a given data set. If the number of parameters is the same for both models, it is fair to compare the likelihoods directly: the model with the higher likelihood is better suited to describe the process. If one of the two models has more parameters, direct comparison is not appropriate anymore. In general it is easier to fit a model with more parameters to given data than a model with fewer parameters. On the other hand, a model with more parameters might only address random features of the data. As J. von Neumann used to say, "with four parameters I can fit an elephant and with five I can make him wiggle his trunk" [50]. Thus, we need to use a criterion that is suited to compare models with different number of parameters. Information theory (see Burnham and Anderson [34]) provides a formula for the distance (Kullback Liebler distance) from a candidate model to a "true" model. Akaike extended this idea to provide an unbiased approximation to the distance, which can be applied to experimental data.

The AIC (Akaike information criterion; the abbreviation was introduced by Akaike himself and originally meant "an information criterion" [2]) is the most frequently used comparison criterion. The AIC is defined as

$$AIC = 2\,\mathcal{LL}(\hat{p}) - 2\,n_p, \tag{7.19}$$

where $\mathcal{LL}(\hat{p})$ is the maximum log-likelihood and $n_p$ is the number of parameters for a given model. The larger the AIC, the better the model. It is based on asymptotic arguments (like the algorithm to obtain approximative confidence intervals from the log-likelihood) and thus has to be handled with care and interpreted carefully. If the number of data points, $N$, is small (rule of thumb $N \leq 40$), then the following *corrected AIC* should be used:

$$AICc = 2\,\mathcal{LL}(\hat{p}) - 2\,n_p \frac{N}{N - n_p - 1}. \tag{7.20}$$

However, the idea of the AIC may be easily understood: Asymptotically, in a perfect fit, each data point adds to the log-likelihood on average one unit. On the other hand, generically, we are able to explain one data point per parameter, even if the model has nothing to do with the data. Thus, we should subtract the number of parameters from the optimal log-likelihood. The factor 2 comes from asymptotic expansions [2]. To a certain degree, the AIC measures the performance of a model independently of the number of its parameters. If we compare models with the same number of parameters, the advice of the AIC and that of the likelihood coincide. If the number of parameters is different, the comparison of the AICs may give a different result than a direct comparison of the likelihoods.

**Example 7.3.1:** *Salmonella.* The mutagenic effect of quinoline on *Salmonella TA*98 is investigated in [27, 147]. (See Table 7.3.) For each dose of quinoline, three independent experiments are performed. A population of Salmonella is exposed to the substance, and the number of mutated colonies is determined. We use count data. The appropriate stochastic model for counts is the Poisson distribution. We need to connect the expected value $\mu$ of this Poisson distribution, that is, the expected number of mutated colonies, with the dose $c$ of quinoline. Without further theoretical considerations, it is possible to write down two plausible models.

The first model (called model 1 hereafter) is based on the fact that any effect is, at first approximation, a linear effect,

$$\mu^{(1)}(c) = \mu_0 + \mu_1\, c, \tag{7.21}$$

with two constant parameters $\mu_0$ and $\mu_1$. The second one (model 2) takes into account that the effect of the substance will change logarithmically with the dose; that is, $\mu(c)$ is more or less a function of $\ln(c)$. Since the logarithm is not defined for $c = 0$, we take for the second model

$$\mu^{(2)}(c) = \nu_0 + \nu_1\, \ln(c + \nu_2), \tag{7.22}$$

with three parameters $\nu_0$, $\nu_1$, and $\nu_2$. The likelihood of finding the natural number $x$, if $x$ is a realization of a Poisson distribution with expectation $\mu$, reads $\mu^x e^{-\mu}/x!$. The joint

**Table 7.3.** *Number of mutated colonies in response to the mutagenic substance quinoline. For each dose, this number is determined for three independent plates (data from [27, 147]).*

| Dose ($\mu$g per plate) | 0 | 10 | 33 | 100 | 333 | 1000 |
|---|---|---|---|---|---|---|
| Plate 1 | 15 | 16 | 16 | 27 | 33 | 20 |
| Plate 2 | 21 | 18 | 26 | 41 | 38 | 27 |
| Plate 3 | 29 | 21 | 33 | 60 | 41 | 42 |

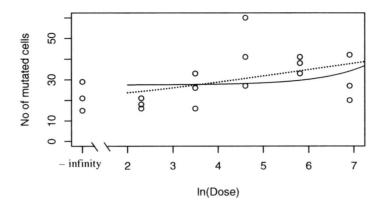

**Figure 7.6.** *Fit of model and data for the salmonella experiment. The solid line corresponds to the fit of model 1; the dashed line corresponds to model 2.*

likelihood for all data $y_{i,j}$ ($i = 1, \ldots, 3$ denotes the plate, and $j = 1, \ldots, 6$ the dose $c_j$) reads

$$\mathcal{L}_{(*)} = \prod_{i,j} \frac{1}{(y_{i,j})!} (\mu^{(*)}(c_i))^{y_{i,j}} e^{-\mu^{(*)}(c_i)}, \tag{7.23}$$

where $* \in \{1, 2\}$ for model 1 or model 2; the log-likelihood is

$$\mathcal{LL}_{(*)} = -\sum_{i,j} \ln((y_{i,j})!) + \sum_{i,j} y_{i,j} \ln(\mu^{(*)}(c_i)) - \sum_{i,j} \mu^{(*)}(c_i). \tag{7.24}$$

Fitting the two models by maximizing $\mathcal{LL}_{(*)}$ yields

| Model | Parameter | | | Log-likelihood |
|---|---|---|---|---|
| Model 1 | $(\mu_0, \mu_1)$ | $=$ | $(27.473, 0.00666)$ | $-24.7$ |
| Model 2 | $(\nu_0, \nu_1, \nu_2)$ | $=$ | $(16.12, 3.12, 3.9)$ | $-16.05$ |

The corresponding fits are shown in Figure 7.6. This table shows that model 2 fits the data better (log-likelihood $-16.05$) than model 1 (log-likelihood $-24.7$). However, the second model has one parameter more than the first one, and thus it is not appropriate to compare the likelihood functions directly. We compute the AIC,

$$\text{AIC}_{(1)} = 2(-24.7 - 2) = -53.4, \qquad \text{AIC}_{(2)} = 2(-16.05 - 3) = -38.1.$$

The AIC values suggest that model 2 is superior to model 1. In other words, it is worth the effort to use the additional parameter in the second model.

## 7.3.2   Likelihood Ratio Test for Nested Models

The AIC can only suggest which model to use. A statistical test is better to interpret. In the classical setting, we derive a general model, where our model candidates are special cases of this general model for certain parameter choices. Typically, we want to distinguish between a simple model and an extended model; we may introduce a parameter $p_0$ that becomes zero for the small model and nonzero for the extended one. Specifically, let the model with parameter $p_0 = 0$ be denoted by $M(0)$, and the extended model by $M(p_0)$. To compare models $M(0)$ and $M(p_0)$, we use a statistical test with the null-hypothesis

$$H_0 : p_0 = 0, \tag{7.25}$$

against the hypothesis

$$H_1 : p_0 \neq 0. \tag{7.26}$$

To apply the likelihood ratio test (LRT), we have to assume that

- the sample size is large enough (rule of thumb $N \geq 25$), and

- if model $M(p_0)$ is fitted to the data, then the parameter $p_0$ is normally distributed.

Under these assumptions, the test may be constructed as follows: Given the data, for each of the models $M(0)$ and $M(p_0)$ we calculate the likelihood and the log-likelihood. It turns out (see, e.g., Berger, Casella, and Berger [19]) that the ratio of the likelihoods $\mathcal{L}(M(p_0))/\mathcal{L}(M(0))$ is $\chi^2$-distributed. Even more, the value for $\lambda$, defined as

$$\lambda := -2(\mathcal{LL}(M(p_0)) - \mathcal{LL}(M(0))), \tag{7.27}$$

is $\chi^2$ distributed. Hence we use the $\chi^2$-distribution for our test. Given significance level $\alpha$ (typically 0.01 or 0.05), we check if $p_0 = 0$ is in the confidence interval, by comparing $\lambda$ to the $\chi^2_{1,\gamma}$-value, where the confidence level is $\gamma = 1 - \alpha$ and the degree of freedom corresponds to the difference in the number of parameters of the models $M(0)$ and $M(p_0)$. For calculation of a confidence interval we use equation (7.6) with $\mathcal{LL}(\pi)$. If this confidence interval includes 0, we cannot reject the null-hypothesis and thus we keep the simple model $M(0)$. If the confidence interval does not contain zero, we find that we need the extended model $M(p_0)$. In that case, we can use the value $\lambda$ from (7.27) and compare it to the $\chi^2_{1,\gamma}$-distribution. If $\lambda$ falls to the right of $\chi^2_{1,\gamma}$, then the null-hypothesis is rejected. For $\lambda = 6.76$ (Table 7.4), the null-hypothesis is rejected at significance level $\alpha = 0.05$. The $P$-value, or probability that the null-hypothesis is rejected erroneously, is given by the area under the $\chi^2_1$ curve to the right of $\lambda$ (Figure 7.4). For $\lambda = 6.76$, $P < 0.01$ (Table 7.4). The $P$-value gives the probability that the simpler model $M(0)$ is erroneously discarded.

We study this method in the following example.

**Example 7.3.2:** *Cell Growth.* To illustrate the use of AIC, AICc, and LRT, we will fit three models to the cell growth data of Gause [63], using least squares (7.10). The data are given in Project 10 in Chapter 9 and are more fully discussed in Section 10.1. For now we do not worry about where the models are from or how they are derived. We just postulate that each of the following models might be appropriate to describe the observed data. Let $f(t)$ denote the population size of *Paramecium caudatum*. The *logistic equation* is

$$f' = rf \left( 1 - \frac{f}{K} \right); \tag{7.28}$$

the Gompertz model reads

$$f' = rf \ln \left( \frac{f}{K} \right); \tag{7.29}$$

**Table 7.4.** *Analysis of the fits of logistic (dashed), Gompertz (dotted), and Bernoulli (solid) models to Gause's* Paramecium caudatum *population growth data, using the method of least squares. The fits are shown graphically in Figure 7.7.*

| Model | RSS | $\hat{\sigma}^2$ | $\mathcal{LL}$ | AICc | $\lambda$ | P |
|---|---|---|---|---|---|---|
| Logistic | 4838 | 201.6 | $-63.7$ | $-131.9$ | 6.76 | 0.01 |
| Gompertz | 11837 | 493.2 | $-74.4$ | $-153.4$ | — | — |
| Bernoulli | 3650 | 152.1 | $-60.3$ | $-127.8$ | — | — |

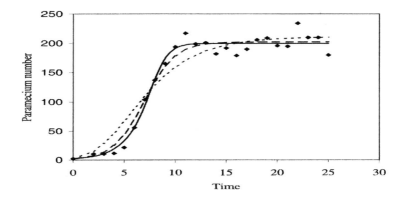

**Figure 7.7.** *Least-squares fit of logistic (dashed), Gompertz (dotted), and Bernoulli (solid) models to Gause's* Paramecium caudatum *data (see Project 10 in Chapter 9), using least squares. The fits are analyzed in Table 7.4. The best fit comes from the three-parameter Bernoulli model, the second best from the two-parameter logistic model (which is equivalent to the Bernoulli model with $\theta = 1$), and the worst from the Gompertz model.*

and the Bernoulli equation reads

$$f' = rf\left(1 - \left(\frac{f}{K}\right)^{\theta}\right). \tag{7.30}$$

The logistic and the Gompertzian model have two parameters each, $r$ and $K$, whereas the Bernoulli model has three parameters, $r$, $K$, and $\theta$. If we set $\theta = 1$ the Bernoulli model, then the logistic model follows. Hence the logistic model is a special case of the Bernoulli model, which means that we can use the LRT to compare the logistic and the Bernoulli models. In Figure 7.7, we show the best fit to the data of Gause for each of the models.

The residual sum of squares (RSS) is used to estimate the variance of the error $\hat{\sigma}^2 = RSS/N$, where $N$ is the number of data points. From this, the log-likelihood is calculated using $\mathcal{LL} = -N\hat{\sigma}^2/2$.

In Table 7.4, we show for each model the RSS, the estimated variance ($\hat{\sigma}^2$), the log-likelihood, and the corrected AICc. The AIC, corrected for the small number of points (7.20), indicates that the Bernoulli model is the best from an information theory perspective.

Because the logistic model is nested within the Bernoulli model by choosing $\theta = 1$, the LRT can be used to test whether the logistic equation should be rejected in favor of the more complex Bernoulli model. The value $\lambda = 6.76$ (equation (7.27)) indicates that the probability of erroneously rejecting the simpler logistic model in favor of the more complex Bernoulli model is approximately one in a hundred ($P = 0.01$). (See also Figure 7.3.) In other words, the Bernoulli model is strongly supported through the LRT as well as using the AICc.

The logistic equation has an analytical solution as given in Exercise 3.9.18. The Gompertz and Bernoulli equations can also be solved analytically. You may want to try this or, alternatively, refer to Thieme [156].


## 7.3.3  Cross Validation

An appealing method for comparing models is based on their ability to predict data that have not been used to fit the parameters. If we consider two models and compare their likelihood function directly, we may incorrectly prefer one model over another, because one of the models may fit random structures in the data (that vanish if we repeat the experiment). This is especially true if we compare models with a different number of parameters. This so-called overfitting (which is one of the dangers in highly structured models) cannot be recognized on the basis of the likelihood only. Cross validation avoids this problem by splitting the data set into a training set and a test set. Using the training data, the model parameters are fit (e.g., by means of a maximum likelihood procedure). These parameters are then used to predict the data of the test set. The likelihood for the two models under consideration can then be compared directly. A given set of data is often split several times in different ways. For example, we define the first data point as a test set, and predict this data point by using all other data for fitting. Then we leave out the second data point, the third data point, etc. In this way, we do not bias our data selection for fitting or for testing, since all are used for fitting and all are used for testing. We obtain $n$ likelihood values for $n$ data points. We are even able to determine which data points support which of the models, a fact that may lead to further insight. For further discussion of cross validation, we refer the reader to Haefner [77].

**Example (salmonella, continued).** In order to evaluate the two salmonella models, we perform cross validation. As discussed, there are several designs possible for choosing training and test data sets. In the present case, we decide to use the data for a certain dose (three data points) as the test set, and the remaining data as the training set. Hence, we perform six maximum likelihood estimations and evaluations for the test data points. The result is shown in Figure 7.8. We find that model 2 is not able to predict the data for zero dose at all, since the prediction of the mean value at a dose of zero is negative. For all other dosages, both models perform quite similarly; the second model is slightly better, especially at high dosages because it yields larger likelihood values. The sum of the log-likelihoods for the test data (for dosages $\geq 10\mu g$) is $-95.3$ for model 1 and $-39.7$ for model 2. Also, the cross validation tells us that the second model is superior to the first model. However, inspection of Figure 7.8 shows that the difference is not always large. If we find additional criteria (expert opinions, theoretical foundations of models), we should give these criteria a stronger weight than the results of AIC and cross validation.

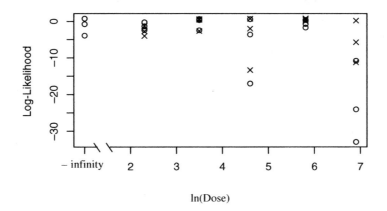

**Figure 7.8.** *Cross validation for the Salmonella example, log-likelihood of test data points. For any given dose there are three replicate plates (Table 7.3). Circles correspond to model 1 and crosses to model 2. Note that model 2 cannot predict the test points where there is no quinoline added.*

## 7.4 Optimization Algorithms

We have seen before that parameter estimation is simply an optimization problem once the likelihood has been determined. It is often impossible to derive the solution for these optimization problems analytically. We will not consider the optimization method explicitly in this text, since it forms a field of research on its own. We refer the interested reader to [58, 59].

If the function that is to be maximized is smooth, one may use the gradient in order to determine the direction of the steepest descent/ascent. The two most popular algorithms that are based on this concept are the *Newton method* and the *hill-climbing algorithm* [58, 59]. If the function is not smooth or there are a lot of local maxima, then stochastic algorithms may perform better than deterministic ones. Here, *simulated annealing* [163], the *genetic algorithm* [88], and the *generic algorithm* [163] are the best known examples.

We consider a simple form of the hill-climbing algorithm and of the generic algorithm in order to understand the constructions as well as gain valuable practical tools for our optimization. However, parameter estimation is an art, and there are no general concepts that always work. You will find that you learn a lot about your data as well as about your model if you try to fit parameters.

### 7.4.1  Algorithms

#### (a) Hill-Climbing Algorithm

Consider a function $f : \mathbb{R}^n \to \mathbb{R}$ that is to be maximized. At any location $p \in \mathbb{R}^n$, the vector $\nabla f(p)$ points in the direction of increasing values of $f$. If we are at a local maximum, then $\nabla f(p) = 0$. Thus, if we constantly change $p$ in the direction of $\nabla f(p)$, we expect to find a maximum at $\nabla f(\hat{p}) = 0$. Hence we solve

$$\dot{p} = \nabla f(p)$$

until we reach a stationary point in order to obtain local maxima. The ODE will quickly find reliable solutions of the maximization problem if there is only one global maximum (best case) or only a few maxima that are all acceptable. However, if there are many (low) local maxima, the algorithm is likely to fail. In this case, we may use a lot of randomly chosen initial values and take the best local maximum determined. This latter variant is quite close to the stochastic generic algorithm considered below.

### (b) Generic Stochastic Algorithm

The idea of the generic algorithm is almost trivial. As before, we start at a certain location $p$ and go into the direction that increases $f$. The hill-climbing algorithm suggests using the gradient of $f$ in order to obtain this direction. Instead, the generic stochastic algorithm tests randomly chosen values $q$ in the neighborhood of $p$ until a better point has been determined, that is, until $f(q) > f(p)$. In the first steps of this algorithm, the initial phase, we allow for a large diameter of the neighborhood in order to avoid local maxima. We then slowly decrease the diameter of the neighborhood in order to force convergence.

This algorithm is an example of an inhomogeneous Markov chain (see Section 5.2.2) that—hopefully—converges in a probabilistic sense to the global maximum of the function. This algorithm only needs a function $f$ that is pointwise defined, while the hill-climbing algorithm needs differentiability of $f$. Furthermore, it has fewer problems with local maxima since there is the possibility of jumping from one local maximum to another one. However, once we are close to the global maximum (and consider a smooth function $f$), the hill-climbing algorithm is preferable since the convergence is faster.

## 7.4.2  Positivity

Very often the relevant parameters in a model represent rates. Therefore, we may require positivity for some components of our parameter vector $p$, that is, $p_k \geq 0$ for $k \in K \subset \{1, \ldots, n\}$. If the function $f$ assumes its maximum for large, positive values $p$, this constraint is easily satisfied. This may be not the case if $p_k$ is quite small, or if the "true" parameter is even zero. Random structures of the data may yield a negative optimum that is not acceptable anymore (perhaps the model cannot be evaluated anymore if, for example, a function $\log(p)$ is involved, etc.). We need to prevent the optimum from becoming negative.

For the hill-climbing algorithm, the remedy is as follows. In order to prevent the trajectory of the ODE $\dot{p} = \nabla f(p)$ from leaving the positive cone of $\mathbb{R}^n$, the vector $\nabla f(p)$ needs to point inward if a component of $p$ becomes negative. We may ensure this by using the penalized function

$$\tilde{f}(p) = f(p) + g(p),$$

where

$$g(p) = -\sum_{i \in K} e^{-p_i/\varepsilon}$$

is the penalty term that becomes large if $p$ tries to leave the positive cone. If $p_k \gg 0$, $k \in K$, then $g(p)$ and its derivatives are almost zero, $\|\nabla g(p)\| \approx 0$. Only if a component $p_k, k \in K$, tends to zero does $\|\nabla g(p)\| \sim 1/\varepsilon$ become large.

For the generic algorithm, it is quite straightforward to prevent negativity of the solution. We discard all test points that do not fulfill the positivity condition. Then our optimum will have the required properties.

## 7.5 What Did We Learn?

First, we need to specify a model. How to perform this task is the main theme of the present book. Once we do have the model, we are interested in rigorous analysis in order to obtain an idea about the behavior we may expect. Often, this task cannot be performed completely; however, simulations and numerical analysis will guide us to a certain degree. These results may lead to an initial, qualitative bridge between experiment and model. We may, for example, check if periodic behavior that was observed in nature can be explained by the model. A stronger bridge is a (semi)quantitative analysis of data. In this case, we fit parameters of a model in order to recover and predict experimental results in a quantitative way.

At the center of all parameter fitting is a quality measure for a parameter set with respect to experimental data. The better a given parameter set fits the data, the larger this quality measure. To derive this quality measure, we assume that the data are typical, that is, the probability of finding these data is large. In general, a model predicts the probability of given data as a function of parameters. The function that maps parameters into the probability of data is called the likelihood function. The idea is to find the parameter set that maximizes the likelihood function, the so-called maximum likelihood estimation.

First we need to know the likelihood function. In stochastic models, the likelihood is a canonical outcome of the model; for deterministic models (where no stochasticity is involved), we have to add a stochastic component, which is mostly the measurement error. The deterministic model predicts the expected values for a measurement; the actual data point is distributed according to a normal distribution with a certain variance around this expected value. If the measurement errors are independent for all data points with constant variance, this approach leads to the usual least-squares method.

Next we maximize the likelihood. In almost all cases, this task can only be performed numerically. Optimization algorithms that are able to deal with high dimensions (a lot of parameters) and a rough fitness landscape (a lot of nonlinearity in the models leading to many local maxima) are required. It seems that stochastic algorithms, like the generic algorithm, are well suited to performing this task.

Once we know the likelihood function and the maximum likelihood estimator, some diagnostic tools are provided by the asymptotic theory (large number of data). One may derive confidence intervals, which tell us something about the uncertainty of our estimation. These confidence intervals are closely related to tests, so that we are able to ask whether the data support the hypothesis that a parameter of the model is nonzero (this is the most common question in practical applications). The answer is "yes" if the confidence interval does not include zero. We may also compare models either by testing, in case they are nested, or by the more heuristical approaches like AIC or cross validation. Model comparison is of special interest if one extends simple toy models to complex and more realistic quantitative models. In a toy model, many effects are neglected. One may add one effect after the other, always ensuring by model comparison tools that the extension of the model is supported by the data.

However, as a rule, in case of doubt, we should trust our intuition instead of formal model comparison tools, especially if we have experience with experiments, models, and data.

## 7.6  Further Reading

A nice, basic introduction to the ideas of likelihood, maximum likelihood estimators, and confidence intervals can be found in the book by Adler [1]. An introduction to the Bayesian approach is given in the book by Gilks, Richardson, and Spiegelhalter [67]. In several short articles, the idea of Bayes statistics, related numerical algorithms, problems of model comparison, and applications are described. It is a pleasure to browse through this book. A source of interesting examples for statistical tests and parameter estimation are the collections [147, 148].

A general introduction to identifiability of parameters and observability of the state for compartmental models, with an emphasis on linear models, is the book by Godfrey [69]. In this book, applications from biology (metabolism, pharmakocinetics, and ecology) are also outlined. Especially for PET, these aspects are addressed in [74].

The background of the AIC (historical as well as mathematical) is described in the article of Akaike [2]. Two articles in [67] are also concerned with model comparison and describe several methods (with practical applications). The book by Burnham and Anderson [34] provides a comprehensive introduction to information theory, with many examples.

There are a lot of books concerned with optimization theory. An informal introduction for the deterministic algorithms can be found in the books of Fletcher [58, 59]. A more mathematically rigorous exposition is given in Hiriat-Urruty [85]. Both books also address several methods for incorporating constraints like positivity. The classical book for genetic algorithms is Holland [88]. This book is a rather intuitive introduction to stochastic optimization algorithms and emphasizes looking at a problem that parallels the theory for biological evolution. For algorithms like the generic algorithm and simulated annealing one may consider the book [163], which also includes many more references.

## 7.7  Exercises for Parameter Estimation

**Exercise 7.7.1: Maximum likelihood estimation for binomial distribution.** *Consider the classical experiment of tossing a coin. We toss a coin n times and obtain tails k times. Determine the maximum likelihood estimator for the model, that is, the maximum of the function*

$$g(p) = \ln \left[ \left( \begin{array}{c} k \\ n \end{array} \right) p^k (1 - p)^{n-k} \right].$$

**Exercise 7.7.2: Maximum likelihood estimation for exponential distribution.**

(a) *Consider data points $x_1, \ldots, x_n$, which are n independent realizations of an exponentially distributed random variable X; that is, for any $0 \le a < b$ we find*

$$P(a \le X \le b) = \int_a^b \lambda e^{-\lambda x} \, dx.$$

*Derive the likelihood for the data points $x_i$, the log-likelihood, and the maximum likelihood estimator for the parameter $\lambda$.*

(b) *What is the interpretation of $\lambda$? How can we interpret the maximum likelihood estimator?*

(c) *Assume that you know the infectious period for four cases of measles (dummy data):*

$$3\ Days, 5\ Days, 4\ Days, 7\ Days.$$

*Compute the recovery rate, assuming an exponential distribution of the infectious period. Is this assumption appropriate? If we compute this rate in order to use it in an ODE describing the dynamics of measles, is this assumption crucial?*

**Exercise 7.7.3: Hill-climbing algorithm.**

(a) *Consider a function $f \in C^1(\mathbb{R}, \mathbb{R})$ that has a unique maximum. Find an example where the hill-climbing algorithm fails to determine this maximum.*

(b) *Consider a function $f \in C^1(\mathbb{R}^n, \mathbb{R})$ that has a unique maximum and $\nabla f = 0$ only at the location of this maximum. Show that the hill-climbing algorithm never fails in this situation.*

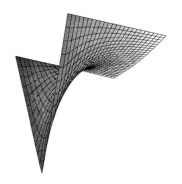

# Part II
# Self-Guided Computer
# Tutorial

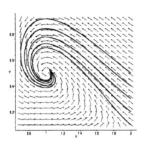

# Chapter 8

# Maple Course

## 8.1 First Steps in Maple

In this part of the book we show how to use a computer software package (such as Maple) to simulate the various model types that were introduced in Part I, on theory. Moreover, we show how to use the models for data fitting and parameter estimation. The computer course is designed so that it can be used by students who have no computational experience at all, as well as by students who are already familiar with a computer software package (such as Maple, *Mathematica*, MATLAB, C++, or similar).

On Unix machines, Maple is started with the command:

```
xmaple
```

In Windows, Maple is started by clicking the following menu items:

> Start
> Programs
> Maple 9
> Maple 9

(note that you may have a different version of Maple, such as Release 8). A large Maple window will appear. Inside this window is another, smaller window, called a Maple worksheet. All Maple commands are entered in the worksheet.

### 8.1.1 Constants and Functions

**Constants**

Try the first example:

```
> a := 5;
> b := 7;
> a*b;
```

You should see

```
> a := 5;
```

$$a := 5$$

```
> b := 7;
```

$$b := 7$$

```
> a*b;
```

$$35$$

The meaning of these lines should be apparent. The first command gives the value 5 to the constant named $a$ and the second line gives the value 7 to the constant named $b$. The third line multiplies the two constants. Each command is followed by a semicolon. If the semicolon is missing, then the program expects more input, since it assumes that the command is not complete.

The ":=" command assigns a value permanently, whereas the "=" command assigns a value temporarily. Try the following:

```
> c = 3.2;
> a*c;
```

You will get:

```
> c = 3.2;
```

$$c = 3.2$$

```
> a*c;
```

$$5\,c$$

Although we set $c$ to be 3.2, it appears that Maple forgot about it and just used the symbol $c$. Here we encounter the fact that Maple is a computer-algebra package, which calculates with both numbers and symbols.

## Functions

We will now define our first function (note that Maple is very particular about the syntax of its commands and the punctuation; however, it is not very particular about spacing within commands, and you may find it helpful to insert spaces to make the commands more readable, or you may eliminate spaces for more efficient typing):

```
> f := x -> a*exp(b*x);
```

$$f := x \to a\ \mathbf{e}^{(bx)}$$

We defined a function $f(x)$, where the argument $x$ has been omitted on the left-hand side. We use "->" to indicate that $x$ is mapped to $a\exp(bx)$. Note that it is important to include the multiplication signs in the Maple commands. What happens when you omit them?

Now we evaluate $f$ at $x = 3$:

```
> f(3);
```

$$5\,\mathbf{e}^{21}$$

Note that $a$ and $b$ have been replaced by the corresponding values defined above.

Maple will recognize an expression as a function *only* if you specify the independent variable. Thus, if you simply enter into Maple

```
> f1 := a*exp(b*x);
```

then you cannot evaluate $f_1$ at any point, say $x = 3$, by entering

> f1(3);

A way you could still get a value for $f_1$ at $x = 3$ is to use the subs command:

> subs(x=3, f1);

$$5 \, e^{21}$$

This is a little tedious, and it is better to use the first method described above.

Functions of more than one variable are defined similarly. For example, $g(x, a, b) = ae^{bx}$ is entered into Maple as

> g := (x, a, b) -> a*exp(b*x);

$$g := (x, a, b) \to a \, e^{(b \, x)}$$

## The evalf Command

Certainly, $f(3) = 5e^{21}$ is correct, but we would also like to have a decimal. We force Maple to evaluate a floating point number with the evalf command:

> evalf( f(3) );

$$0.6594078670 \quad 10^{10}$$

## Plotting Functions

We would like to plot the function $f$ defined above:

> plot( f(x), x=0..0.5 );

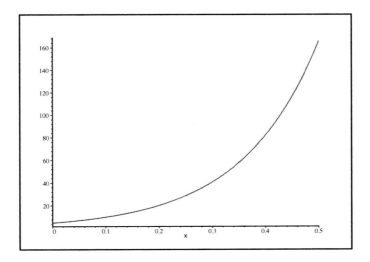

The first argument of the plot command is the function to be plotted, and the second argument specifies the domain.

**Exercise 8.1.1.**

(a) *Define a function $h(t) = 5\sin(ct)$. Don't forget about the implied multiplications!*

(b) *Choose $c = 1.3$ and evaluate h at $t = 0.0$, $t = 1.0$, and $t = 3.2$.*

(c) *Plot the function h for $t \in [0, 5]$.*

## Getting Help

For any command, you can use the help manual, included with the Maple program, to get additional information. For example, to find out more about the `plot` command, type

```
> ?  plot
```

Of course, the "?" command is of little use when you don't know the name of the command for which you need help. At this time, you should explore the general help menu, obtained by clicking on the help button in the top right corner of the main Maple window. Perhaps one of the most useful menu items is "Topic Search". Select it, and type `plot`, and see what comes up. Also note the topic index at the top of any Maple help window. If you have trouble finding what you're looking for via the "Topic Search", try navigating the help index.

## Saving Your Worksheet

To save the worksheet you are working in, click on "File", and then on "Save". Once you have saved your worksheet, a more convenient way of saving it at any later time is by holding the "Control" key and typing "s" on the keyboard.

## Executing the Whole Worksheet

If, in your worksheet, you had defined a := 5, and then at the bottom of your worksheet after reopening Maple you enter a, you will get

```
> a;
```

$$a$$

Maple clears its internal memory after closing the worksheet, so you would have to either re-execute the command where you had defined $a$ to be 5, or (a more practical thing to do) you could execute the whole worksheet by clicking on "Edit", "Execute", then "Worksheet". Now everything is as it was before you closed the worksheet.

## Clearing Maple Memory

It will be helpful to know that there are several commands that can be used to clear Maple's internal memory, either partially or completely. To clear a specific variable, use the `unassign` command (the quotes around the variable name are important):

```
> d := 1;
```

$$d := 1$$

```
> d;
```

$$1$$

```
> unassign( 'd' );
> d;
```

$$d$$

You can also use the `unassign` command to clear several variables at once (for example, `unassign( 'a', 'b' )`). To clear Maple's entire internal memory, use the `restart` command. After entering this command, Maple acts as if you had just started. It is good practice to start each new worksheet with the `restart` command. That way, when you re-execute the entire worksheet at a later time, Maple's memory will be cleared first, and Maple won't be confused by previous definitions and declarations.

## 8.1.2 Working with Data Sets

In most applications, it is necessary to work with experimental data. Moreover, data can be analyzed using mathematical models. To illustrate this procedure, we use the following example, listing the mass and size of 10 brown trout (*Salmo trutta forma fario*) in Table 8.1.

### Lists and the `seq` Command

Each experimental measurement consists of two numbers, mass and size. We will define two lists to save these measurements, one for the mass and one for size. A list in Maple is an ordered set of elements enclosed in square brackets, and the elements in the list can be numbers, lists themselves, and so on. We begin by defining the list of masses:

```
> mass := [31,45,52,79,122,154,184,210,263,360];
```

$$mass := [31, 45, 52, 79, 122, 154, 184, 210, 263, 360]$$

**Table 8.1.** *Brown trout data.*

| Mass (in grams) | Size (in mm) |
|:---:|:---:|
| 31 | 140 |
| 45 | 160 |
| 52 | 180 |
| 79 | 200 |
| 122 | 220 |
| 154 | 240 |
| 184 | 260 |
| 210 | 280 |
| 263 | 300 |
| 360 | 320 |

We can extract specific elements from the list by specifying their position in the list. For example:

```
> mass[5];
```

$$122$$

We continue by defining the list of sizes. Of course, we could use a command similar to the one above. But it should be obvious that the size of the fish increases by 20 from fish to fish, with the size of the $i$th fish given by $120 + i * 20$. We can take advantage of this observation, and use the seq command to assign the list of sizes in a clever way, as follows:

```
> size := [ seq( 120+i*20, i=1..10 ) ];
```

$$[140, 160, 180, 200, 220, 240, 260, 280, 300, 320]$$

The seq command is used to construct a sequence of values according to some given rule, specified in the first argument, here the formula for the size of the $i$th fish. The second argument specifies that only 10 values are needed, obtained by letting $i$ be the integers from 1 to 10 in turn. Since the seq command is surrounded by square brackets, the sequence is placed in a list, which is the data structure we need for the remainder of the data analysis.

### Plotting with Lists

We now use the lists we have just defined to show the brown trout data graphically. We begin by creating a new list containing the coordinates of the points to be plotted, with each coordinate in the form [mass[i],size[i]]:

```
> L := [ seq( [mass[i],size[i]], i=1..10 ) ];
```

$$L := [[31, 140], [45, 160], [52, 180], [79, 200], [122, 220], \ldots, [360, 320]]$$

To show the data graphically, we again use the plot command, but now with a different set of arguments than before:

```
> plot( L, style=point, symbol=circle );
```

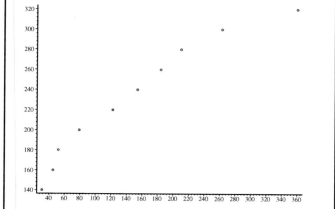

**Exercise 8.1.2.** *Research the* plot *command and remake the plot, this time labeling the axes and adding a title. Note that if you label your axes* mass *and* size, *you will run into trouble, because these words have already been defined as arrays. It is always good practice to enclose your labels in double quotes; that is, use* ``mass'' *and* ``size''.

### Data Transformations

The above plot supports the idea of a power law of the form

$$\text{size} = a \, \text{mass}^b$$

to describe the brown trout data. We would like to find $a$ and $b$ (the fitting parameters). If we take the logarithm of the above formula,

$$\ln(\text{size}) = \ln(a) + b * \ln(\text{mass}),$$

then ln(size) is a linear function of ln(mass), and $a$ and $b$ can be found easily from the $y$-intercept and the slope of the function. We first transform the brown trout data to a logarithmic scale:

```
> log_m := [ seq( evalf(log(mass[i])), i=1..10 )
  ];
```

$$log\_m := [3.433987204, 3.806662490, 3.951243719, \ldots, 5.886104031]$$

```
> log_s := [ seq( evalf(log(size[i])), i=1..10 )
  ]:
```

Here, we used a colon (":") to finish the last command instead of a semicolon (";"). The colon means that the result of this command will not be printed on the screen. If at a later time we are interested in seeing the value of the list log_s, we can use the print command:

```
> print( log_s );
```

$$[4.941642423, 5.075173815, 5.192956851, \ldots, 5.768320996]$$

In the next section, we use the lists containing the transformed data to determine the values of the fitting parameters $a$ and $b$ by linear regression.

## 8.1.3   Linear Regression

We expect the transformed data to have a linear relationship. Let's have a look:

```
> log_L := [ seq( [log_m[i],log_s[i]], i=1..10 )
  ]:
> plot( log_L, style=point, symbol=circle );
```

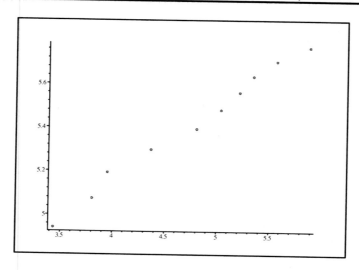

From the graph, it appears that indeed there is a linear relationship between the mass and the size of the fish. We now try to find the regression line, which is the straight line that best fits the data. To learn more about regression, see Devore and Peck [44].

## Loading Maple Library Packages

Statistical functions are available in Maple. However, they are not activated right away. We have to include the appropriate library package:

```
> with( stats );
```

$$[anova, \; describe, \; fit, \; importdata, \; random, \; statevalf, \; ...]$$

When we activate the statistics package, Maple shows which additional functions are available. Other useful packages are

- `linalg`: contains linear algebra functions

- `plots` and `plottools`: these packages contain a variety of plotting functions and tools

- `DEtools`: contains functions that help you work with differential equations

To learn more about these packages, click on the Help button in your Maple window, select the "Using Help" menu item, and navigate to "Mathematics . . . /Packages . . ." in the resulting help window.

**Fitting with the Least-Squares Method**

We use the `fit` command to find the regression line:

```
> fit[ leastsquare[ [x,y], y=a1*x+b1,
    {a1,b1} ] ]( [log_m,log_s] );
```

$$y = 0.3350651881\,x + 3.816026381$$

The `fit` command performs some sort of linear or nonlinear regression. The argument `leastsquare` means that the sum of the squares of the pointwise distance between the curve and the data should be minimized. The argument `[x,y]` specifies the variables of the fitting function. We expect a linear relationship. Hence, we use a general linear hfunction $y = a_1^* x + b_1$, where $a_1$ and $b_1$ have to be determined by the fit. The argument `{a1,b1}` specifies the unknown parameters. Finally, in the parentheses at the end of the command, we specify the two data sets which have to be fitted. Maple finds that the best fit is achieved when $a_1 = 0.3350651881$ and $b_1 = 3.816026381$.

We could also try a quadratic fit. If we assume that there is a quadratic relation, then we write:

```
> fit[ leastsquare[ [x,y], y=a1*x*x+b1*x+c1,
    {a1,b1,c1} ] ]( [log_m,log_s] );
```

$$y = -0.009928883861\,x^2 + 0.4272857455\,x + 3.607969257$$

As you can see, the coefficient of $x^2$ is very small compared to the other coefficients. This indicates that a linear fit is better.

**Checking the Fit**

Now we would like to plot the regression line. We need to convert the results of the fit to a function. We do this in two steps. First, we assign the result of the fit to `fitresult`:

```
> fitresult := fit[ leastsquare[ [x,y],
    y=a1*x+b1,
    {a1,b1} ] ]( [log_m,log_s] );
```

$$\textit{fitresult} := y = 0.3350651881\,x + 3.816026381$$

Next, we assign a function `linfit` with the `rhs` command:

```
> linfit := rhs( fitresult );
```

$$\textit{linfit} := 0.3350651881\,x + 3.816026381$$

Note that Maple understands `linfit` to be a function, but not `fitresult`. Now let's plot the fitted function together with the brown trout data:

```
> plot( [log_L,linfit], x=3.2..6,
    style=[point,line],
    symbol=circle );
```

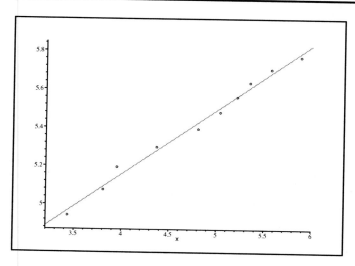

That's looking very good! Finally, we will check the fit with the original data set on the nonlogarithmic scale. To do so, we need to transform `linfit` to the nonlogarithmic scale, which we accomplish with the following three steps:

```
> g1 := exp( linfit );
```

$$g1 := e^{(0.3350651881\,x+3.816026381)}$$

```
> g2 := subs( x=log(m), g1 );
```

$$g2 := e^{(0.3350651881\,ln(m)+3.816026381)}$$

```
> s := simplify( g2 );
```

$$s := 45.42335413\,m^{\left[\frac{3350651881}{10000000000}\right]}$$

We used the `subs` command to replace `x` with `log(m)` in the function `g1`, where `m` represents the mass of the trout, and we used the `simplify` command to find an easier way of presenting the solution. Sometimes the `simplify` command produces an easier expression, so it is a useful tool to have available (however, in some cases the command may not help much).

Now let's see how our function for the size as a function of mass, `s(m)`, fits the original brown trout data (contained in the list `L`) on the nonlogarithmic scale:

```
> plot( [L,s], m=30..360, style=[point,line],

  symbol=circle );
```

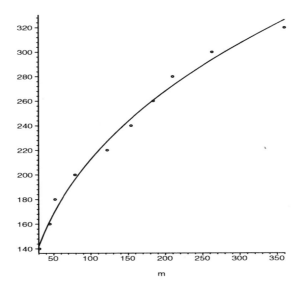

This also is looking very good!

**Exercise 8.1.3: World population 1850–1997.** *The world population from* 1850 *to* 1997 *is given in Table* 8.2.

(a) *Define lists containing the data.*

(b) *Plot the data (world population versus year).*

(c) *Transform the data set for the world population to a logarithmic scale, and show your results in a graph.*

(d) *Fit the data of the world population. Show the fit on both the logarithmic and the nonlogarithmic scale. What type of function did you use to fit the data?*

(e) *Try fitting the data directly with a quadratic and a cubic function. Which fit do you think is best? If the simplified version of the cubic function does not plot, then try it without the* simplify *command.*

**Table 8.2.** *World population data.*

| Year | Size of population (in millions) |
|------|----------------------------------|
| 1850 | 1200 |
| 1940 | 2249 |
| 1950 | 2509 |
| 1960 | 3010 |
| 1970 | 3700 |
| 1985 | 4800 |
| 1997 | 5848.7 |

## 8.2   Discrete Dynamical Systems: The Ricker Model

We briefly discussed the Ricker model in Section 2.2.4 and saw that use of the model is appropriate for describing populations with nonoverlapping generations. We determined the fixed points of the model, as well as their stability, and alluded to the fact that the Ricker model can exhibit complex dynamics, such as cycles and chaos, for certain choices of the model parameters. We will now use Maple to conduct a thorough investigation of the dynamics of this model.

We begin with simplifying (2.24) by letting $a = e^r$, $b = r/k$, and $\bar{x}_n = bx_n$. After dropping the overbars, we obtain the simplified Ricker model,

$$x_{n+1} = ax_n e^{-x_n}. \tag{8.1}$$

Of interest for the remainder of this section is the behavior of (8.1) and its dependence on the value of the model parameter $a$. Although the restriction $r > 0$ for the original Ricker model implies $a > 1$, we will study (8.1) in more generality, and allow $a > 0$.

Let $x^*$ be a fixed point of (8.1); that is, $x^*$ satisfies

$$x^* = f(x^*) = ax^* e^{-x^*}.$$

Although it is easy to solve this equation for $x^*$ by hand, we will use Maple so that we can learn the solve command:

```
> f := x -> a*x*exp(-x):
> solve( f(x)=x, x );
```

$$0, -\ln\left(\frac{1}{a}\right)$$

The arguments of the solve command are, first, the equation and, second, the quantity we want to find. In this case, we obtain two fixed points: $x_1^* = 0$ and $x_2^* = \ln(a)$. The trivial fixed point $x_1^* = 0$ describes a population which is extinct. Note that this fixed point exists for all values of the model parameter $a$. The nontrivial fixed point, $x_2^* = \ln(a)$, exists only for $a > 1$ (this is consistent with our earlier observation that $r > 0$ implied $a > 1$).

To determine the stability of the fixed points, we need

$$f'(x) = ae^{-x}[1 - x],$$

so that

$$f'(x_1^*) = f'(0) = a,$$
$$f'(x_2^*) = f'(\ln(a)) = 1 - \ln(a).$$

**Exercise 8.2.1.**

(a) *At what value of $a$ does the stability of the trivial fixed point, $x_1^* = 0$, change?*

(b) *Plot the function $g(a) = 1 - \ln(a)$. For which values of $a$ is $|g(a)| < 1$ ($|g(a)| > 1$)? That is, when is the nontrivial fixed point stable (unstable)?*

(c) *Sketch (by hand) a partial bifurcation diagram for the simplified Ricker model, (8.1), as we did in Section 2.2.3 for the rescaled logistic map (see Figure 2.7).*

In terms of the simplified Ricker model, we say that there are bifurcations at $a = 1$ and $a = e^2$. The bifurcation at $a = 1$ is called a *transcritical bifurcation*. We defer discussion of the bifurcation at $a = e^2$ to later.

We are interested in plotting solution trajectories for various values of the model parameter $a$. For a given value of $a$, suppose that we wish to iterate the map 20 times and plot the iterates as a function of the iteration number. This means that we need to create a list with 21 coordinates of the form $[i, x_i]$ (the 20 iterates plus the initial condition). Since we need to keep track of the current iterate to create the next, writing an appropriate `seq` command is a bit tricky. Perhaps an easier way to create the list of coordinates is to use a `for` statement and build up the list recursively with the `op` command. We take a small detour now to learn about these two commands separately.

### The `for` Statement

The `for` statement is a type of repetition statement. It provides the ability to execute a command or a sequence of commands repeatedly. The sequence of the commands to be executed repeatedly is listed between a `do` command and an `od` command (`od` is `do` spelled backwards). The remainder of the statement specifies the number of times that the sequence of commands needs to be executed. Before executing the command below, note that we have stretched the statement over four lines and added indentation for readability. If you would like to do the same (without getting warning messages), you can press the "Shift" and "Enter" keys. Maple will give a new line to type on, but it will treat it as the same line of input. The "Shift + Enter" trick is very useful when editing procedures, for example, to insert new lines. You do not need to stretch the statement over four lines (you can write the entire statement on one line if you wish). Try the following now:

```
> for i from 0 by 2 to 20 do
>     print(i);
>     print(i+1);
> od;
```

The variable `i` counts the number of times the sequence of commands is executed. Here, Maple is instructed to start with $i = 0$ (`from 0`), increase $i$ by 2 (`by 2`) at the end of each repetition, and terminate the repetition as soon as $i > 20$ (`to 20`). Did the command give the result you expected?

### The `op` Command and Recursive Definition of Lists

The `op` command removes the outer square brackets of a list:

```
> L1 := [ 1, [1,2], [1,2,3], 2 ];
```

$$L1 := [1, [1, 2], [1, 2, 3], 2]$$

```
> op(L1);
```

$$1, [1, 2], [1, 2, 3], 2$$

We use this to our advantage to insert elements in a list. For example, the following inserts [33,12,-3] into L1 (at the end) to create a new list L2:

```
> L2  :=  [ op(L1),  [33,12,-3]  ];
```

$$L2 := [1, [1, 2], [1, 2, 3], 2, [33, 12, -3]]$$

To create a list recursively, we start with an empty list (L:=[]), and use a for command to insert elements into the list one at a time. For example,

```
> L  :=  []:
> for i from 1 to 5 do
>     L  :=  [ op(L),  [i,i*i]  ];
> od;
```

$$L := [[1, 1]]$$

$$L := [[1, 1], [2, 4]]$$

$$L := [[1, 1], [2, 4], [3, 9]]$$

$$L := [[1, 1], [2, 4], [3, 9], [4, 16]]$$

$$L := [[1, 1], [2, 4], [3, 9], [4, 16], [5, 25]]$$

Note that our for loop repeatedly redefines the *same* variable, L, recursively. In detail, here's what happened. Initially, Maple has the information that L is the empty list, L = []. The command op(L) returns the empty list with the brackets removed. Hence, when Maple starts the for loop, at $i = 1$, it uses this information, and

```
L=[op(L),[1,1*1]]  =  [[1,  1]]
```

Thus, after the first iteration of the for loop, Maple is at the stage where L=[[1,1]], and op(L) returns [1,1]. That is, the second time through the loop, we get

```
L=[op(L),[2,2*2]]
=[[1,1],[2,4]]
```

Maple continues in this fashion until it reaches the final iterate ($i = 5$ in this case), and then the final list, [[1,1],[2,4],[3,9],[4,16],[5,25]], is our list L.

## Plotting a Trajectory

We now return to the Ricker model and use the new Maple commands to create a list of coordinates of a trajectory, let's say for $a = 0.8$. (From the linear stability analysis, we know that the trivial fixed point $x_1^* = 0$ is stable in this case, and the nontrivial fixed point $x_2^* = \ln(a)$ does not exist yet.) We begin by specifying the value of the model parameter $a$

and setting the initial condition $x_0$ (the zeroth iterate):

```
> a := 0.8:   iter := 1.0:
```

Next, we create a list of coordinate points, using the iteration number i as the $x$-coordinate in $[i, x_i]$, and the corresponding iterate iter as the $y$-coordinate in $[i, x_i]$:

```
> L := [ [0,iter] ]:
> for i from 1 to 20 do
>    iter := evalf( f(iter) );
>    L := [ op(L), [i,iter] ];
> od:
```

Note that we have used the evalf command to force a numerical evaluation of the iterates instead of a symbolic evaluation (this speeds up the calculation significantly).

Finally, we plot the list of coordinates:

```
> plot( L, style=point, symbol=box );
```

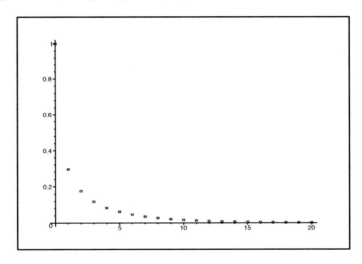

As you can see, with $a = 0.8$, the population dies out, at least with the initial condition $x_0 = 1.0$. The numerical result is consistent with the results of our linear stability analysis. You can check that the population dies out with other initial conditions as well by making an appropriate change on the line specifying the initial condition and pressing the return key a few times to re-execute that line and the following lines.

## 8.2.1   Procedures in Maple

We're interested in seeing the behavior of the model for different values of the parameter $a$. We can continue making appropriate changes in the lines we already have on the screen. Instead of changing the value of $a$ over and over, it is more elegant to define a procedure which plots the trajectory for a given value of $a$. Here's how we define such a procedure (note that Maple ignores lines that begin with "#", so we can use these to include comments in our Maple program):

```
> # define a procedure
> plot_trajectory := proc( a )
>    # definition of local variables
>    # (valid only in this procedure)
>    local iter, i, L;
>    # initialize the first iterate
>    iter := 1;
>    # initialize the list where we
>    # collect coordinate points
>    L := [ [0,iter] ];
>    # algorithm to calculate the trajectory
>    # for the given value of a
>    for i from 1 to 30 do
>       iter := evalf( a * iter * exp(-iter) );
>       L := [ op(L), [i,iter] ];
>    od:
>    # plot the trajectory
>    plot( L, style=point, symbol=box );
> end;
```

We can now easily plot trajectories for different values of the parameters $a$:

```
> plot_trajectory(0.8);
> plot_trajectory(1.0);
> plot_trajectory(5.0);
> plot_trajectory(8.0);
> plot_trajectory(13.0);
> plot_trajectory(14.5);
> plot_trajectory(20.0);
```

You should observe that the qualitative behavior of the Ricker model changes drastically when the value of the parameter $a$ is changed, as should be expected from the results of the linear stability analysis. Before we investigate interesting types of behavior, we will dissect the above Maple code.

1. First, we give the name of the procedure, which is followed by ":=".

2. The keyword `proc` indicates that we are going to define a procedure. In parentheses, we give a list of input parameters, which are separated by commas.

3. Next, we define local variables. The values for these variables are only known to this particular procedure and they cannot be used outside of this procedure.

4. Then the Maple commands follow, and define the action of the procedure.

5. The `end` command ends the definition of a procedure.

**Exercise 8.2.2.** *As our procedure stands right now, the initial condition of the iteration is set by the programmer (you) within the procedure. Modify the procedure so that the initial condition for the iteration also can be specified by the user. You should test your procedure with a variety of initial conditions.*

Maple has many other commands to help with programming. Go to the help menu, then click on "Programming" and "Flow Control" to find out about some more of them. Exercise 8.2.3 introduces you to `if` statements.

**Exercise 8.2.3.** *The "signum" function, which is thought of the "sign" of a number, can be defined as*

$$\text{sgn}(x) := \begin{cases} -1 & \text{if } x < 0, \\ 0 & \text{if } x = 0, \\ 1 & \text{if } x > 0. \end{cases}$$

*Write a Maple procedure to define the signum function in this way. Note that Maple's own name for the signum function is* `signum` *so you must give it a different name.*

### 8.2.2   Feigenbaum Diagram and Bifurcation Analysis

We would like to understand the changes in the qualitative behavior of the Ricker model. We will focus on the steady-state behavior of the model. We ask the following question. For a given value of the model parameter $a$, what is the steady-state behavior of the model? We use the numerical capabilities of Maple to help us answer this question and create a Feigenbaum diagram (also known as an orbital bifurcation diagram), as we did for the rescaled logistic equation in Section 2.2.3 (see Figure 2.13).

To accomplish this, we do the following. For each value of $a$ of interest, we ask Maple to iterate the model a large number of times so that we can be sure that we have reached steady state. Then we throw out most of the iterations and save only the last few. Finally, we plot the iterations that we kept ($a$ is on the $x$-axis, and the value of the population at steady state is on the $y$-axis). If the model converges to a fixed point for a particular value of $a$, then the points for that value of $a$ will all be plotted on top of each other. If the model converges to an orbit of period 2, then there will be two distinct points for that value of $a$, and so on. You will create the Feigenbaum diagram for the Ricker model in two steps via the following two exercises.

**Exercise 8.2.4.** *Define a procedure which iterates the Ricker model a total of* 600 *times for a particular value of a. The arguments of the procedure should be the parameter a and the initial condition $x_0$. Your procedure should return a list (e.g.,* `mylist`*) that contains the coordinates for the points that will appear in the Feigenbaum diagram. Note that you do not need to create coordinates for the first* 500 *iterates, only for the last* 100 *iterates. Hint: Each coordinate in the list should be of the form* `[a,iter]`*, not* `[i,iter]`*. To return the list, the last command in your procedure before the* `end` *command should be* `mylist;`*.*

**Exercise 8.2.5.**

(a) *Define a big list, say* `biglist`*, which collects all coordinates for the Feigenbaum diagram for values of a from* 0 *to* 10 *in steps of* 1. *For each value of a, use the procedure from the previous exercise.*

(b) *Plot* `biglist` *to view the Feigenbaum diagram. If you're confident of your results, repeat the exercise with smaller steps of a to obtain more details in the diagram (be careful not to make the steps too small, though, or you will have to wait a long time to view the result).*

(c) *Recall that the linear stability analysis of Exercise* 8.2.1 *predicted a* transcritical *bifurcation at a* = 1 *and another bifurcation at a* = $e^2$. *How are the bifurcations manifested in the Feigenbaum diagram(s) you produced? What kind of bifurcation occurs at a* = $e^2$?

The diagram you produced in the previous exercise shows a typical route to chaos, namely, the period-doubling route. For small values of $a$, we find one stable fixed point. As $a$ increases, the fixed point loses its stability at a *period-doubling bifurcation*, and we obtain a stable orbit of period 2 instead. As $a$ increases further, the period is doubled again to 4 and further to 8 and so on. We will analyze this process in more detail below.

Recall that a 2-cycle is defined by the values $u$ and $v$ with

$$u = f(v), \qquad v = f(u).$$

If we apply $f$ twice, we get

$$u = f(f(u)), \qquad v = f(f(v)).$$

This suggests that we should consider the second-iterate function $f(f(.))$ instead of $f(.)$, since fixed points of this function correspond to a 2-cycle.

We will find the iterates visited during the 2-cycle for $a = 8$ and verify that they correspond to the fixed points of the second-iterate function, which we define as follows:

```
> f2 := x -> f( f(x) );
```

$$f2 := x \rightarrow f(f(x))$$

**Exercise 8.2.6.**

(a) *Estimate the values of the iterates visited during the 2-cycle observed for a* = 8 *from the bifurcation diagram.*

(b) *Obtain accurate values of these iterates by using your* `plot_trajectory` *procedure.*

(c) *Plot f(f(x)) for a* = 8, *together with the diagonal line y* = *x, and verify that two of the fixed points of the second-iterate function correspond to this same 2-cycle.*

In the above exercise, you should have found that the higher iterate of the 2-cycle lies between 2.6 and 3. We can solve for this iterate, which is one of the fixed points of the second-iterate map, as follows:

```
> solution := fsolve( subs(a=8, f2(x)=x), x,
    2.6..3 );
```

$$solution := 2.772588722$$

Note that `fsolve` finds only one solution at a time.

**Exercise 8.2.7.** *Find the other nonzero solution we expected, and verify that these two nonzero solutions together correspond to the 2-cycle at $a = 8$ (verify that $f(u) = v$ and $f(v) = u$, where $u$ and $v$ are the solutions found).*

Note that you can find the two iterates of the 2-cycle analytically, by using the `solve` command instead of the `fsolve` command, but Maple gives a result in terms of the `RootOf` command. In this case, you must use the `evalf` command to get the numerical answer.

## 8.2.3   Application of the Ricker Model to *Vespula vulgaris*

The common wasp (*Vespula vulgaris*) was introduced into New Zealand sometime in the 1980s. They are aggressive to native insects and compete for resources with native birds. Barlow, Moller, and Beggs [13] observed the insect population over five years in seven locations. They counted the nests per hectare (1 hectare = 2.47 acres) each spring from 1988 to 1992. The data is shown in Table 8.3.

**Exercise 8.2.8.** *In this exercise, we use the Ricker model, (2.24), written in the following form:*

$$x_{n+1} = ae^{-bx_n}x_n.$$

(a) *Fit the Ricker model to the data from Site 1. That is, find the values of the parameters $a$ and $b$ that best fit the data set. Hint: Show that the Ricker model can be transformed so that there is a linear relationship between the quantities $\ln(x_{n+1}/x_n)$ and $x_n$, as follows:*

$$\ln\left(\frac{x_{n+1}}{x_n}\right) = \ln(a) - bx_n.$$

**Table 8.3.** *Data for the common wasp* Vespula vulgaris, *from* [13].

|        | 1988 | 1989 | 1990 | 1991 | 1992 |
|--------|------|------|------|------|------|
| Site 1 | 8.6  | 31.1 | 7.0  | 11.7 | 10.2 |
| Site 2 | 2.7  | 6.9  | 3.3  | 4.4  | 3.1  |
| Site 3 | 10.5 | 15.8 | 8.2  | 11.6 | 12.1 |
| Site 4 | 1.0  | 1.9  | 6.0  | 1.0  | 1.0  |
| Site 5 | 16.0 | 11.8 | 10.0 | 15.7 | 19.9 |
| Site 6 | 18.5 | 32.9 | 17.1 | 13.6 | 13.0 |
| Site 7 | 20.5 | 19.8 | 12.9 | 15.7 | 10.6 |

*Hence, you can find values for a and b from a regression line for*

$$\left( x_n, \ln \left( \frac{x_{n+1}}{x_n} \right) \right), \qquad t = 1988, \dots, 1991.$$

(b) *Check the fit for the data two ways: (1) in a plot of* $\ln (x_{n+1}/x_n)$ *versus* $x_n$, *and (2) in a plot of* $x_{n+1}$ *versus* $x_n$.

(c) *What behavior is predicted for the wasp population at Site 1, based on the results of Exercise 8.2.5?*

## 8.3   Stochastic Models with Maple

Maple has several features for simulating stochastic processes. In particular, it has random number generators. To get Maple to pick a random *integer* between 1 and 6 inclusive, use the `rand` command:

```
> die := rand(1..6);
```

Maple responds with a bunch of computerspeak that defines `die` to be a function with no arguments. To get a random number, or roll the die, type

```
> die();
```

**Exercise 8.3.1.** *Roll your computer die several times. What would constitute a fair die?*

To get Maple to pick a random number between 0 and 1, we need the `random` command in the statistics package. We define a function named `number` that generates the random numbers:

```
> with(stats):
> number := x -> random[ uniform[0,1] ](1);
```

As before, `number` is a function that requires no arguments (x can be considered to be a dummy variable). To get a random number, type

```
> number();
```

**Exercise 8.3.2.** *Generate several of these random numbers. Why do you think the word* `uniform` *appears as one of the arguments in the definition of the* `number` *function?*

Let's try something a little more complicated. We consider an individual jumping equal-sized distances along a horizontal line. At each time step, the individual must make a decision to jump to the right or the left. Suppose that, at each time step, the probability an individual jumps to the right is $R$ and the probability the individual jumps to the left is $L = 1 - R$. Furthermore, we assume that the event of jumping to the right is given by a Bernoulli random variable with probability $p = R$. Recall that a Bernoulli random variable is a discrete random variable that takes on the value 1 with probability $p$ and the value 0 with probability $1 - p$.

**Exercise 8.3.3.** *To make life more interesting, Maple prefers to consider the Bernoulli random variable to be a special case of the binomial random variable. What is the special case, and why does this work?*

Try the following:

```
> R := 0.5:
> rnum := x -> random[ binomiald[1,R] ](1);
> [ seq( rnum(), i=1..20 ) ];
```

As you can see, our function `rnum` randomly returns a 1 with probability *R* and a 0 otherwise, as desired. If we can make a transformation so that we either obtain a 1 with probability *R* and a −1 otherwise, then we can use the result to keep track of the location of the individual. We will call the transformed function `jump`. It is defined as follows:

```
> jump := x -> 2*random[ binomiald[1,R] ](1) - 1;
```

**Exercise 8.3.4.** *Explain the transformation, and evaluate the* `jump` *function a few times. Did it work?*

One way to simulate a sequence of jumps at successive time steps is to first create a function, say `newloc`, whose argument is the location of the individual before a jump and whose value is the location of the individual after a jump, as follows:

```
> newloc := x -> x + jump();
```

By recursive evaluation of `newloc`, we simulate the random walk made by the individual. We let the individual begin at location 5, and then we let the individual make 10 jumps, as follows:

```
> loc := 5;
> for i from 1 to 10 do
>    loc := newloc( loc );
> od;
```

**Exercise 8.3.5.** *Let's visualize the random walk:*

(a) *Create a procedure called* `plot_sim` *that plots location versus time for an individual who makes successive random jumps. Each jump is one unit to the right with probability R, and otherwise one unit to the left. The arguments of* `plot_sim` *should be R, the probability of jumping one unit to the right, a, the initial location of the individual, and* `numjumps`, *the number of jumps the individual makes. Of course, you should incorporate the idea behind the function* `newloc` *defined above.*

(b) *Determine the effect of varying the value of R. Run your procedure several times, keeping the number of jumps at* `numjumps=25`, *and the initial location constant at* `a=0`, *but varying the value of R. Explain the results (it is a good idea to run the simulation several times for each value of R).*

We won't always need to generate a graph of the random walk. We will define a function that returns the location of the individual after `numjumps` steps directly, and try it out for numjumps=25, as follows:

```
> finalloc := (x,numjumps) ->
> (newloc@@numjumps)(x);
> finalloc(0,25);
```

Here "@@" describes functional composition (to find out more about this, use the Maple help feature (?@@)).

We can now evaluate the function finalloc many times so as to empirically deduce the probabilities associated with different endpoints after a fixed number of jumps (25 is a nice number to start with), always starting at the same location. We evaluate the function finalloc(0,25) 500 times, and save the results in a list using the seq command (this will take Maple a while, so be prepared to be patient):

```
> M := 500:
> loclist := [ seq( finalloc(0,25), i=1..M ) ]:
```

To generate a histogram of the results, we use the histogram command in the statplots package (note that we assign the plot to graph1, and then use the display command to view the plot; this may seem inefficient, but this allows us to use the plot again later, as you will see soon):

```
> with( statplots );
> graph1 := histogram( loclist ):
> with( plots ):
> display( graph1 );
```

**Exercise 8.3.6.** *In Chapter 5, we learned that if the random walk is unbiased ($R = 0.5$), the distribution of individuals after a large number of time steps should be approximated by a Gaussian distribution with mean $\mu = 0$ and variance $\sigma^2$ equal to the number of time steps (or the number of jumps, numjumps). We will test the theory here.*

(a) *Use appropriate Maple commands to evaluate the mean and variance of loclist and compare the results with theory.*

(b) *Create a new plot of the appropriate Gaussian (using the theoretical values for the mean and variance) and assign this plot to the variable graph2.*

(c) *Plot graph1 and graph2 on the same set of axes with the following command:*

```
> display( graph1, graph2 );
```

*How good is the approximation?*

(d) *Repeat the above with values of numjumps larger and smaller than 25, and compare.*

**Exercise 8.3.7.** *Theory also tells us that the mean-squared displacement of individuals undergoing the random walk processes should increase linearly with time if the walk is unbiased ($R = 0.5$). Let $x_i(t)$ be the location of individual $i$ at time $t$. Then test the theoretical prediction by using Maple to plot $(\sum_{i=1}^{M}(x_i(t)^2))/M$ versus $t$ for $t = 1, 2, 3, 4, 5$. Test your code with a small value for $M$. When your code works, you will want to use a relatively large value for $M$ (and be patient!). What is the slope of the line? Explain the value of the slope using the theory discussed in Chapter 5.*

# 8.4   ODEs: Applications to an Epidemic Model and a Predator–Prey Model

In a previous section, we studied a model that describes population growth in terms of discrete-time intervals. However, in some cases it is important to know the state of the system at any time. This can be achieved using differential equations. In this section, we focus on ODEs. Recall that an autonomous ODE can be written as

$$\frac{d}{dt}x(t) = f(x(t)),$$

where $t \in \mathbb{R}$, $x \in \mathbb{R}^n$, and $f : \mathbb{R}^n \to \mathbb{R}^n$. The left-hand side, $dx(t)/dt$, is the rate of change of the state variable $x(t)$, and the right-hand side, $f(x(t))$, summarizes all factors which cause a change in $x(t)$ (e.g., birth, death, creation, removal, etc.). We will investigate two specific models, one describing the time course of an infection in a population, and the second describing a simple predator–prey system.

## 8.4.1   The SIR Model of Kermack and McKendrick

To obtain the basic epidemic model of Kermack and McKendrick, we split the population into a class $S$ of susceptible individuals, a class $I$ of infective individuals, and a class $R$ of recovered or deceased individuals. First, we consider the transition from class $S$ to class $I$. Not every encounter between a susceptible and an infective individual leads to infection of the susceptible. We consider a small time step $\Delta t$, and we introduce the parameter $\beta$, which measures the average number of effective contacts per unit time per infective individual (an effective contact is one in which the infection is transmitted from an infective to a susceptible individual). When an infection is successful, the newly infected individual is removed from the class of susceptibles and added to the class of infectives. Thus, in the small time step $\Delta t$, the change in the number of susceptible individuals, $\Delta S$, is

$$\Delta S = -\beta \Delta t S I,$$

and the change in the number of infective individuals, $\Delta I$, is

$$\Delta I = \beta \Delta t S I.$$

Dividing both sides of the equations by $\Delta t$ and taking the limit as $\Delta t \to 0$ gives

$$\dot{S} = -\beta S I,$$
$$\dot{I} = \beta S I.$$

Next, we consider the transition from class $I$ to class $R$. Depending on the disease, infectives either recover (here, we assume that individuals recover with permanent immunity to the disease) or they die. Both cases lead to the same model. We assume that the rate of recovery is $\alpha$. That is, in a time step of $\Delta t$, the number of individuals that undergo the transition from class $I$ to class $R$ is $\alpha \Delta t I$. After the limiting process, the full model then reads

$$\dot{S} = -\beta S I,$$
$$\dot{I} = \beta S I - \alpha I,$$
$$\dot{R} = \alpha I.$$

Note that $R$ is decoupled from the rest of the system (once a solution for $I$ is known, $R$ is known as well, and $R$ does not feed back onto the equations for $S$ and $I$). So it is sufficient to study the first two equations of the model. We won't be able to find an explicit solution, giving $S$, $I$, and $R$ as functions of time $t$. A little later, we will learn how to find solutions numerically. Before we do that, we will look for solutions in the form $I(S)$, that is, $I$ as a function of $S$. We have

$$\frac{dI}{dS} = \frac{dI}{dt}\frac{dt}{dS} = \frac{dI}{dt}\left(\frac{dS}{dt}\right)^{-1} = \frac{\beta SI - \alpha I}{-\beta SI} = -1 + \frac{\alpha}{\beta S}.$$

We can solve this equation by hand, but we choose here to use Maple. First, we tell Maple about the differential equation. We begin by defining $f$ as the right-hand side of the differential equation:

```
> f := x -> -1 + alpha/(beta*x);
```

$$f(x) := x \to -1 + \frac{\alpha}{\beta x}$$

Then we set up the differential equation (we will use small letters instead of capital letters for $S$ and $I$ in our Maple commands, since Maple interprets I as the imaginary unit $i$, where $i^2 = -1$):

```
> eqn := diff( i(s), s) = f(s);
```

$$eqn := \frac{d}{ds}i(s) = -1 + \frac{\alpha}{\beta s}$$

Note that the `diff` command has been used to specify the derivative.

We solve our equation for $I$ as a function of $S$ with the `dsolve` command:

```
> dsolve( eqn, i(s) );
```

$$i(s) = \frac{\alpha \ln(s)}{\beta} - s + \_C1$$

Since we did not specify any initial conditions, the solution includes a constant of integration, here called "$\_C1$" by Maple.

**Exercise 8.4.1.** *Let $I_0$ denote a number of newly infected individuals in an otherwise susceptible population $S_0$.*

(a) *Determine the constant $C1$ such that $I(S_0) = I_0$.*

(b) *Create a Maple function for $I(S)$ with the appropriate constant of integration.*

(c) *Choose some values for the parameters $\alpha$, $\beta$, $I_0$, and $S_0$, and plot the function $I(S)$ (e.g., choose $\alpha = 0.04$, $\beta = 0.0002$, $I_0 = 10$, $S_0 = 990$). What is an appropriate domain for your graph (think about the maximum and minimum number of susceptible individuals during the infection)? Does the infection progress as you expected?*

We would really like to find solutions of the model as a function of time. As mentioned before, we won't be able to do so explicitly. However, we can solve the system numerically. We assign some values for $\alpha$ and $\beta$ and define the system of differential equations:

```
> alpha := 0.04:  beta := 0.0002:
> eq1 := diff(s(t),t) = -beta*s(t)*i(t):
> eq2 := diff(i(t),t) = beta*s(t)*i(t) -
  alpha*i(t):
```

We solve the system with the dsolve command:

```
> numsol := dsolve( {eq1, eq2, s(0)=990.0,
  i(0)=10},
  {s(t),i(t)}, type=numeric, output=listprocedure
  ):
```

Finally, we plot the solution with the odeplot command, found in the plots package. We show the solution two different ways (you should recognize the graph obtained the second way):

```
> with(plots);
> odeplot( numsol, [[t,s(t)],[t,i(t)]], 0..100 );
> odeplot( numsol, [s(t),i(t)], 0..100 );
```

**Exercise 8.4.2.** *Experiment with different values of the model parameters and initial conditions. Try several cases, and verify that there is an epidemic outbreak when $R_0 = S_0\beta/\alpha > 1$ and no outbreak when $R_0 < 1$.*

## 8.4.2  A Predator–Prey Model

We denote the size of the prey population at time $t$ by $x(t)$ and the size of the predator population by $y(t)$. We assume that in the absence of a predator the prey population approaches its carrying capacity as modeled by the logistic law (Verhulst's growth model),

$$\dot{x} = ax(1 - x/K),$$

where $a > 0$ is the per capita growth rate and $K > 0$ is the carrying capacity. We also assume that the predator population cannot survive without prey, and model this with an exponential decay equation,

$$\dot{y} = -dy.$$

For each encounter of a predator with a prey there is a certain probability that the prey will be eaten. We apply the Law of Mass Action and represent removal of prey by predators as $-bxy$, where $b > 0$ is a rate constant. Several successful hunts by the predator will result in the production of offspring. This is modeled with a term $cxy$, where $c > 0$ is a reproduction rate. In general, $b \neq c$ (why?). We obtain the following predator–prey model:

$$\dot{x} = ax\left(1 - \frac{x}{K}\right) - bxy,$$

$$\dot{y} = cxy - dy.$$

We nondimensionalize this model by letting $\tau = at$, $\kappa = \frac{c}{d}K$, $g = \frac{d}{a}$, $u(\tau) = \frac{c}{d}x(\frac{\tau}{a})$, and $v(\tau) = \frac{b}{a}y(\frac{\tau}{a})$, to get

$$\dot{u} = u\left(1 - \frac{u}{\kappa}\right) - uv,$$

$$\dot{v} = g(u - 1)v.$$

The steady states of the system satisfy the following algebraic system:

$$u\left(1 - \frac{u}{\kappa}\right) - uv = 0,$$
$$g(u - 1)v = 0.$$

We solve for the steady states with the `solve` command, for the specific case $\kappa = 2$ and $g = 1$:

```
> kappa := 2:   g := 1:
> solve( {u*(1-u/kappa)-u*v=0,  g*(u-1)*v=0},
  {u,v});
```

$$\{v = 0, u = 0\}, \{v = 0, u = 2\}, \left\{u = 1, v = \frac{1}{2}\right\}$$

Maple finds three steady states. We wish to determine the stability of each of the steady states. For this, we need the Jacobian matrix of the right-hand side of the system of equations, and it is

$$Df(u, v) := \begin{pmatrix} 1 - 2\dfrac{u}{\kappa} - v & -u \\ gv & g(u - 1) \end{pmatrix}.$$

We need to evaluate the Jacobian matrix at each of the three steady states and then determine its eigenvalues to deduce their stability. We will use Maple to do this as well. We need to use the package `linalg`:

```
> with(linalg):
```

The `jacobian` command gives us exactly what we are looking for (only now the actual values of $\kappa$ and $g$ are included):

```
> desys := vector( [u*(1-u/kappa)-u*v, g*(u-1)*v]
  ):
> jacobian( desys, [u,v] );
```

$$\begin{bmatrix} 1 - u - v & -u \\ v & u - 1 \end{bmatrix}$$

Unfortunately, we have to jump through some hoops to get Maple to understand the above as a function of $u$ and $v$, as follows:

```
> temp := (u,v) -> jacobian( desys, [u,v] ):
> Df := (s,t) -> subs( u=s, v=t, temp(u,v) ):
> Df(u,v);
```

$$\begin{bmatrix} 1 - u - v & -u \\ v & u - 1 \end{bmatrix}$$

We define the three matrices m1, m2, and m3 to be the Jacobian matrix evaluated at the three steady states, respectively:

```
> m1 := Df(0, 0);
```

$$m1 := \begin{bmatrix} 1 & 0 \\ 0 & -1 \end{bmatrix}$$

```
> m2 := Df(2, 0);
```

$$m2 := \begin{bmatrix} -1 & -2 \\ 0 & 1 \end{bmatrix}$$

```
> m3 := Df(1, 0.5);
```

$$m3 := \begin{bmatrix} -0.5 & -1 \\ 0.5 & 0 \end{bmatrix}$$

We proceed to find the eigenvalues with the `eigenvals` command:

```
> ev1 := eigenvals(m1);
> ev2 := eigenvals(m2);
> ev3 := eigenvals(m3);
```

**Exercise 8.4.3.** *Verify that the steady states $(0, 0)$ and $(2, 0)$ are saddles, and that $(1, 0.5)$ is a stable spiral.*

Note that if you also are interested in eigenvectors for each eigenvalue, then the `eigenvects` command will be useful.

We can go further and visualize the complete phase portrait. For this, we need the `DEtools` package for differential equations:

```
> with(DEtools);
```

First, we define the differential equations:

```
> de1 := diff(u(t),t) = u(t)*(1-u(t)/kappa)-u(t)*v(t);
> de2 := diff(v(t),t) = g*(u(t)-1)*v(t);
```

We now can graph the vector field defined by the above differential equations using the `dfieldplot` command:

```
> dfieldplot( [de1,de2], [u(t),v(t)],
    t=0..1, u=0..2, v=0..0.8 );
```

Finally, we specify a list of initial conditions with the `seq` command. Each initial condition is in the form $[t_0, u(t_0), v(t_0)]$. The corresponding trajectories can then be drawn in the phase plane with the `phaseportrait` command:

```
> initcond := seq( [0,2,i*0.1], i=1..5);
> phaseportrait( [de1,de2], [u(t),v(t)], t=0..10,
    {initcond}, stepsize=0.1, linecolor=black );
```

## 8.5   PDEs: An Age-Structured Model

In this section, we consider the age-structured model developed in Section 4.2.

**Exercise 8.5.1.** *For most populations, newborn individuals are not immediately capable of reproduction. So it is natural to expect the lower limit of integration in (4.4) to be a number bigger than 0. Similarly, for some species, the female population stops reproduction after a certain age. Discuss the limits of integration in this context. Why can we integrate from 0 to ∞?*

Maple is not capable of handling PDEs, except the most basic ones. However, this need not deter us, as we can discretize our model and use Maple to help out with a numerical approximation to the solution.

For ease of discussion, assume discrete time steps of one year and let

$$u_i^j := \text{number of individuals with age } i \text{ at the beginning of year } j.$$

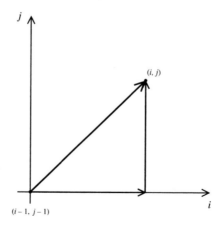

**Figure 8.1.** *Schematic for the discrete approximations of the time and age partial derivatives.*

We choose the discrete versions of the two derivatives in our model according to the schematic shown in Figure 8.1:

$$\frac{\partial u(t,a)}{\partial t} = u_i^j - u_i^{j-1},$$

$$\frac{\partial u(t,a)}{\partial a} = u_i^{j-1} - u_{i-1}^{j-1},$$

so that the discrete version of our model becomes

$$u_i^j = u_{i-1}^{j-1} - \mu_i u_{i-1}^{j-1} \text{ for } i \geq 1,$$

$$u_0^j = \sum_{i=0}^{n} b_i u_i^{j-1}.$$

Note that we have accounted for a maximum age of $n$ (for human populations, we can safely take $n = 100$).

### Arrays versus Lists

We are going to develop a numerical simulation of the discrete model. For programming purposes, it will be convenient to use arrays instead of lists to keep track of the population in each of the age classes, as well as the death rates, etc. Arrays are similar to lists, in that they are an ordered set of elements. However, indexing for arrays is more flexible than for lists. To illustrate the difference, we first create both a list and an array with the same six elements. Here's the list:

```
> L := [ seq( i*i, i=0..5 ) ];
```

$$L := [0, 1, 4, 9, 16, 25]$$

And here's the array:

```
> A := array(0..5,[]):
> for i from 0 to 5 do
>    A[i] := i*i:
> od;
```

Now compare L[0] and A[0]:

```
> L[0];
Error, invalid subscript selector
> A[0];
```

$$0$$

The latter is as expected, and as we wish, since the array is indexed from 0 to 5. If we had wanted to, we could have indexed the array from 10 to 15 (in this case, to extract the first element of the array, we would need to use A[10]). Try it! In the case of the list, the first element of the list must be extracted with L[1], even though we created the elements in the list by letting i run from 0 to 5.

For the discrete age-structured model, the age classes run from 0 to $n$, and so arrays with indices that run from 0 to $n$ will be most convenient.

### Creating the Simulation for the Age-Structured Model

For our numerical simulation, we begin with a population that is equally distributed over all age classes. We set up an array for $u_i^j$ and assign a population of 1 to each age class. (Obviously, we cannot have a fractional number of individuals, so you should scale these numbers; i.e., $u_i^j = 1$ means 1 million individuals in the corresponding age class, for example.)

```
> pop := array(0..100,[]):
> for i from 0 to 100 do pop[i] := 1.0; od:
```

Next, we set up an array containing the birth rate for each of the age classes. For the time being, we assume a birth rate of 7% between the ages of 20 and 35:

```
> birth := array(0..100,[]):
> for i from 0 to 100 do birth[i] := 0.0; od:
> for i from 20 to 35 do birth[i] := 0.07; od:
```

Similarly, we set up an array containing the survival rate for each of the age classes. We assume a death rate of 1% for each age class, equivalent to a survival rate of 99%:

```
> survival := array(0..100, []):
> for i from 0 to 100 do survival[i] := 0.99; od:
```

To let our population evolve over time, we define a procedure that calculates the population in year $j + 1$ (new_pop) from a population in year $j$ (last_pop). We will make use of the sum command. The syntax of this command is straightforward, but you should note the quotation marks:

```
> evolve := proc( last_pop )
>    # local variables
>    local new_pop, i;
>    # define new_pop
>    new_pop := array(0..100,[]);
>    # determine the fraction of the population
>    # that survives
>    for i from 1 to 100 do
>       new_pop[i] := survival[i]*last_pop[i-1];
>    od:
>    # determine the newborns (age class 0)
>    new_pop[0] := sum( 'birth[i]*last_pop[i]',
     'i'=0..100 );
>    # return the new population
>    RETURN( new_pop );
> end;
```

Finally, we define a procedure which lets us plot the number of individuals in each age class of a population:

```
> with( plots ) :
> plotpop := proc( some_pop )
>   # local variables
>   local pop_list, i;
>   # create the list of coordinates to be
    plotted from some_pop
>   pop_list := [ seq( [i,some_pop[i]], i=0..100
    ) ];
>   # and create the plot
>   plot( pop_list, view=[0..100,0..2],
    style=point,
    symbol=circle );
> end;
```

**Running the Simulation**

Now we have defined all that we need: an array to contain the number of individuals in each age class in a population, arrays for the birth and survival rates, a numerical algorithm that updates the number of individuals in each age class, and a procedure to plot the result. Let's start the simulation.

First, we plot the initial population:

```
> plotpop( pop );
```

Now we calculate the population after one year:

```
> pop := evolve( pop );
```

Let's see what happened:

```
> plotpop( pop );
```

We can let the population progress several years with the following recursion:

```
> for i from 1 to 10 do
>   pop := evolve( pop );
> od:
> plotpop( pop );
```

**Exercise 8.5.2.** *What do you observe? Describe what happens to the population as time progresses.*

**Exercise 8.5.3.** *Modify the birth and death rates and study the behavior of the population over time (you will need to re-initialize the population each time you specify new birth and death rates).*

## 8.6   Stochastic Models: Common Colds in Households

In Section 5.7.1, we developed a discrete Markov chain model for the outbreak and transmission of a cold in a household. In this section, we show how to implement the model
developed so far. First we define the transition matrix $P$. We translate the formulas from
above into a program. Let $N$ be the family size, and the matrix $P_{(.,.),(.,.)}$ will be called
transition.

```
> # Family size
> N := 5;
> # Transition array, P(i,j)(k,l) =
  transition[i,j,k,l]
> transition := array(0..N,0..N,0..N,0..N);
```

We first initialize the matrix transition with 0 and then we use formulas (5.66) and
(5.67).

```
> # Initialize everything with zero
> for n1 from 0 by 1 to N do
>    for n2 from 0 by 1 to N do
>     for n3 from 0 by 1 to N do
>      for n4 from 0 by 1 to N do
>       transition[n1,n2,n3,n4] := 0;
>      od:
>     od:
>    od:
> od:
> # First type of transition:  infections.
> for i from 1 by 1 to N-1 do
>    for j from 0 by 1 to N-i-1 do
>      transition[i,j,i+1,j] :=
  R0*(N-i-j)/(R0*(N-i-j)+1);
>    od:
> od:
> # Second type of transition:  recovery.
> for i from 1 by 1 to N do
>    for j from 0 by 1 to N-i do
>     transition[i,j,i-1,j+1] := 1/(R0*(N-i-j)+1);
>    od:
> od:
> # Third type of transition:  absorbing states.
> for j from 0 by 1 to N do
>    transition[0,j,0,j] := 1;
> od:
```

Next, we choose the parameter $R_0$ and define the state variable $q_{i,j}(n)$, that is, the probability
distribution over the states. For the Maple code we call it probs and start in the state $(1, 0)$.

```
> R0 := 0.154;
> probs := array(0..N,0..N);
> for n1 from 0 by 1 to N do
>    for n2 from 0 by 1 to N do
>      probs[n1,n2] := 0;
>    od:
> od:
> probs[1,0] := 1;
```

The last step is to define a procedure that iterates our state from one event to the next event.

```
> iter := proc( probs_in )
>    # define local state variable
>    local probs_new, n1, n2, n3, n4;
>
>    # initialize the local state
>    probs_new := array(0..N,0..N);
>    for n1 from 0 by 1 to N do
>      for n2 from 0 by 1 to N do
>        probs_new[n1,n2] := 0;
>      od;
>    od:
>
>    # now, iterate once
>    for n1 from 0 by 1 to N do
>      for n2 from 0 by 1 to N do
>        for n3 from 0 by 1 to N do
>          for n4 from 0 by 1 to N do
>            probs_new[n3,n4] := probs_new[n3,n4]
>            + probs_in[n1,n2]*transition[n1,n2,n3,n4];
>          od:
>        od:
>        od:
>      od:
>
>    # return the new state
>    RETURN( probs_new );
> end;
```

Finally, we have to iterate over many events until all probability mass is contained in the absorbing states (in other words, the sum over the probabilities of the absorbing states equals 1). How often do we have to iterate? This is just the length of the longest path from state $(1, 0)$ into the states $(0, j)$ along the arrows of Figure 5.10. The longest path ends up in $(N, 0)$, and all paths connecting $(0, 1)$ with $(0, N)$ have length $2N - 1$ (check!).

```
> # now iterate 2*N-1 times:
> for k from 1 by 1 to (2*N-1) do
>    probs := iter(probs);
> od:
> # Plot the final size distribution
> L := [];
> for n1 from 1 by 1 to N do
>    L := [op(L),[n1,probs[0,n1]]];
> od:
> with(plots);
> myplot1 := plot(L, style=point, symbol =
  circle):
> display(myplot1);
```

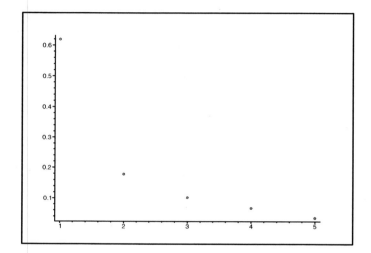

Given the parameter $R_0$, we are able to obtain the picture of the final size epidemics.

**Exercise 8.6.1.**

(a) *Vary $R_0$. How does the picture change? Can you explain the changes?*

(b) *Vary the household size $N$. What happens?*

## 8.6.1  Application to Data

We now apply our theory to data that are given in Table 5.3. We have to come up with an estimate for $R_0$. One possibility would be to do a least-squares fit: vary $R_0$ until the error between data and theoretical prediction is minimal. Using statistical tools, one may refine this approach, taking an appropriate variance structure into account. However, we use a shortcut. Let us consider the probability of finding exactly one infected person during the epidemic within our family. Since there is exactly one path from state $(1, 0)$ to state $(0, 1)$, the probability for this absorbing state is already given after one iteration, that is, by the element $P_{(1,0),(0,1)}$. Since $P_{(1,0),(0,1)} = 1/(R_0\,(N-1)+1)$,

$$R_0 = \frac{1}{(N-1)} \left( \frac{1}{\text{Prob. for one infected}} - 1 \right).$$

Furthermore, we have a simple estimate for the probability of finding exactly one infected person during the epidemic. If we let $F_i$ be the number of families with $i$ infected persons, then this probability is approximately $F_1/F$ with $F = \sum_{i=1}^{N} F_i$. Hence, $R_0$ may be estimated by $\hat{R}_0$,

$$\hat{R}_0 := \frac{1}{(N-1)} \left( \frac{F}{F_1} - 1 \right).$$

For our data we obtain $\hat{R}_0 = 0.154$, the numerical value we used in our program in order to calculate the final size distribution. The plot myplot1 already contains the distribution for the appropriate parameter. We add the empirical distribution to this plot.

```
> # Data
> final := [112,35,17,11,6];
> total := final[1]+final[2]+final[3]+final[4]+final[5];
>
> # empirical final size distribution
> efsd := [];
> for n1 from 1 by 1 to 5 do
>    efsd := [op(efsd),[n1,evalf(final[n1]/total)]];
> od:
>
> myplot2 := plot(efsd, style=point, symbol=cross):
> display(myplot1, myplot2);
```

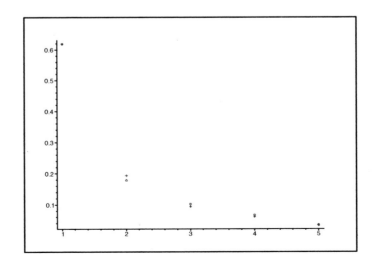

We find that data and theoretical prediction agree surprisingly well. In order to judge the agreement in more detail one has to use statistical methods (see, e.g., [1, 15]), and this is beyond of the scope of this book.

**Exercise 8.6.2.** *Heasman and Reid [83] divided all households (with five members) into three classes: overcrowded, crowded, and uncrowded. The data of Table 5.3 represents overcrowded households. In Table 8.4 the data for crowded and uncrowded households are given. Estimate $R_0$ and plot the theoretical and empirical final size distribution. Is $R_0$ different for the three types of households? Can you explain the results?*

**Table 8.4.** *Frequencies for the size of outbreaks for crowded and uncrowded households. Data are taken from [15].*

| Final infected | Number of families (crowded) | Number of families (uncrowded) |
|:---:|:---:|:---:|
| 1 | 155 | 156 |
| 2 | 41 | 55 |
| 3 | 24 | 19 |
| 4 | 15 | 10 |
| 5 | 6 | 2 |

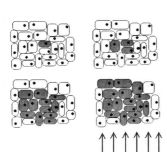

# Part III

# Projects

# Chapter 9
# Project Descriptions

In this chapter we present a collection of open-ended problems in mathematical biology. Many of these problems have not previously been studied with a mathematical model.

## 9.1 Epidemic Models

### Project 1: HIV

Table 9.1 shows data, taken from [42], on the HIV epidemic in Cuba from 1986 until 2000. Design a model which describes the epidemic spread of HIV in Cuba, and fit the data in Table 9.1. Which are the relevant parameters of your model? Try to introduce control mechanisms to lower the number of AIDS cases. Compare your control mechanism with the data of the given time period. You need to look into facts about HIV transmission and Cuba's control strategy via the Internet or in appropriate textbooks.

### Project 2: Smallpox

An outbreak of smallpox in Abakaliki in southeastern Nigeria in 1967 was reported by Bailey and Thomas [11]. People living there belong to a religious group that is quite isolated and declines vaccination. Overall, there were 30 cases of infection in a population of 120 individuals. The time (in days) between newly reported pox cases is given in the following sequence:

$$13, 7, 2, 3, 0, 0, 1, 4, 5, 3, 2, 0, 2, 0, 5, 3, 1, 4, 0, 1, 1, 1, 2, 0, 1, 5, 0, 5, 5.$$

Develop a model which describes these data and analyze the epidemic outbreak.

**Table 9.1.** *HIV data from Cuba 1986–2000 (data from* [42]*).*

| Year | HIV cases | AIDS cases | Death from AIDS |
|------|-----------|------------|------------------|
| 1986 | 99        | 5          | 2                |
| 1987 | 75        | 11         | 4                |
| 1988 | 93        | 14         | 6                |
| 1989 | 121       | 13         | 5                |
| 1990 | 140       | 28         | 23               |
| 1991 | 183       | 37         | 17               |
| 1992 | 175       | 71         | 32               |
| 1993 | 102       | 82         | 59               |
| 1994 | 122       | 102        | 62               |
| 1995 | 124       | 116        | 80               |
| 1996 | 234       | 99         | 92               |
| 1997 | 364       | 121        | 99               |
| 1998 | 362       | 150        | 98               |
| 1999 | 493       | 176        | 122              |
| 2000 | 545       | 251        | 142              |

## Project 3: Influenza

In the *British Medical Journal* in 1978, a report was published with detailed statistics of a
flu epidemic in a British boarding school [5]. The school had 733 pupils [6], all but 30 of
whom were boarders, and all boys. Of these, 512 were confined to bed during the epidemic,
which lasted from January 22nd to February 4th, 1978. It seems that one infected boy
initiated the epidemic. The school dealt with the epidemic by putting a boy to bed as soon
as it was discovered that he was infected. Detailed data are shown in Table 9.2. Model the
epidemic outbreak.

## Project 4: Yellow Fever in Senegal in 2002

Yellow fever (YF) is a viral hemorrhagic fever transmitted by infected mosquitoes. YF is
spread into human populations in three stages:

1. **Sylvatic** (or jungle). YF occurs in tropical rain forests where mosquitoes, which feed
   on infected monkeys, pass the virus to humans who work in the forest.

2. **Intermediate.** YF occurs as infected individuals bring the disease into rural villages,
   where it is spread by mosquitoes among humans (and also monkeys).

3. **Urban.** YF occurs as soon as an infected individual enters urban areas. This can lead
   to an explosive epidemic in densely inhabited regions. Domestic mosquitoes carry
   the virus from person to person.

**Table 9.2.** *Influenza in a boarding school (data from [6]).*

| Date | In bed | Convalescent |
|------|--------|--------------|
| Jan. 22nd | 3 | 0 |
| Jan. 23rd | 8 | 0 |
| Jan. 24th | 26 | 0 |
| Jan. 25th | 76 | 0 |
| Jan. 26th | 225 | 9 |
| Jan. 27th | 298 | 17 |
| Jan. 28th | 258 | 105 |
| Jan. 29th | 233 | 162 |
| Jan. 30th | 189 | 176 |
| Jan. 31st | 128 | 166 |
| Feb. 1st | 68 | 150 |
| Feb. 2nd | 29 | 85 |
| Feb. 3rd | 14 | 47 |
| Feb. 4th | 4 | 20 |

**Table 9.3.** *Yellow Fever in Senegal,* 2002 *(data from the disease outbreak news archives of the WHO* [167]*).*

| Report date | Cases (total) | Deaths (total) |
|-------------|---------------|----------------|
| Jan. 18th | 18 | 0 |
| Oct. 4th | 12 | 0 |
| Oct. 11th | 15 | 2 |
| Oct. 17th | 18 | 2 |
| Oct. 24th | 41 | 4 |
| Oct. 31st | 45 | 4 |
| Nov. 20th | 57 | 10 |
| Nov. 28th | 60 | 11 |

The epidemic can be controlled by vaccination. The YF vaccine is safe and effective, and provides immunity within one week in 95% of those vaccinated.

Table 9.3 shows a data set of YF cases and YF deaths during an outbreak in Senegal in 2002, collected from the Internet archives of the World Health Organization (WHO) [167]. As soon as the virus was identified, a vaccination program was started (Oct. 1, 2002). On Oct. 11, 2002, the disease was reported in Touba, a city of 800,000 residents. More information can be found on the WHO website [167].

1. Develop a model for the three stages of YF as outlined above.

2. Include a fourth stage which describes vaccination in urban areas.

3. Fit your model to the data.

4. What would have happened without vaccination?

5. Would you expect the disease to die out, or to become persistent?

## Project 5:  Cholera in South Africa 2000–2001

The seventh cholera pandemic began in Indonesia in 1961. Over the following 40 years, the virus *Vibric cholerae* O1 spread around the world, mainly into underdeveloped countries. In South Africa, the cholera epidemic arrived in mid-August 2000. In Table 9.4, we show data on the number of cholera cases and cholera-related death cases, taken from the disease outbreak news archives of the WHO [167].

Large cholera outbreaks are usually related to contaminated water. The cholera virus is present in brackish water through algae blossom and through human feces. Only 10–20% of infected individuals suffer from severe symptoms. Many individuals do not show symptoms at all, but their feces are infectious. Cholera is a serious disease since the progress of symptoms can be very fast if not treated.

**Table 9.4.** *Cholera outbreak in South Africa 2000–2001 (data from the WHO archives [167]). The number highlighted with an asterisk was later corrected by the WHO to 31; to show that data can be inconsistent, the original number is given here.*

| Date | Cases (total) | Deaths (total) |
|------|---------------|----------------|
| Oct. 13, 2000 | 2175 | 22 |
| Oct. 18, 2000 | 3075 | 26 |
| Oct. 19, 2000 | 3279 | 27 |
| Oct. 26, 2000 | 3806 | 33* |
| Nov. 02, 2000 | 4270 | 32 |
| Nov. 09, 2000 | 4583 | 33 |
| Nov. 19, 2000 | 5285 | 35 |
| Nov. 27, 2000 | 5876 | 35 |
| Dec. 05, 2000 | 6548 | 35 |
| Dec. 19, 2000 | 8137 | 41 |
| Dec. 29, 2000 | 11183 | 51 |
| Jan. 09, 2001 | 15983 | 60 |
| Jan. 14, 2001 | 19499 | 66 |
| Jan. 25, 2001 | 27431 | 74 |
| Feb. 04, 2001 | 37204 | 85 |
| Feb. 14, 2001 | 48647 | 108 |
| Feb. 22, 2001 | 56092 | 120 |
| Mar. 03, 2001 | 62607 | 131 |
| Mar. 14, 2001 | 69761 | 139 |
| Mar. 28, 2001 | 78140 | 163 |
| Apr. 16, 2001 | 86107 | 181 |

The WHO recommends four major control mechanisms, namely,

1. hygienic disposal of human feces,

2. adequate supply of safe drinking water,

3. good food hygiene and cooking, and

4. washing hands after defecation and before meals.

More information about this disease, control mechanisms, and vaccination can be found at the website of the WHO [167].

Develop a model for the outbreak of cholera in South Africa.

1. First, model the epidemic without any control mechanism.

2. Then include the recommended control mechanisms in the model and see if you can obtain a better fit to the data.

3. Use your model to determine which of the above control mechanisms is most effective.

4. Can you predict the further development of the disease, provided that all control measures are in place?

## Project 6: SARS Outbreak

Detailed data on confirmed SARS (Severe Acute Respiratory Syndrome) cases in Canada in 2003 are given in Table 9.5.

Construct a model that can be used to determine the number of current new SARS cases, deaths, and recoveries each day. Start by assuming a "closed" population with an initial infection of one person. Further assume that there is no quarantining of SARS patients, and that no measures are taken to reduce the likelihood of infection from one individual to another.

Consider deterministic and stochastic versions of this model. Try running the stochastic models on the computer. What level of variability is associated with the predictions?

Can the data be used to estimate parameters in the model? If so, use the data to do this.

One key disease control goal is to eradicate the outbreak of SARS through quarantining and preventative measures. Assess the effectiveness of these control measures on the disease dynamics. If you have time, consider the case where the population is no longer "closed," but where new infections can be imported and exported.

## Project 7: Paths of an Epidemic

In Section 5.7.1, we considered a model for the final size of an epidemic of the common cold in a household. The data collected by Heasman and Reid in [83] are even more detailed. We call the person who got the disease first (i.e., from someone outside of the family) the

**Table 9.5.** *SARS outbreak in Canada, 2003. Every case is listed on the date the patient showed the first symptoms, and classified by probable way of infection: T = travel, F = household, H = health care setting, O = others. Data from a graphic from the Health Canada webpages* [82].

| Date | T | F | H | O |
|---|---|---|---|---|
| Feb. 23 | 1 | 0 | 0 | 0 |
| Feb. 26 | 0 | 1 | 0 | 0 |
| Feb. 28 | 1 | 0 | 0 | 0 |
| Mar. 3 | 0 | 1 | 0 | 0 |
| Mar. 5 | 0 | 1 | 0 | 0 |
| Mar. 7 | 0 | 1 | 0 | 0 |
| Mar. 9 | 0 | 0 | 1 | 0 |
| Mar. 10 | 0 | 0 | 1 | 0 |
| Mar. 12 | 0 | 0 | 2 | 0 |
| Mar. 13 | 1 | 0 | 2 | 0 |
| Mar. 15 | 0 | 0 | 2 | 0 |
| Mar. 16 | 0 | 1 | 1 | 0 |
| Mar. 17 | 0 | 0 | 5 | 0 |
| Mar. 18 | 0 | 0 | 7 | 0 |
| Mar. 19 | 0 | 0 | 9 | 0 |
| Mar. 20 | 0 | 0 | 7 | 0 |
| Mar. 21 | 0 | 0 | 3 | 0 |
| Mar. 22 | 0 | 0 | 4 | 0 |
| Mar. 23 | 0 | 0 | 1 | 0 |
| Mar. 24 | 1 | 0 | 3 | 0 |
| Mar. 25 | 0 | 1 | 4 | 0 |
| Mar. 26 | 0 | 4 | 3 | 0 |
| Mar. 27 | 0 | 3 | 4 | 0 |
| Mar. 28 | 0 | 2 | 5 | 0 |
| Mar. 29 | 0 | 4 | 3 | 0 |
| Mar. 30 | 0 | 2 | 1 | 0 |

| Date | T | F | H | O |
|---|---|---|---|---|
| Mar. 31 | 0 | 0 | 2 | 1 |
| Apr. 1 | 1 | 3 | 1 | 0 |
| Apr. 2 | 0 | 2 | 1 | 0 |
| Apr. 3 | 0 | 2 | 1 | 2 |
| Apr. 4 | 0 | 1 | 6 | 0 |
| Apr. 5 | 0 | 3 | 0 | 1 |
| Apr. 6 | 0 | 2 | 1 | 1 |
| Apr. 7 | 0 | 1 | 0 | 1 |
| Apr. 8 | 0 | 1 | 2 | 0 |
| Apr. 9 | 0 | 1 | 0 | 0 |
| Apr. 10 | 0 | 0 | 1 | 0 |
| Apr. 11 | 0 | 1 | 0 | 0 |
| Apr. 14 | 0 | 1 | 1 | 0 |
| Apr. 15 | 0 | 0 | 1 | 0 |
| Apr. 16 | 0 | 0 | 3 | 0 |
| Apr. 17 | 0 | 0 | 2 | 0 |
| Apr. 19 | 0 | 0 | 1 | 0 |
| Apr. 22 | 0 | 0 | 1 | 0 |
| Apr. 25 | 0 | 0 | 1 | 0 |
| Apr. 29 | 0 | 0 | 1 | 0 |
| Apr. 30 | 0 | 0 | 1 | 0 |
| May 1 | 0 | 0 | 1 | 0 |
| May 3 | 0 | 0 | 1 | 0 |
| May 7 | 0 | 0 | 1 | 0 |
| May 9 | 0 | 0 | 2 | 0 |
| May 11 | 0 | 0 | 3 | 0 |

| Date | T | F | H | O |
|---|---|---|---|---|
| May 12 | 0 | 0 | 2 | 0 |
| May 13 | 0 | 0 | 2 | 0 |
| May 14 | 0 | 0 | 1 | 0 |
| May 15 | 0 | 0 | 2 | 0 |
| May 16 | 0 | 0 | 3 | 0 |
| May 17 | 0 | 0 | 4 | 0 |
| May 18 | 0 | 0 | 5 | 0 |
| May 19 | 0 | 0 | 5 | 0 |
| May 20 | 0 | 0 | 4 | 0 |
| May 21 | 0 | 1 | 5 | 0 |
| May 22 | 0 | 0 | 6 | 0 |
| May 23 | 0 | 0 | 4 | 0 |
| May 24 | 0 | 0 | 8 | 0 |
| May 25 | 0 | 0 | 5 | 0 |
| May 26 | 0 | 0 | 6 | 0 |
| May 27 | 0 | 0 | 7 | 0 |
| May 28 | 0 | 0 | 5 | 0 |
| May 29 | 0 | 1 | 8 | 0 |
| May 30 | 0 | 0 | 1 | 0 |
| May 31 | 0 | 1 | 1 | 0 |
| Jun. 1 | 0 | 1 | 2 | 0 |
| Jun. 2 | 0 | 0 | 1 | 0 |
| Jun. 3 | 0 | 0 | 1 | 0 |
| Jun. 4 | 0 | 0 | 2 | 0 |
| Jun. 8 | 0 | 2 | 0 | 0 |
| Jun. 12 | 0 | 1 | 0 | 0 |

*first generation.* The persons who got the disease directly from this primary infected are called the *second generation.* The second generation infects the third generation and so on. It is very difficult to identify explicitly the members of each infected generation (see, e.g., [15, 83]). However, it is possible to estimate the number of infected individuals in each generation. The data shown in Table 9.6 list the number of infecteds in each generation for 181 families. The table shows all possible ways in which the epidemic can spread through a household of five members. For each possible path, the number of families is given where this path has been identified. Find a model that can describe these numbers and fit the model parameters.

**Table 9.6.** *Data for paths of an epidemic. Shown are all possible infection paths for households of five members, and the corresponding number of households where this particular path has been identified. (Data from [15].)*

| 1st Gen. | 2nd Gen. | 3rd Gen. | 4th Gen. | 5th Gen. | Number of families |
|----------|----------|----------|----------|----------|--------------------|
| 1 |   |   |   |   | 413 |
| 1 | 1 |   |   |   | 131 |
| 1 | 1 | 1 |   |   | 36 |
| 1 | 2 |   |   |   | 24 |
| 1 | 1 | 1 | 1 |   | 14 |
| 1 | 1 | 2 |   |   | 8 |
| 1 | 2 | 1 |   |   | 11 |
| 1 | 3 |   |   |   | 3 |
| 1 | 1 | 1 | 1 | 1 | 4 |
| 1 | 1 | 1 | 2 |   | 2 |
| 1 | 1 | 2 | 1 |   | 2 |
| 1 | 1 | 3 |   |   | 2 |
| 1 | 2 | 1 | 1 |   | 3 |
| 1 | 2 | 2 |   |   | 1 |
| 1 | 3 | 1 |   |   | 0 |
| 1 | 4 |   |   |   | 0 |

## 9.2   Population Dynamics

### Project 8: Models for Extinction

Many populations are endangered and are on the verge of extinction. While some go extinct, others recover. The goal in this project is to apply stochastic birth-death models to modeling populations on the verge of extinction. These models are described in Section 5.6.2.

In this project, we ask you to analyze some brand new data on Swedish wolf populations. Table 9.7 shows wolf population data in Sweden from 1980 to 2001 (data from Liberg [108]). Prior to 1991, the Swedish wolf population was small and remained steady. Since 1991, a significant increase in the population has been observed.

Here is some basic wolf biology: reproductive units are packs; one reproduction event typically results in a litter of pups; there is rarely more than one reproduction event per pack; and not all wolves are within packs (some may be "lone").

One theory is that this isolated Swedish wolf population suffered from "inbreeding depression" due to genetic similarity of individuals in the population, and that this ended with the emigration of a single Russian wolf to the Swedish population in about 1991. Assume that this is the case. Use the data to calculate the birth and death rates prior to 1991 and after 1991. Based on these, calculate the mean time to extinction, and mean and variance in population size as a function of time before and after 1991.

**Table 9.7.** *The number of wolves in the Scandinavian wolf population 1980–1981 to 2000–2001 (data from* [108]*). The table shows the minimum, maximum, and mean total number of wolves during each winter.*

| Year winter | Total (min) | Total (max) | Total (mean) |
|---|---|---|---|
| 1980–1981 | 2 | 6 | 4 |
| 1981–1982 | 3 | 3 | 3 |
| 1982–1983 | 3 | 3 | 3 |
| 1983–1984 | 8 | 8 | 8 |
| 1984–1985 | 6 | 6 | 6 |
| 1985–1986 | 7 | 7 | 7 |
| 1986–1987 | 5 | 5 | 5 |
| 1987–1988 | 6 | 6 | 6 |
| 1988–1989 | 10 | 10 | 10 |
| 1989–1990 | 8 | 8 | 8 |
| 1990–1991 | 8 | 8 | 8 |
| 1991–1992 | 16 | 18 | 17 |
| 1992–1993 | 19 | 22 | 20 |
| 1993–1994 | 16 | 30 | 28 |
| 1994–1995 | 29 | 39 | 34 |
| 1995–1996 | 34 | 45 | 39 |
| 1996–1997 | 41 | 57 | 49 |
| 1997–1998 | 50 | 72 | 61 |
| 1998–1999 | 62 | 78 | 70 |
| 1999–2000 | 67 | 81 | 74 |
| 2000–2001 | 87 | 97 | 92 |

Detailed analyses of populations at risk can be quite complex and involve detailed mathematical models. For an example applied to grizzly bears, see the article by Boyce et al. [24].

## Project 9: Growth of Cell Populations

In an article by Baker et al. [12], the growth of yeast cells is reported. The authors consider an experiment with *Schizosaccharomyces pombe*, where over a period of 8 hours the population size is measured in cells/mL every half hour. The experimental data are shown in Table 9.8.

Plot the data, and show that they cannot be explained reasonably with an exponential growth model.

The reason is as follows. Initially, the cells are synchronized, that is, they are at the same stage in their cell cycle, and the cells divide at approximately the same time. However, there are some variations in the time of cell division. Some cells divide a little bit earlier or later than others. Eventually, this destroys the synchronization and cells proliferate randomly.

**Table 9.8.** *Experimental data for* Schizosaccharomyces pombe *taken from* [12].

| Time (hrs) | 0.0 | 0.5 | 1.0 | 1.5 | 2.0 | 2.5 | 3.0 | 3.5 | 4.0 |
|---|---|---|---|---|---|---|---|---|---|
| Cells/mL | 114 | 116 | 114 | 108 | 112 | 107 | 108 | 128 | 169 |

| Time (hrs) | 4.5 | 5.0 | 5.5 | 6.0 | 6.5 | 7.0 | 7.5 | 8.0 | |
|---|---|---|---|---|---|---|---|---|---|
| Cells/mL | 201 | 212 | 214 | 245 | 262 | 297 | 314 | 340 | |

Develop a model which accounts for these effects and which reproduces and explains the data.

## Project 10: Cell Competition

*A possible solution of this project is discussed in detail in Section* 10.1.

In [63], Gause reported data for the following experiment. In two containers containing the same growth medium, populations of *Paramecium caudatum* and *Paramecium aurelia* are grown. The two populations are measured once per day. Also, in a larger container, the two populations are mixed and grown together. In this situation, the two populations compete for the same resources. Again, the populations are measured once per day. Gause's data are shown in Table 9.9.

Develop a model of the competition, and fit it to the given data. What does your model predict about the long-term viability of the populations (will both populations survive, or will one population become extinct)?

## Project 11: Fairy Rings

On many lawns in Alberta (and elsewhere), one can observe fairy rings. These are concentric rings (or parts of rings) of dead grass, or grass of a different color. For many residents as well as for golf-course owners, fairy rings are considered to be a lawn disease. Fairy rings are caused by a fungus which lives in the soil and spreads radially. As it grows and branches, the fungus forms a dense mesh of hyphae, so that water can no longer penetrate to the roots of the grass. Hence the good grass dies, and other wild grass types dominate. In a famous paper by Schantz and Peimeisel [139], detailed observations on fairy rings can be found. If the paper by Schantz and Peimeisel is not available at your institution, you may want to use Internet resources to learn more about basic features of fairy rings.

In this project, we would like to design a simple model that can describe the growth and branching of the fungus. It is recommended to study the situation in one spatial dimension first.

Once your model fungus grows well, you could either (i) generalize the model to two-dimensional spread, or (ii) include a grass population which competes for resources with the fairy ring fungus, or several grass species that compete with each other. If time allows, think of control mechanisms which might slow down or stop the fungus growth.

**Table 9.9.** *Paramecium competition data (from Gause [63]). Mean density is measured in individuals per* 0.5 cm³.

| Day # | Mean density | | | |
|---|---|---|---|---|
| | In isolation | | In competition | |
| | *P. aurelia* | *P. caudatum* | *P. aurelia* | *P. caudatum* |
| 0 | 2 | 2 | 2 | 2 |
| 1 | — | — | — | — |
| 2 | 14 | 10 | 10 | 10 |
| 3 | 34 | 10 | 21 | 11 |
| 4 | 56 | 11 | 58 | 29 |
| 5 | 94 | 21 | 92 | 50 |
| 6 | 189 | 56 | 202 | 88 |
| 7 | 266 | 104 | 163 | 102 |
| 8 | 330 | 137 | 221 | 124 |
| 9 | 416 | 165 | 293 | 93 |
| 10 | 507 | 194 | 236 | 80 |
| 11 | 580 | 217 | 303 | 66 |
| 12 | 610 | 199 | 302 | 83 |
| 13 | 513 | 201 | 340 | 55 |
| 14 | 593 | 182 | 387 | 67 |
| 15 | 557 | 192 | 335 | 52 |
| 16 | 560 | 179 | 363 | 55 |
| 17 | 522 | 190 | 323 | 40 |
| 18 | 565 | 206 | 358 | 48 |
| 19 | 517 | 209 | 308 | 47 |
| 20 | 500 | 196 | 350 | 50 |
| 21 | 585 | 195 | 330 | 40 |
| 22 | 500 | 234 | 350 | 20 |
| 23 | 495 | 210 | 350 | 20 |
| 24 | 525 | 210 | 330 | 35 |
| 25 | 510 | 180 | 350 | 20 |

# Project 12: Optimal Spatial Foraging

Individuals tend to move so as to optimize resource intake. The study of this process is generally called optimal foraging theory. In this problem, we would like you to investigate a simple theoretical model for foraging.

Consider a pond surrounded by a ring of moisture-loving plants. The plants have varying quality. In a given time step, flightless insects living on a given plant have the opportunity to remain on the same plant, move to the plant to the right, or move to the plant to the left. The pond stops insects from jumping to plants other than their immediate neighbors. Assume that each plant has a fixed quality, but that the quality varies from plant to plant.

Consider the case where each insect bases its movement decision only upon the quality of the current plant from which it may move. Write a master equation describing the possible ways to get to plant $i$ in one time step. Note that the probabilities of not moving, jumping to the right, or jumping to the left will depend upon the fixed local plant quality levels.

Simulate this model. What is the outcome? Do the insects congregate on high-quality plants? Now try the model with different behavioral rules (you choose the rules). What is the outcome?

As a challenge, you may want to extend this model in any number of possible ways. Three possibilities are given here:

1. Consider the model with plant dynamics included. Assume that herbivory by an insect on the plant will reduce the plant quality. What does your model show?

2. Consider the effect of bird predation on the insect populations.

3. Analyze the random walk process by taking the limit where the space and time steps become small, converting the random walk to a PDE model. This will involve some adeptness with Taylor series. If you choose to do this, analyze the steady-state (time-independent) solutions to the PDE model. What do they look like?

### Project 13: Mass March of Termites into the Deadly Trap

Typical carnivore pitcher plants feed on any insect which happens to slip into the pitcher. This catching strategy leads to a relatively constant supply of nutrients. As reported by Merbach et al. [118], the pitcher plant *Nepenthes albomarginata* has developed a different, unique strategy to catch certain termites in huge amounts. On the outside of the pitcher, just below the peristome, *N. albomarginata* has a rim of white hairs. These hairs can be eaten by termites (e.g., *Hospitalitermes bicolor*). If an individual termite encounters this hairy rim, it will feed on it happily, but it will not slip into the pitcher. The termite gets away with it just to inform other termites about its find. Subsequently, termites will return by the thousands to benefit from the hairy meal on the outside rim of the pitcher. The masses of termites still arriving push the first individuals into the pitcher. This way, a single pitcher can catch 100 to 1000 termites per meal.

Model this particular catching strategy. When the model runs successfully, modify the trapping strategy of your model plant, and see if you can increase the rate of capturing termites by the plant. Compare the strategy of *N. albomarginata* with other strategies, such as an olfactory attractor with a slippery rim and no hairs.

## 9.3 Models for Spatial Spread

### Project 14: The Chemotactic Paradox

*A possible solution of this project is discussed in detail in Section* 10.2.

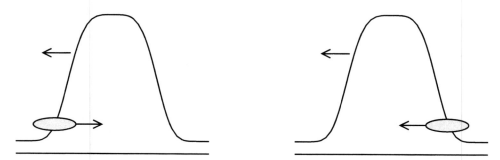

**Figure 9.1.** *Sketch of the chemotactic paradox.*

A variety of mechanisms have evolved by which living systems sense the environment in which they reside and respond to signals they detect, often by changing their patterns of movement. The movement response can entail changing the speed of movement and the frequency of turning, which is called *kinesis*, it may involve direct movement, which is called *taxis*, or it may involve a combination of these. Taxis and kinesis may be characterized as positive or negative, depending on whether they lead to accumulation at high or low concentrations of an external stimulus. Typical stimuli for microorganisms are light, gravitation, pressure, or chemical signals. Tactic and kinetic responses both involve the detection of the external signal and transduction of this signal into an internal signal that triggers the response.

An example of chemotaxis occurs in *Dictyostelium discoideum*, where individuals aggregate in response to a signal from "organizer" cells. Individual cells relay the signal to their neighbors, thereby causing an outward moving traveling wave of the chemical through a *Dictyostelium* population. We are interested in the movement of a single cell as it responds to the moving signal pulse.

As a model case, we assume that the wave is moving in one spatial dimension from right to left, as shown in Figure 9.1. A cell at the wave front senses an increasing signal concentration and moves forward (to the right), opposite to the direction of the wave. As soon as the wave back passes the cell, the cell senses a negative gradient of the signal concentration. Hence it should move backwards (to the left), which is in the direction of the wave. Overall, the cell would spend more time in the wave back than in the wave front, which should give a net displacement to the left. In experiments, however, cells move to the right only. This is called the *chemotactic paradox*. Formulate a hypothesis to resolve the paradox and investigate your hypothesis with the use of a mathematical model. Can you resolve the paradox?

## Project 15: Movement of Flagellated Bacteria

Flagellated bacteria swim in a manner that depends on the size and shape of the body of the cell and the number and distribution of the flagella. The common intestinal organism *Escherichia coli* looks like a small cocktail sausage. It has a rod-shaped body about $10^{-4}$ cm in

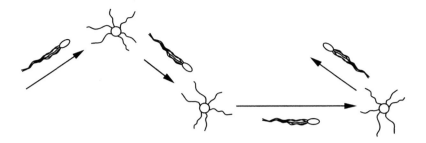

**Figure 9.2.** *Schematic of the random walk of* Escherichia coli.

diameter and $2 \times 10^{-4}$ cm in length. Approximately six flagellar filaments emerge at random points on the sides of the body and extend about three body lengths into the surrounding medium. When these flagella turn counterclockwise, they form a synchronous bundle that pushes the body steadily forward; the cell is said to "run." When they turn clockwise, the bundle comes apart and the flagella turns independently, moving the cell body this way and that in a highly erratic manner; the cell is said to "tumble." These modes alternate, and the cell executes a three-dimensional random walk, as illustrated in Figure 9.2. An impressive collection of movies of moving cells can be found on the webpages of Howard Berg of the Rowland Institute at Harvard. The motor mechanisms are described in Armitage [7].

Design a model to describe the movement of an individual bacterium. Once your simulated bacterium runs well, study a whole population of bacteria. Plot the mean-squared displacement as a function of time. In Section 5.4.2 on random walks, we found that for a diffusion process, the mean-squared displacement grows linearly with time. What do you find with the random walk of the bacteria, especially for small time intervals?

Put a source of a chemical signal at the origin, for example, by setting a signal concentration

$$S(x, y) = \alpha e^{-\frac{x^2+y^2}{\beta}},$$

with appropriate parameters $\alpha$ and $\beta$. Bias your random walk such that the bacterium chooses directions toward $(0, 0)$ more often.

If time allows, let your source do a random walk too and let it be chased by a bacterium.

## Project 16: Movement of Amoebae

In [146], Soll describes the orientation of a slime amoeba (*Dictyostelium discoideum*) in response to an external chemical signal. In Figure 9.3, reproduced from the paper, it is shown how a cell rotates its body axis in the direction of a chemical gradient which is pointing upwards. The individual cell measures the signal concentration along its body surface. It elongates its surface along one side and contracts it along the other sides. This eventually leads to polarization along the chemical gradient. Develop a model that reproduces this polarization behavior.

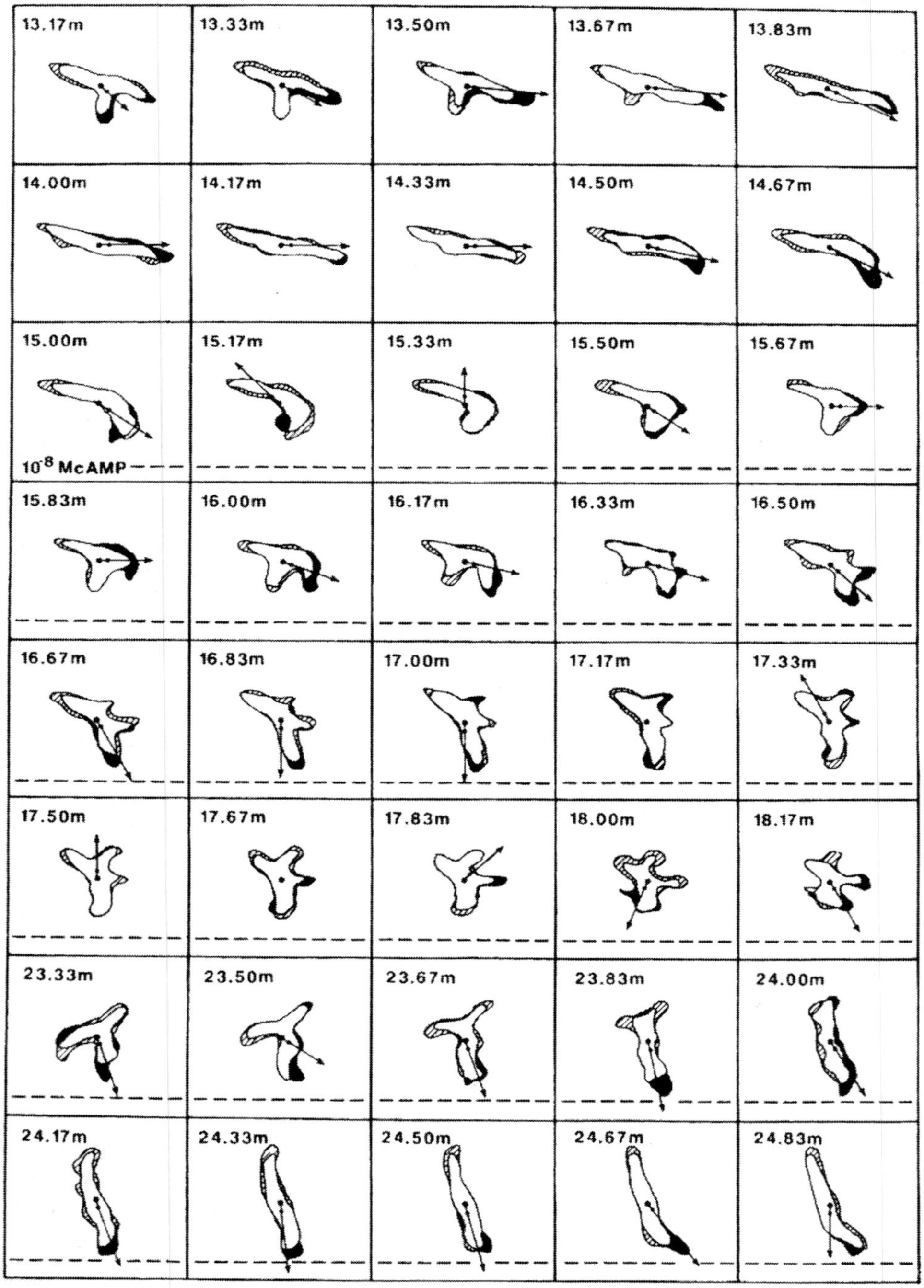

**Figure 9.3.** *Illustration of the movement of amoebae.  Black regions indicate areas of extension, striped regions indicate areas of retraction, and the arrow indicates the orientation of the cell.  (Figure 4 of Soll [146], reprinted with permission from Kluwer Academic/Plenum Publishers.)*

## Project 17: Home Ranges

*For this project, Sections 8.1–8.3 of Okubo and Levin [127] are needed.*

Many animals have home ranges or territories. These include wolves, coyotes, badgers, squirrels, birds, and lizards. The goal of this project is to develop a spatially explicit model for home-range movements of an animal. Individuals typically cue their movement behavior based on sound, familiar and foreign scent marks, prey density, and familiarity with a particular region. Some animals use one cue over another. For example, birds primarily use sound (bird calls), while wolves primarily use scent marks.

Read Sections 8.1–8.3 from the book by Okubo and Levin [127]. Simulate the Holgate model (cases 1 and 2) for an individual's movement. Repeat this for a large number of individuals (realizations of the stochastic process). What do the spatial distributions of large numbers of individuals look like in each of the two cases? Formulate a third case where the bias does not vary with distance. Simulate this case. Now let $\epsilon = k\Delta x$, where $\Delta x$ is the spacing between grid points. Derive a PDE model for this third case by taking the "diffusion limit" discussed in chapter 5. Calculate the steady-state distribution for this PDE model, compare with simulation results, and plot the results.

Develop more realistic models for animal movement. You may want to start by reading "An Olfactory Orientation Model for Mammals' Movements in Their Home Ranges" by Benhamou [18]. Simulate the animal movement using random numbers drawn from a Gaussian and the algorithm given on page 382 of [18]. Print out a movement path similar to Figure 1, page 382. Modify the animal movement simulator as described on the top of page 383. Print out a space-use pattern for a 10,000-step path as shown in Figure 2, page 383. Discuss the algorithm and output.

Design your own model for home-range or territorial movement behavior. Territories involve interactions between adjacent home-range holders through signaling (e.g., scent-marking), and thus require two or more individuals interacting in space. Choose an animal, and tailor a territorial model to fit the biology. Simulate the model. Which rules give rise to stable territories?

## Project 18: Re-invasion of Otters to California's Coast

One example of a biological invasion is the spread of a re-invading sea otter species off the coast of California. It was thought to be extinct until a relict population was found off Point Sur in 1914. Under protection from hunting, it grew and spread spatially, and now it calls much of the west coast of North America its home. Details on the early spatial spread are given in Figure 9.4 and Table 9.10 (from [110]). Most of the sea otter activity occurs within 1 km of the coast, and so the spread can be thought to be linear (up and down the coast).

Plot the distance spread versus time in northward and southward directions. Also plot the total range radius versus time. Discuss why spread may be different in north and southward directions. Derive a mathematical model for the spread process. You may want to research the life history of sea otters so as to make your model realistic.

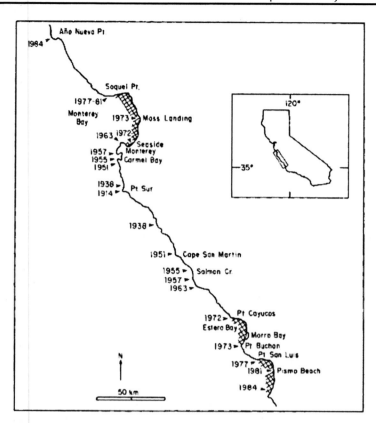

**Figure 9.4.** *Graphical illustration of the range expansion of sea otters. (Figure* 1
*of Lubina and Levin* [110], *reprinted with permission from Chicago University Press.)*

## 9.4   Physiology

### Project 19:  Pupil Control System

The pupil is the opening in the middle of the eye through which light enters the eye.  In
many animals, including humans, involuntary contraction and dilation of the pupil regulates
the intensity of light entering the eye.  The pupil will contract under bright light conditions,
while it will dilate under low light conditions.

Now suppose that you shine a tiny spot of light onto the eye, always in the same
location, and that the spot initially is on the edge of the pupil.  At first, the pupil will contract
in response to the spot of light.  After the contraction, the light no longer enters the pupil,
so it will dilate.  After the dilation, the light again enters the pupil, so it will contract again,
and so forth.  An oscillation has been generated, for example, as shown in Figure 9.5.

Develop a model to reproduce the phenomenon.  Begin by developing a model of the
light intensity as a function of the pupil radius.  Then add in negative feedback.  Can you
obtain oscillations?  Investigate the incorporation of a delay representing the time it takes
for the eye to respond to a change in light intensity.

**Table 9.10.** *Experimental data on the range expansion of sea otters. (Table 1 of Lubina and Levin [110], reprinted with permission from Chicago University Press.)*

RANGE EXPANSION AND POPULATION SIZE OF THE CALIFORNIA SEA OTTER
ALONG THE CALIFORNIA COAST (IN km)

| | EXTENT OF RANGE INCREASE | | ESTIMATED TOTAL RANGE | POPULATION SIZE |
|---|---|---|---|---|
| YEAR | North | South | | |
| 1914 | ? | ? | (11) | (50) |
| 1938 | 11 | (21) | 43 | 310 |
| 1947 | 8 | 23 | 74 | 530 |
| 1950 | 2 | 13 | 89 | 660 |
| 1955 | 3 | 16 | 108 | 800 |
| 1957 | 11 | 6 | 125 | 880 |
| 1959 | 6 | 6 | 137 | 1050 |
| 1963 | 5 | 10 | 152 | 1190 |
| 1966 | 0 | 6 | 158 | 1260 |
| 1969 | 6 | 13 | 177 | 1390 |
| 1972 | 0 | 15 | 192 | 1530 |
| 1973 | 23 | 29 | 244 | 1720 |
| 1974 | 6 | 5 | 255 | 1730 |
| 1975 | 8 | 0 | 263 | ? |
| 1976 | 10 | 6 | 279 | 1789 |
| 1977 | 8 | 6 | 293 | ? |
| 1978 | 0 | 0 | 293 | ? |
| 1979 | 0 | 6 | 299 | (1443) |
| 1980 | 0 | 13 | 312 | ? |
| 1982 | 0 | 0 | 312 | 1338 |
| 1983 | 26 | 15 | 353 | 1226 |
| 1984 | 0 | 0 | 353 | 1203 |
| 1986 | ? | ? | ? | 1400 |

NOTE.—The extent of the range is determined by the linear distance along the coastline between the outermost main raft of otters at the population boundaries. Point Sur was used as the location of the division between the northern and southern populations. The total estimated population size was based on aerial and shore counts. Parentheses indicate that the estimate was considered unreliable; a question mark means that no estimate was made.

SOURCES.—E. Ebert, pers. comm.; Riedman and Estes, MS; Estes, unpubl. data.

## Project 20: Modeling of Heart Beats

In this project (inspired by Chapter 1.13 in [1]), you are asked to investigate the production of heart beats.

A mathematician's view of the apparatus for beating of the heart is shown in Figure 9.6. The sinoatrial (SA) node is the pacemaker. Its function is to send signals at regular intervals

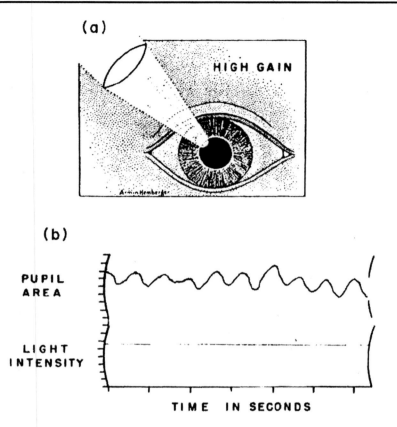

**Figure 9.5.** (a) *Technique used for pupil stimulation. Light here is focused on the border of iris and pupil. Small movements of the iris result in large changes in light intensity of the retina.* (b) *Example of spontaneous high gain oscillations in pupil area obtained with constant light stimulus using high gain operating condition illustrated in* (a). *(Figures 11 and 12 in Section II of Stark [149], reprinted with permission from Kluwer/Plenum Press.)*

to the atrioventrical (AV) node. Upon receipt of a signal from the SA node, the AV node checks the condition of the heart and decides whether to tell the heart to contract or not.

For a simple model of the heart, it is sufficient to describe the behavior of the AV node. The AV node uses an electrical potential to keep track of the condition of the heart. In particular, this potential decreases exponentially during the time between signals from the SA node. When the AV node receives a signal from the SA node, one of two things happens. If the potential is too high, it means that the heart is not yet ready to contract again, and the AV node ignores the signal. Otherwise, the AV node tells the heart to contract. The contraction of the heart causes the potential of the AV node to increase (for simplicity, you may assume that the increase is a constant).

Develop a model describing the electrical potential of the AV node. Under what conditions does your model produce regular heart beats?

Investigate the production of irregular beating patterns by modifying parameters in your model. Two patterns of clinical interest are second-degree block and the Wenckebach

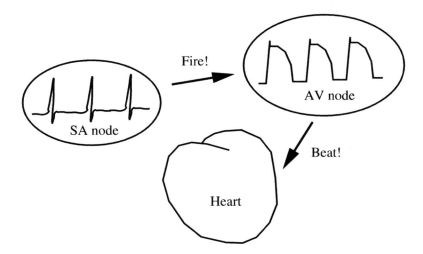

**Figure 9.6.** *A schematic of the heart beat control mechanism. Figure adapted from Adler* [1].

phenomenon. Second-degree block refers to the situation in which the heart skips every other beat (i.e., the AV node blocks every other signal from the SA node). The Wenckebach phenomenon refers to the situation in which the heart skips a beat every now and then, while it beats normally most of the time. Can your model produce other beating patterns?

## Project 21: Ocular Dominance Columns

Visual information is transmitted via the optic nerve to the visual cortex. Scientists studying the visual cortex of cats and monkeys discovered columns (bands or stripes) of neurons that selectively respond to visual information from one eye or the other. The bands are interlaced, as shown in Figure 9.7.

Hubel et al. [94] suggested that the columns are formed through a competition process during the first several months after birth. Neurons in the visual cortex have a number of synapses receiving inputs from the eyes. A synapse is associated either with the right eye or the left eye. Initially, all neurons are binocular, that is, it has both right- and left-eye synapses, and the synapses are intermixed randomly. During development, synapses can switch allegiance from one eye to the other, as a result of competition. Swindale [153] demonstrated that ocular dominance patterns can be generated by assuming that interactions between right- and left-eye synapses follow two simple rules:

(1) Local interactions (within a region 200 $\mu$m in diameter) are stimulatory (for example, in a region where right-eye synapses dominate, there will be an increase in the number of right-eye synapses at the expense of left-eye synapses);

(2) Interactions over larger distances (200–600 $\mu$m) between opposite-eye synapses are inhibitory (for example, in an annular ring surrounding a region where right-eye

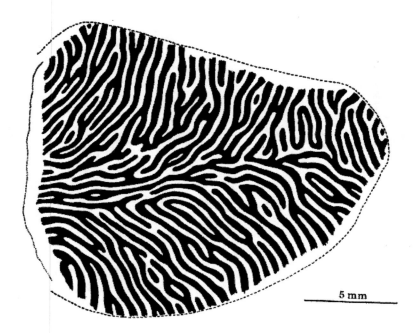

**Figure 9.7.** *Ocular dominance stripes in an area of the visual cortex of a macaque monkey. Black represents neurons receiving input from one eye; white represents neurons receiving input from the other eye. (Figure 24 (a) from Hubel and Wiesel [93], reprinted with permission by the Royal Society of London.)*

synapses dominate, there will be an increase in the number of left-eye synapses at the expense of right-eye synapses).

Develop a cellular automaton simulation that implements the assumptions mentioned above. Initially, you may assume that the number of synapses remains fixed.

Investigate the effect of growth of the visual cortex during development of the ocular dominance stripes. How does this affect the pattern produced?

Investigate the effect of monocular deprivation (restricting the input from one eye for a period of time during development). Are there times at which monocular deprivation causes a change in the eventual pattern?

## Project 22: The Sound of Many Hands Clapping

A popular topic in mathematical biology is the self-organization of many individual units, whether they be cells or people, to produce patterns in space and/or time.

An interesting example of self-organization is rhythmic applause produced by audiences in concert halls. An audience indicates its appreciation for a performance by the strength and nature of its applause. In [124], Neda et al. note that several transitions between fast, incoherent clapping and slower, synchronized clapping may occur. Apparently, an audience has a desire for both synchronization and high noise intensity. However, synchronization can only occur with a low clapping frequency, corresponding to a low noise intensity. When members of the audience increase their clapping frequency to increase the noise intensity, synchronization is lost.

Develop a model that reproduces the phenomenon of self-organization by clapping audiences.

Experimental findings on the clapping phenomenon can be found in Neda et al. [124]. For some modeling ideas, the work by Strogatz on the synchronization of male fireflies may be helpful (see [152] for an introduction).

## Project 23: Tumor Growth and Radiotherapy

The development of cancer is a multistage process involving multiple genetic events. Damage to DNA can cause mutations. Some mutations alter the function of tumor suppressor genes, such as P53, which controls cell growth. Such mutated cells grow faster than normal cells.

Develop a model that describes cancer growth in an environment with limited resources (such as oxygen). Start with one cell which has undergone a mutation for faster growth (initiated cell). The initiated cell grows and proliferates. In successive generations, more and more mutations occur and some cells become more and more aggressive. The aggressive cells grow faster, but they need more resources. Assume that all cells, cancerous and healthy, compete for the same resources.

Model the different mutation stages. Does your model show that the tumor grows unlimited, or will the growth come to a halt due to limited resources?

If time allows, add *radiotherapy* to your model (see Figure 9.8). Assume that an ionizing beam is used periodically, which (i) kills cells and (ii) enhances mutations. Unfortunately, radiation also affects normal cells. What happens? Can you design a good treatment plan? According to your model, how often would you radiate and in which time periods to optimize treatment and to minimize side effects on normal tissue? For further reading on cancer in general, we recommend the "World Cancer Report" by Stewart and Kleihues [150].

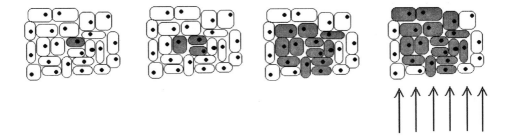

**Figure 9.8.** *Schematic of tumor progression and radiation treatment.*

**Figure 9.9.** *Shells of* Olivia *and* Conus *species.*

## Project 24:  Mollusk Patterns

Many mollusks show very interesting patterns on their shells. Figure 9.9 shows examples of
*Olivia spec.* and *Conus spec.* Since these shells grow gradually on the outer edge only and
since the patterns do not change later, it can be seen as the time record of a one-dimensional
pattern-producing system.

These shell patterns are very similar to some patterns produced by some simple Wolfram automata (discussed in Section 6.1.1). Therefore, they might be modeled with cellular
automata as done by Kusch and Markus [103].

Find rules for a one-dimensional cellular automaton that produces patterns as shown in
Figure 9.9. You may also look for pictures of other *Olivia* and *Conus* species and reproduce
their patterns. If you look closely at these pictures, you see that real shell patterns are never
as perfect as patterns produced by simulations. Introduce stochasticity in your automaton
to generate more realistic patterns.

Please note that if two patterns look alike, it does not necessarily mean that they are
produced by the same mechanism. You may compare your model with the reaction-diffusion
models in [117].

## Project 25:  Run-Bike-Fun

A "Run-Bike-Fun" sports event takes place every year in a small university town in Germany.
Each participating team consists of two people. Both people have to complete a 15 km course
through a combination of running and cycling. Each team has one bicycle. Only one person
is allowed to ride the bicycle at any one time, but team members can switch between running
and cycling as often as they wish. The first team with both partners at the finish line wins.

At the beginning of the race, one person starts riding the bicycle, and the other starts running. After some time, the cyclist gets off the bicycle, puts it down, and starts running. When the other runner reaches the bicycle, he/she picks it up and starts cycling.

What is the optimal switching strategy? At which locations along the course should the switch(es) occur?

You may wish to begin by assuming that it takes no time to get on/off the bicycle, that both team members are $x$ times faster at cycling compared to running, and that people run/cycle with constant velocity. Based on your own experience, estimate the value of $x$. When/where should you switch?

In reality, people get tired. How might you describe that? Would you use the same description for running and cycling? How does this affect the optimal strategy? Also, switching between cycling and running takes time. How does this affect the optimal strategy? What if two people with different abilities form a team?

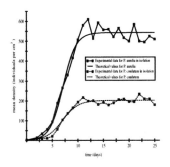

# Chapter 10
# Solved Projects

## 10.1  Cell Competition

In this section we present and discuss a solution for the cell competition project (Project 10 in Chapter 9). The model is based on the work of Renouf and Choczynska [136].

We analyze the data gathered by Gause [63] and summarized in Table 9.9. Populations of *Paramecium aurelia* and *Paramecium caudatum* were grown in isolation and in competition over a period of 25 days. Nutrient levels were kept constant, resulting in both intra- and interspecific competition. The data from Table 9.9 are plotted in Figure 10.1.

When *P. aurelia* and *P. caudatum* are grown in isolation, it appears that both populations reach a steady state over the course of the experiment. When the two populations are grown in competition, the data suggests that *P. aurelia* reach a steady state, albeit at a lower level than in isolation. *P. caudatum* seem to head either towards extinction or towards a steady state as well, also at a lower level than in isolation. A question of interest for this project is to find out which of these situations is most likely for the given data.

### 10.1.1  *Paramecium caudatum* in Isolation

We model the growth of an isolated population with the logistic equation (encountered in Section 3.1 as (3.6))

$$\frac{dx}{dt} = rx \left( 1 - \frac{x}{K} \right), \tag{10.1}$$

where $x(t)$ is the mean density (in individuals per 0.5 cm$^3$) at time $t$ (in days), $r$ is the instantaneous rate of increase (births/deaths), and $K$ is the carrying capacity per 0.5 cm$^3$. We assume constant $K$ and $r$, linear density dependence, no time lags, no migration, no age structure, and limited resources.

We define and solve (10.1) in Maple, as shown in Section 8.4.1. Note that we have taken the initial condition to be $x(0) = 2$, in accordance with the data from Table 9.9.

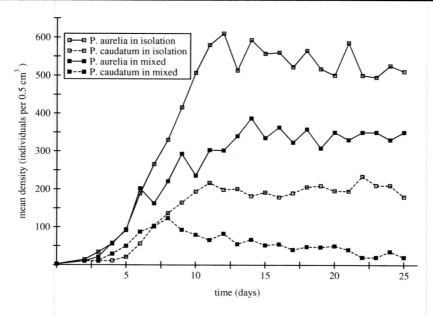

**Figure 10.1.** *Growth of* P. aurelia *and* P. caudatum. *Data taken from Table* 9.9.

```
> eq:=diff(x(t),t)=r*x(t)*(1-(x(t)/K));
```

$$eq := \frac{\partial}{\partial t}x(t) = r\,x(t)\left(1 - \frac{x(t)}{K}\right)$$

```
> sol:=rhs(dsolve({eq, x(0)=2}, {x(t)}));
```

$$sol := \frac{K}{1 + \frac{1}{2}\mathbf{e}^{(-r\,t)}(K - 2)}$$

We are interested in the values of $r$ and $K$ so that the solution $x(t)$ of (10.1) best fits the experimental data. We show how to determine $r$ and $K$ with a simple iterative process for *P. caudatum*, and leave the determination of $r$ and $K$ for *P. aurelia* as an exercise to the reader.

First, we obtain a good fit visually by trial and error. With $r = 0.65$ and $K = 200$, the fit is reasonable, as can be seen by executing the following set of Maple commands:

```
> r:=0.65:  K:=200:
> caudatum:=[[0,2], [2,10], [3,10],..., [25,180]]:
> data:=plot(caudatum, style=point):
> theory:=plot(sol, t=0..25):
> with(plots):
> display(data, theory, labels=["time steps",
  pop]);
```

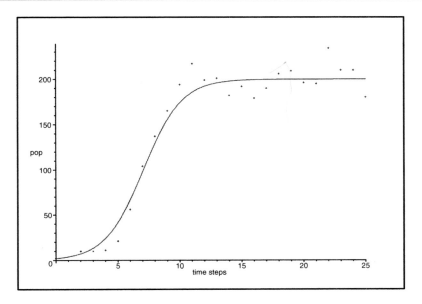

**Figure 10.2.** *The plot of population versus time steps. The model solution,* sol, *is shown as a solid curve; the experimental data points are shown as dots.*

The list, caudatum, is the list of data points for *P. caudatum* in isolation from Table 9.9. Defining $E_i$ to be the mean density of population at time $t = i$, for $i = 0, 1, \ldots, 25$, the entries in this list are then of the form $[i, E_i]$. Note that we have created two plots, but assigned the results to the data and theory variables, and superimposed these two plots with the display command. The result is shown in Figure 10.2. As can be seen, the fit is quite reasonable, but we can do better. To obtain the *best* fit, we use the least-squares method (see p. 209), in which we minimize the sum of the squares of the differences between the theoretical and experimental values at each point. The theoretical value equivalent to $E_i$ is $x(i; r, K)$, the expected mean density of individuals on day $i$ for specified values of $r$ and $K$. Let the sum of squares be

$$D(r, K) = \sum_{i=0}^{25} [x(i; r, K) - E_i]^2.$$

Our task is to find values of $r$ and $K$ so that the sum $D(r, K)$ is minimized.

The following Maple procedure, sum_sq, will be used to determine the sum of squares. The list referred to, caudatum, is the same list as above.

```
> sum_sq:=proc(r,K)
>    local D, d, a, i, p, fa, sola, eqa;
>    eqa:=diff(x(p),p)=r*x(p)*(1-(x(p)/K));
>    sola:=dsolve({eqa, x(0)=2}, {x(p)});
```

```
>    # solves the differential equation
>    fa:=rhs(sola);
>    # initial D with term corresponding to i = 0
>    D:=(evalf(subs(p=0,fa))-2)^2;
>    # update D
>    for i from 2 to 25 do
>       a:=evalf(subs(p=i,fa));
>       d:=evalf((a-(caudatum[i][2]))^2);
>       D:=D+d;
>    od:
>    D;
> end;
```

We proceed by computing the sum of squares over a range of $K$ values, from $K_1$ to $K_2$ in steps of 0.1, for a fixed value of $r$. The following procedure, range, outputs a list of the form $[K, D(r, K)]$, for $K = K_1, \ldots, K_2$:

```
> range:=proc(r,K1,K2)
>    local L, i, s;
>    L:=[];
>    for i from K1 by 0.1 to K2 do
>       s:=sum_sq(r,i);
>       L:=[op(L), [i,s]];
>    od;
>    L;
> end;
```

We now use the range procedure to compute the sum of squares for $r = 0.65$ and values of $K$ between $K_1 = 190$ and $K_2 = 210$, and plot the resulting sum of squares as a function of $K$.

```
> q:=range(0.65, 190, 210);
> plot(q, labels=[K, sum_sq]);
```

The result is shown in Figure 10.3. We can see that the minimum occurs at approximately $K = 203$. By examining the entries in list q, we find that the $K$ value for which $D(r, K)$ is a minimum is $K = 203.2$ (with error less than 0.1). We now fix $K = 203.2$ and use a similar procedure to find the best $r$ value. We find $r = 0.66$. We repeat these two steps to improve $r$ and $K$. The process converges, and we find that the values of $r$ and $K$ that produce the best fit of the model solution to the data are $r = 0.66$ and $K = 202.6$. The reader is encouraged to use the above approach to find the best fit parameters for the data of *P. aurelia*. We find $r = 0.79$ and $K = 543.1$. In Figure 10.4, the best fit solutions are shown with the experimental data.

## 10.1.2   The Two Populations in Competition

Now that we know the growth dynamics of the two species in isolation, we will investigate the growth dynamics in competition. We assume the Lotka–Volterra competition model,

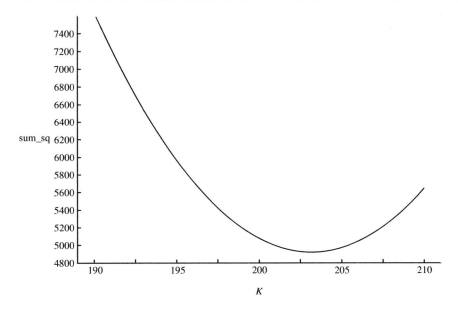

**Figure 10.3.** *The sum of squares, $D(r, K)$, for $r = 0.65$, as a function of $K$. The minimum is attained at $K = 203.2$ (with an error less than 0.1).*

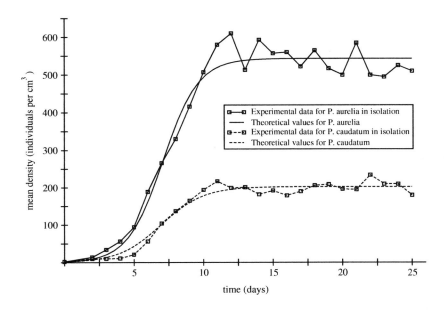

**Figure 10.4.** *Comparison of the best fit solutions of the logistic model, (10.1), and Gause's experimental data for growth of* P. caudatum *and* P. aurelia *in isolation.*

discussed briefly in Section 3.3.2:

$$\frac{dN_1}{dt} = r_1 N_1(t) \left[ \frac{K_1 - N_1(t) - \beta_{12} N_2(t)}{K_1} \right], \tag{10.2}$$

$$\frac{dN_2}{dt} = r_2 N_2(t) \left[ \frac{K_2 - N_2(t) - \beta_{21} N_1(t)}{K_2} \right]. \tag{10.3}$$

Here, $N_i$ represents the population density (in individuals per 0.5 cm$^3$) of species $i$, for $i = 1, 2$. Species 1 is *P. aurelia*, and species 2 is *P. caudatum*. The term $r_i$ represents the instantaneous rate of increase of species $i$, and $K_i$ represents the carrying capacity of species $i$. The parameter $\beta_{12}$ represents the per capita effect of *P. caudatum* on the population growth of *P. aurelia*, and $\beta_{21}$ represents the per capita effect of *P. aurelia* on the population growth of *P. caudatum*. Notice that if the second species is absent, that is, $N_2(t) = 0$, then (10.2) reduces to the logistic equation. This is expected, since the first species will then not be affected by the second species. We already know the values of $r_1, r_2, K_1$, and $K_2$: they are the values we determined in the previous section. Now we are interested in finding best fit values for $\beta_{12}$ and $\beta_{21}$.

We define system (10.2)–(10.3) with Maple:

```
> eqN1:=diff(N1(t),t)
    = r1*N1(t)*(K1 - N1(t) - beta12*N2(t))/K1;
> eqN2:=diff(N2(t),t)
    = r2*N2(t)*(K2 - N2(t) - beta21*N1(t))/K2;
```

$$eq\,N1 := \frac{\partial}{\partial t} N1(t) = \frac{r1\,N1(t)(K1 - N1(t) - \beta12\,N2(t))}{K1}$$

$$eq\,N2 := \frac{\partial}{\partial t} N2(t) = \frac{r2\,N2(t)(K2 - N2(t) - \beta21\,N1(t))}{K2}$$

To obtain the solution, we specify the values of the model parameters and solve the system numerically. As we did in the previous section for $r$ and $K$, we start with guesses for the values of $\beta_{12}$ and $\beta_{21}$ and modify them until we get a good visual fit of the model to the experimental data. Once we have a good visual fit, we improve upon the values of $\beta_{12}$ and $\beta_{21}$ using the same iterative approach as above (the details are left to the reader). We find that the best fit values are $\beta_{12} = 2.17$ and $\beta_{21} = 0.36$:

```
> r1:=0.79; K1:=543.1; r2:=0.66; K2:=202.6;
  beta12:=2.17; beta21:=0.36;
> sol:=dsolve({eqN1,eqN2,N1(0)=2,N2(0)=2},
  {N1(t),N2(t)}, type=numeric,
  method=classical[rk4], output=listprocedure):
```

Superimposing plots of the experimental data and the best fit solution as before, we obtain the graph shown in Figure 10.5. Agreement between the data and the model is good. From Figure 10.5, it appears that *P. caudatum* is heading towards either extinction or a small steady-state population. Consequently, *P. aurelia* would experience no or less interspecific competition, and it would grow towards its carrying capacity in isolation or close to it. To determine which of these situations is predicted by the model, we can run the simulation beyond 25 days. We will do so in Section 10.1.4. In the meantime, we will proceed with

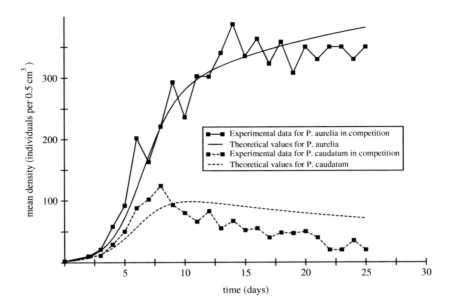

mean density (individuals per 0.5 cm$^3$)

time (days)

**Figure 10.5.** *Comparison of the best fit solution to the competition model and Gause's experimental data for growth of* P. caudatum *and* P. aurelia *in competition.*

a detailed phase-plane analysis of the competition model in the next section. The analysis will lead to a valuable insight about the strength of our conclusions.

### 10.1.3   Phase-Plane Analysis of the Competition Model

**Nullclines and Steady States**

We begin by finding the nullclines of the model, (10.2)–(10.3). The nullclines for *P. aurelia* are obtained by setting (10.2) to zero, yielding two straight lines, namely, the trivial nullcline

$$N_1 = 0, \qquad\qquad\qquad (10.4)$$

and the nontrivial nullcline

$$K_1 - N_1 - \beta_{12}N_2 = 0. \qquad\qquad\qquad (10.5)$$

When plotted in the $(N_1, N_2)$ phase plane, the trivial nullcline lies along the $N_2$-axis. The nontrivial nullcline has slope $-1/\beta_{12}$ and the intercepts are located at $(K_1, 0)$ and $(0, K_1/\beta_{12})$. Similarly, the nullclines for *P. caudatum* are obtained by setting (10.3) to zero, yielding the trivial nullcline

$$N_2 = 0 \qquad\qquad\qquad (10.6)$$

and the nontrivial nullcline

$$K_2 - N_2 - \beta_{21}N_1 = 0. \qquad\qquad\qquad (10.7)$$

The trivial nullcline lies along the $N_1$ axis. The nontrivial nullcline has slope $-\beta_{21}$ and the intercepts are located at $(K_2/\beta_{21}, 0)$ and $(0, K_2)$.

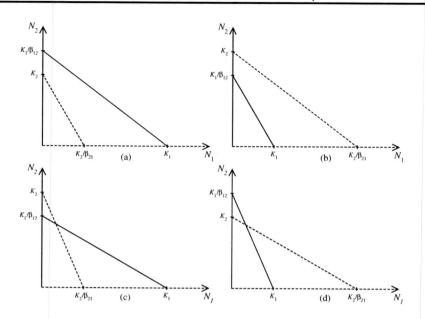

**Figure 10.6.** *Four essentially different cases for the cell competition model, (10.2)–(10.3), based on the relative positions of the $N_1$- and $N_2$-intercepts of the nontrivial null-clines. Solid lines represent the $N_1$ nullclines given in (10.4) and (10.5); dashed lines represent the $N_2$ nullclines given in (10.6) and (10.7).  (a) Case I: $K_2/\beta_{21} < K_1$ and $K_1/\beta_{12} > K_2$; (b) case II: $K_2/\beta_{21} > K_1$ and $K_1/\beta_{12} < K_2$; (c) case III: $K_2/\beta_{21} < K_1$ and $K_1/\beta_{12} < K_2$; (d) case IV: $K_2/\beta_{21} > K_1$ and $K_1/\beta_{12} > K_2$.*

**Table 10.1.** *Summary of the cases shown in Figure 10.6.*

|  | Case I | Case II | Case III | Case IV |
|---|---|---|---|---|
| $\beta_{21}K_1 - K_2$ | $> 0$ | $< 0$ | $> 0$ | $< 0$ |
| $\beta_{12}K_2 - K_1$ | $< 0$ | $> 0$ | $> 0$ | $< 0$ |
| $\beta_{12}\beta_{21} - 1$ | $> 0$ | $< 0$ | $> 0$ | $< 0$ |

Based on the relative positions of the $N_1$- and $N_2$-intercepts of the nontrivial nullclines, we distinguish four different cases, as shown in Figure 10.6 and summarized in Table 10.1.

Steady states occur at the intersection of a nullcline for $N_1$ and a nullcline for $N_2$. In Figure 10.6, we look for intersections of a solid curve and a dashed curve. We see that the following three steady states always exist, independent of the choice of model parameters:

$$S_1 : (N_1, N_2) = (0, 0),$$
$$S_2 : (N_1, N_2) = (K_1, 0),$$
$$S_3 : (N_1, N_2) = (0, K_2).$$

Steady state $S_1$ is not so interesting, since it represents the situation in which both populations have gone extinct. Steady state $S_2$ represents the situation in which species 2 has gone extinct, and species 1 has reached its carrying capacity, $K_1$. Similarly, steady state $S_3$ represents the situation in which species 1 has gone extinct, and species 2 has reached its carrying capacity, $K_2$. Note that $S_2$ and $S_3$ represent the steady states of the populations in isolation.

A fourth steady state may or may not exist, depending on whether (10.5) and (10.7) intersect in the first quadrant or not. When it exists, as in cases III and IV, it is given by

$$S_4 : (N_1, N_2) = \left( \frac{\beta_{12} K_2 - K_1}{\beta_{12} \beta_{21} - 1}, \frac{\beta_{21} K_1 - K_2}{\beta_{12} \beta_{21} - 1} \right).$$

Steady state $S_4$ is of particular interest. If it exists and is stable, it represents coexistence of the two populations.

## Linear Stability Analysis

As shown in Section 3.4.2, we can determine the stability of a steady state by using the Jacobian matrix. For our system, (10.2)–(10.3), the Jacobian matrix is

$$Df(N_1, N_2) = \begin{pmatrix} r_1 - \frac{2r_1 N_1}{K_1} - \frac{r_1 \beta_{12} N_2}{K_1} & \frac{-r_1 \beta_{12} N_1}{K_1} \\ \frac{-r_2 \beta_{21} N_2}{K_2} & r_2 - \frac{2r_2 N_2}{K_2} - \frac{r_2 \beta_{21} N_1}{K_2} \end{pmatrix}.$$

Evaluating $Df(N_1, N_2)$ at steady state $S_1$ gives

$$Df(0, 0) = \begin{pmatrix} r_1 & 0 \\ 0 & r_2 \end{pmatrix}.$$

The eigenvalues of this matrix are $\lambda_1 = r_1 > 0$ and $\lambda_2 = r_2 > 0$, indicating that steady state $S_1$ always is an unstable node.

Evaluating $Df(N_1, N_2)$ at steady state $S_2$ gives

$$Df(K_1, 0) = \begin{pmatrix} -r_1 & -r_1 \beta_{12} \\ 0 & \frac{r_2}{K_2}(K_2 - \beta_{21} K_1) \end{pmatrix}.$$

The eigenvalues of this matrix are $\lambda_1 = -r_1 < 0$ and

$$\lambda_2 = \frac{r_2}{K_2}(K_2 - \beta_{21} K_1) \quad \begin{cases} < 0 & \text{in cases I and III,} \\ > 0 & \text{in cases II and IV.} \end{cases}$$

Note that the sign of $\lambda_2$ can be determined easily from the information summarized in Table 10.1. Thus, steady state $S_2$ is a stable node in cases I and III, and a saddle in cases II and IV. Similarly, it can be shown that steady state $S_3$ is a stable node in cases II and III, and a saddle in cases I and IV.

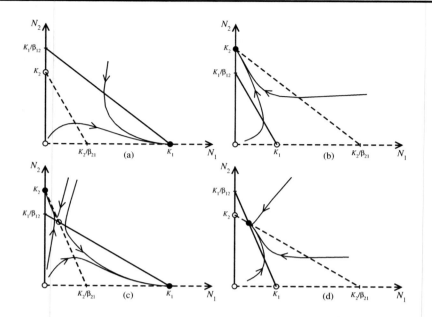

**Figure 10.7.** *Expanded phase plane diagrams, corresponding to Figure* 10.6.
*Filled circles indicate stable steady states; open circles indicate unstable steady states.* (a)
*Case* I; (b) *Case* II; (c) *Case* III; (d) *Case* IV.

Lastly, evaluating $Df(N_1, N_2)$ at steady state $S_4$ gives

$$Df\left(\frac{\beta_{12}K_2 - K_1}{\beta_{12}\beta_{21} - 1}, \frac{\beta_{21}K_1 - K_2}{\beta_{12}\beta_{21} - 1}\right) = \begin{pmatrix} \frac{r_1(K_1 - \beta_{12}K_2)}{K_1(\beta_{12}\beta_{21} - 1)} & \frac{r_1(K_1 - \beta_{12}K_2)\beta_{12}}{K_1(\beta_{12}\beta_{21} - 1)} \\ \frac{r_2(K_2 - \beta_{21}K_1)\beta_{21}}{K_2(\beta_{12}\beta_{21} - 1)} & \frac{r_2(K_2 - \beta_{21}K_1)}{K_2(\beta_{12}\beta_{21} - 1)} \end{pmatrix}.$$

Instead of calculating the eigenvalues of this matrix explicitly, it is more convenient to
determine the stability via the determinant and trace (equation (3.22) in Section 3.4.1). We
have

$$\det Df = \frac{r_1 r_2}{K_1 K_2} \frac{(K_1 - \beta_{12}K_2)(\beta_{21}K_1 - K_2)}{(\beta_{12}\beta_{21} - 1)},$$

$$\operatorname{tr} Df = \frac{r_1(K_1 - \beta_{12}K_2)}{K_1(\beta_{12}\beta_{21} - 1)} + \frac{r_2(K_2 - \beta_{21}K_1)}{K_2(\beta_{12}\beta_{21} - 1)}.$$

Using the information summarized in Table 10.1, we find that in case III, $\det Df < 0$. Thus,
steady state $S_4$ is unstable in this case. In particular, it is a saddle (recall Figure 3.10 from
Section 3.4.1). In case IV, steady state $S_4$ is stable, since $\det Df > 0$ and $\operatorname{tr} Df < 0$.
    Phase planes for the four different cases are shown again in Figure 10.7, this time
with the steady states and their stability superimposed. Typical solution trajectories are
shown as well. Note that there is *bistability* between steady states $S_2$ and $S_3$ in case III
(Figure 10.7 (c)), and the final outcome of the competition depends on the initial condition.

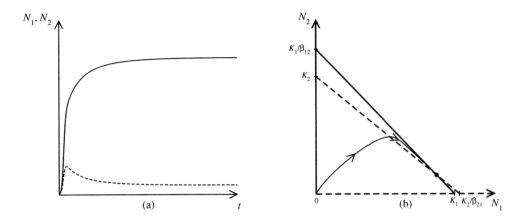

**Figure 10.8.** (a) *Long-term behavior of the best fit solution for the competition model;* (b) *solution trajectory shown in the phase plane.*

### 10.1.4  Model Prediction

We now return to the experimental data of Gause and the best fit solution of the competition model that we found in Section 10.1.2. We determined that the best fit parameters for our model, (10.2)–(10.3), are $r_1 = 0.79$, $K_1 = 543.1$, $r_2 = 0.66$, $K_2 = 202.6$, $\beta_{12} = 2.17$, and $\beta_{21} = 0.36$. From Table 10.1, we can see that case IV holds with these parameter values. That is, the model predicts that the two species can coexist. Indeed, numerical simulation of the model beyond 25 days shows that both populations are predicted to reach a steady state, as shown in Figure 10.8 (a). The solution in the phase plane is shown in Figure 10.8 (b), as well as the nullclines. Note that species 1 approaches a steady state just below its carrying capacity of $K_1 = 543.1$, and species 2 barely survives, approaching a steady state much smaller than its carrying capacity of $K_2 = 202.6$.

In Figure 10.8 (b), we see that the two nontrivial nullclines have very similar slope. This indicates that case IV holds, but just barely. It is not hard to imagine that a slight change in the parameter values will shift and/or tilt the nullclines so that case I holds instead of case IV. In that case, the model predicts that species 1 will drive species 2 to extinction. Thus, although Gause's data set suggests that the two populations in his experiment are able to coexist in competition, there is room to argue this conclusion. This insight follows directly from our detailed analysis of the model in the previous section.

### 10.1.5  An Alternative Hypothesis

In this project, we were able to explain the experiment of Gause successfully by a model that assumed interspecific competition for resources such as nutrients and space. As in every model, we neglected a number of facts, important or not. Therefore, it is necessary to think about alternative hypotheses and to compare different models. Instead of competition for resources, are there other reasons that could underlie limitations in growth?

Consider the production of beer or wine by yeast cells. In these situations, the medium in which the yeast population grows is not renewed, and toxic waste (alcohol) produced by the yeast builds up in the medium. The more sensitive the yeast cells are to alcohol, the sooner they will die, and the less alcohol will be present in the medium after all yeast cells have died (the reason why wine has a higher alcohol content than beer is that the yeast species used to produce wine is less sensitive to alcohol than the species used to produce beer).

The above example leads to the following alternative hypothesis. Suppose that populations of two species are grown in the same medium which is not renewed through the course of the experiment. Consequently, nutrient levels decline, and waste products build up. What happens if the two species have different intrinsic growth rates and different sensitivities to the waste products? Can a model based on this alternative hypothesis explain Gause's experiment?

## 10.2  The Chemotactic Paradox

In this section, we treat the chemotactic paradox (Project 14 in Chapter 9). The model of Section 10.2.1 was derived by Cubbon and Gutermuth [40]. The chemotactic paradox is a classical problem in cell movement and has been studied by a number of researchers. In Section 10.2.2, we relate our model to the literature.

### 10.2.1  A Resolution of the Chemotactic Paradox

Chemotaxis is the response of motile cells in which the direction of movement is affected by the gradient of a diffusible substance. For example, individual cells of the species *Dictyostelium discoideum* aggregate in response to a signal from so-called organizer cells (also called "pacemaker" cells). The chemical signal initially is released by the organizer cells. The signal binds to cell surface receptors on neighboring cells. As a result, neighboring cells change their orientation and move towards the aggregation center. In addition, the neighboring cells are stimulated to release the signal as well. On the level of the entire population of cells, this process causes a signal wave moving outward.

We consider one motile cell and its response to a single-peaked signal wave that is approaching the cell from the right (thus, the signal wave is moving from right to left). If we assume that the cell can sense the local chemical gradient and moves in the direction of a positive gradient, then the cell would move to the right initially as the wave approaches. As soon as the peak of the wave has traveled past the cell, the cell experiences a negative gradient. One would expect the cell to reverse its direction and move to the left. Overall, the cell would show a net displacement to the left. However, it is found experimentally that cells only move to the right or they stop; they do not reverse during aggregation. This is the chemotactic paradox which we will resolve with a mathematical model.

To study this phenomenon, we consider the apparatus which is used by the cell to measure the chemical gradient. Across the cell surface, there are receptor proteins which are able to bind certain chemicals. In the case of *Dictyostelium discoideum*, the chemical signal is cAMP (cyclic adenosine monophosphate). We assume that as soon as the receptors

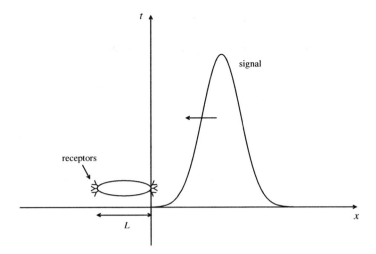

**Figure 10.9.** *Sketch of a cell with receptors on both ends and an incoming traveling signal wave.*

at the front of the cell are occupied, a messenger signal is sent through the cell body which inhibits (closes) the remaining receptors. Thus, the remaining receptors are desensitized, resulting in the cell possessing directional information. We assume that the cell moves in the direction of the occupied receptors.

We begin the model by considering an observer cell located at $x = 0$. Assume that the cell has length $L > 0$. We choose the coordinate system so that the front of the cell always is located at $x = 0$, as shown in Figure 10.9. We use a wave of the chemical signal which has the form of a Gaussian distribution and is moving to the left with unit speed without changing its shape. Let $s(t)$ be the signal concentration at $x = 0$. We model $s(t)$ by

$$s(t) = \frac{1}{\sqrt{2\pi\sigma^2}} \exp\left(-\frac{1}{2}\frac{(t - t_0)^2}{\sigma^2}\right), \tag{10.8}$$

where $\sigma$ is the standard deviation and $t_0$ is the time at which the peak of the wave reaches the front of the cell. Further, we assume that the cell has an equal number of receptors at the front and back. We ignore any receptors on the sides of the cell, since we consider movement in one dimension only. Last but not least, we assume that the number of receptors, $\rho$, is large, so that the use of a model of ODEs is justified.

To model the binding to the receptors, we use a simple reaction as introduced in Section 3.3.1. If we let $r_f$ denote the number of empty front receptors, and $R_f$ the number of occupied front receptors, then

$$r_f + s \underset{\beta}{\overset{\alpha}{\rightleftharpoons}} R_f,$$

with positive binding and dissociation rates, $\alpha$ and $\beta$. Using the Law of Mass Action, we obtain the following model for the front receptors:

$$\frac{d}{dt}r_f(t) = -\alpha r_f(t)s(t) + \beta R_f(t),$$

$$\frac{d}{dt}R_f(t) = \alpha r_f(t)s(t) - \beta R_f(t). \tag{10.9}$$

As mentioned earlier, the total number of front receptors is constant. This fact is reflected in the above equations since $\frac{d}{dt}(r_f + R_f) = 0$. Below, we use this fact to reduce this set of two equations to a single equation.

We have a similar reaction for the receptors $r_b$ and $R_b$ at the back of the cell, namely,

$$r_b + s \underset{b}{\overset{a}{\rightleftharpoons}} R_b.$$

Here, we must account for our assumption that occupied front receptors inhibit the back receptors. We do this by assuming that the reaction rate $a$ is a decreasing function of $R_f$. Specifically, we assume $a = a(R_f) = a_0 e^{-cR_f}$, with $a_0$ and $c$ being positive constants. Since the wave signal $s(t)$ moves with unit speed and the cell has length $L$, the back receptors encounter the same wave profile as the front receptors, but at a later time, $t - \tau$, where $\tau = L$. For dissociation, we choose $b = \beta$. Then the model equations for the receptors at the back of the cell are

$$\frac{d}{dt}r_b(t) = -a(R_f)r_b(t)s(t - \tau) + \beta R_b(t),$$

$$\frac{d}{dt}R_b(t) = a(R_f)r_b(t)s(t - \tau) - \beta R_b(t). \tag{10.10}$$

As before, $r_b + R_b$ is constant.

The system of equations (10.9) and (10.10) forms our basic model. We now use the property of conservation of receptor number to reduce the system of four equations to a system of two equations, tracking only the number of occupied receptors. Since $r_f + R_f = \rho = r_b + R_b$, we can write $r_f = \rho - R_f$ and $r_b = \rho - R_b$. Then (10.9) and (10.10) reduce to

$$\frac{d}{dt}R_f(t) = \alpha(\rho - R_f(t))s(t) - \beta R_f(t),$$

$$\frac{d}{dt}R_b(t) = a_0 e^{-cR_f(t)}(\rho - R_b(t))s(t - \tau) - \beta R_b(t). \tag{10.11}$$

We use the initial conditions $r_f(0) = r_b(0) = \rho$, and will simulate the model up to $t = 20$. We take $\alpha = 1$, $a_0 = 1$, $\beta = 10$, $t_0 = 5$, $\sigma = 1$, and $\tau = L = 1$. We further assume that the total number of receptors on each side is $\rho = 10$. Then we have

$$r_f(t) + R_f(t) = r_b(t) + R_b(t) = 10$$

for all times $t$.

We solve this system of two equations with Maple. In the Maple code, we use the symbols F and B for the occupied front and back receptors, respectively.

```
> n:=20:   t0:=5:   alpha:=1:   beta:=10:   sigma:=1:
  a0:=1:   tau:=1:   k:=20:   c:=100:
> eq1 := diff(F(t),t)=alpha*(1/(sqrt(2*Pi*sigma*sigma))
  *exp((-(t-t0)*(t-t0))/(2*sigma*sigma)))*(k-F(t))-beta*F(t);
```

$$eq1 := \frac{\partial}{\partial t}F(t) = \frac{1}{2}\frac{\sqrt{2}\,\mathbf{e}^{\left(-\frac{1}{2}(t-5)^2\right)}(20 - F(t))}{\sqrt{\pi}} - 10F(t)$$

```
> eq2 := diff(B(t),t)=a0*exp(-c*F(t))*(1/(sqrt(2*Pi*sigma
  *sigma))*exp((-(t-t0-tau)*(t-t0-tau))/(2*sigma*sigma)))
  *(k-B(t))-beta*B(t);
```

$$eq2 := \frac{\partial}{\partial t}B(t) = \frac{1}{2}\frac{\mathbf{e}^{(-100\,F(t))}\sqrt{2}\,\mathbf{e}^{\left(-\frac{1}{2}(t-6)^2\right)}(20 - B(t))}{\sqrt{\pi}} - 10B(t)$$

```
> numsol1 := dsolve({eq1, eq2, F(0) = 0, B(0) = 0},
  {F(t), B(t)}, type=numeric, output=listprocedure):
> with(plots):   with(DEtools):
> odeplot(numsol1, [[t,F(t)],[t,B(t)]],0..20);
```

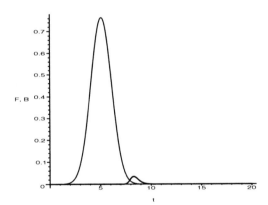

The thin solid curve represents the number of occupied front receptors, and the bold solid curve represents the number of occupied back receptors. Notice that the number of occupied front receptors reflects the shape of the incoming signal wave, while the back receptors are strongly inhibited.

Let the difference in the number of occupied front and back receptors be denoted by $h(t)$,

$$h(t) = R_f(t) - R_b(t).$$

We hypothesize that $h(t)$ is the guiding factor for the cell to decide in which direction to move, with the cell moving to the right (left) whenever $h(t) > 0\,(h(t) < 0)$. Since

$\int_0^{20} h(t)dt > 0$, there are more occupied front receptors than occupied back receptors over the time interval. The following plot of $h(t)$ demonstrates quite clearly that the cell primarily moves to the right, as observed experimentally. Thus, our model reflects a possible resolution of the chemotactic paradox.

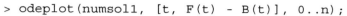

```
> odeplot(numsol1, [t, F(t) - B(t)], 0..n);
```

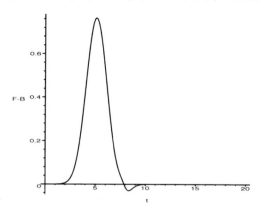

## 10.2.2  Discussion

In 1990, Soll [146] described the chemotactic paradox in some detail. He proposed four mechanisms which may play a role in inhibiting turning by a cell, as follows:

1. an internal excitation-adaptation mechanism could desensitize the cell as a whole;

2. the increasing signal concentration at the wave front decreases the turning rate of a moving cell, while the decreasing signal concentration at the wave back causes the cell to become unresponsive;

3. a decreasing signal concentration may directly inhibit turning;

4. the establishment of cell polarity.

The model presented above falls into the last category. As soon as some receptors are occupied and others are inhibited, the cell is polarized in the direction of the occupied receptors. Another mechanism by which the cell may sense a spatial gradient would be to measure the chemotactic signal concentration along the entire cell surface and somehow compare the concentrations at different locations on its surface. In a recent study by Rappel et al. [134], a simple inhibition mechanism has been introduced which also is able to explain cell polarization. Rappel et al. propose a simple mechanism in which local activation of receptors at the front of the cell generates a second messenger in the interior of the cell. This second messenger diffuses through the interior of the cell and suppresses the activation of the receptors at the back of the cell. Thus, the cell converts the temporal gradient into an initial cellular asymmetry. Rappel et al. suggest that for Dictyostelium discoideum, the internal inhibitor might be cGMP.

A model that fits into Soll's first category, the excitation-adaption mechanism, has been proposed by Höfer et al. [87]. The excitation-adaptation mechanism is important in particular for bacterial chemotaxis, since bacteria are too small to measure spatial concentration differences. They also are too small to polarize. We briefly review the work by Höfer et al. here. It is assumed that active proteins inside the cell can be deactivated during the encounter of the extracellular chemical signal. Höfer et al. propose a simple chemical pathway between an active phase $A$ and a deactivated phase $D$, where the deactivation and the activation rates, $f_+(s)$ and $f_-(s)$, depend on the signal concentration:

$$A \underset{f_-(s)}{\overset{f_+(s)}{\rightleftharpoons}} D.$$

Using the Law of Mass Action, one finds

$$\frac{dA}{dt} = -f_+(s)A + f_-(s)D, \tag{10.12}$$

$$\frac{dD}{dt} = f_+(s)A - f_-(s)D. \tag{10.13}$$

Let $\alpha = \frac{A}{A+D}$ denote the fraction of activated proteins, where $A + D$ is constant. From (10.12), the equation for $\alpha$ is

$$\frac{d\alpha}{dt} = -f_+(s)\alpha + f_-(s)(1 - \alpha). \tag{10.14}$$

A time constant $\tau > 0$ is introduced to indicate that the internal dynamics operate on a different time scale than cell movement, that is,

$$\tau \frac{d}{dt}\alpha(t) = -f_+(s(t))\alpha(t) + f_-(s(t))(1 - \alpha(t)), \tag{10.15}$$

where $s(t)$ is the signal wave as used in the previous section.

To understand the role of $\tau$, we consider the transformation of time. Let $\vartheta = \frac{t}{\tau}$ and $\tilde{\alpha}(\vartheta) = \alpha(\tau\vartheta)$. Then

$$\frac{d}{d\vartheta}\tilde{\alpha}(\vartheta) = \frac{d}{dt}\alpha \cdot \frac{dt}{d\vartheta} = \tau \cdot \frac{d}{dt}\alpha = -f_+\alpha + f_-(1 - \alpha)$$

$$= -f_+\tilde{\alpha} + f_-(1 - \tilde{\alpha}).$$

Thus, on the modified time scale $\frac{t}{\tau}$, we have the original equation (10.14).

Höfer et al. use the following functional forms of $f_+$, $f_-$:

$$f_+(s) = \frac{1 + 0.1s}{1 + s}, \qquad f_-(s) = 2\left(\frac{1 + 0.2s}{1 + 2s}\right).$$

The behavior of (10.15) now can be studied for different values of the *adaptation time*, $\tau$. The adaptation for different values of $\tau$ is shown in Figure 10.10. In each of the figures (a),

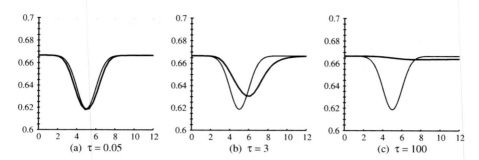

**Figure 10.10.** *Comparison of the cell's sensitivity to a chemical signal $\alpha(t)$ (bold lines) and the incoming signal wave $s(t)$ (thin line) for different values of the adaptation time $\tau$. (a) $\tau = 0.05$; (b) $\tau = 3$; (c) $\tau = 100$.*

(b), and (c), the fraction of activated proteins, $\alpha(t)$, is shown (bold curves) and compared to the incoming signal (thin curves). To facilitate the comparison, the transformation

$$f(t) = 0.6667 - 0.12s(t)$$

is shown instead of $s(t)$. In Figure 10.10 (a), $\tau$ is very small, and we see that $\alpha(t)$ simply follows $s(t)$. Hence the cell is insensitive at the peak of the wave, but becomes sensitive again in the back of the wave. That is, the cell will move to the right in the wave front, and to the left in the wave back. No adaptation occurs that can explain the chemotactic paradox; instead, there is instantaneous adaptation to the local quasi-equilibrium

$$\alpha(t) = \frac{f_-(s(t))}{f_+(s(t)) + f_-(s(t))}.$$

In Figure 10.10 (b), $\tau$ has an intermediate value, and we see that the sensitivity is minimal in the wave back. Thus, an intermediate adaptation time leads to a significant asymmetry between response to wave front and wave back, which leads to positively oriented movement. In Figure 10.10 (c), $\tau$ is very large, hence $\alpha(t)$ shows almost no response to the signal, and the cell does not move at all. In conclusion, an intermediate adaptation response is necessary to obtain oriented movement.

Othmer and Schaap [129] also studied an excitation-adaptation mechanism. Although in real cells many proteins are involved in the internal translation of the external signal, the mechanism can be captured by assuming that two factors are most important, namely, a response factor, $u_1$, and an inhibitor, $u_2$. If $s(t)$ again denotes the chemical signal, then one studies

$$\tau_e \frac{du_1}{dt} = h(s(t)) - (u_1 + u_2), \tag{10.16}$$

$$\tau_a \frac{du_2}{dt} = h(s(t)) - u_2, \tag{10.17}$$

where the function $h(s)$ describes excitation of both $u_1$ and $u_2$, and the factors $\tau_e$ and $\tau_a$ indicate the typical time scales of the excitatory pathway (for $u_1$) and the inhibitory pathway (for $u_2$). As in the model by Höfer et al., the relative size of $\tau_a$ and $\tau_e$ is important for the adaptation effect. The reader is invited to study the model of Othmer and Schaap with $h(s(t)) = H = $ constant and initial data $u_1(0) = u_2(0) = 0$, for the following three cases: (i) $\tau_e \ll \tau_a$, (ii) $\tau_e \gg \tau_a$, and (iii) $\tau_e \approx \tau_a$.

Soll's third mechanism underlying the chemotactic paradox, namely, the dependence on temporal gradients, has recently been considered theoretically by Dolak and Schmeiser [48]. The reader is invited to develop an appropriate model reflecting Soll's second mechanism.

 **Appendix**

# Further Reading

The field of mathematical biology and mathematical modeling of biological systems is growing rapidly. In the mid-1990s, we were looking at a handful of classical texts in mathematical biology. Since then, a plenitude of other excellent textbooks has appeared. Each of them has a different strength and slightly different emphasis. We use this appendix to give a rough overview of those textbooks related to the one you are reading now. Naturally, this overview expresses the personal points of view of the authors.

## Classical Texts

- L. Edelstein-Keshet, *Mathematical Models in Biology* [51].
  This text has become a standard introductory text for beginning graduate students.
- R.M. May, *Theoretical Ecology* [114].
- J. Murray, *Mathematical Biology* I: *An Introduction* [121].
- J. Murray, *Mathematical Biology* II: *Spatial Models and Biomedical Applications* [122].
  The original text by Murray recently has appeared as a 3rd edition, and split into these two volumes. They give an ample and detailed overview of mathematical models for biological systems. The level of exposition is aimed at higher-level graduate students and researchers.
- S.I. Rubinow, *Introduction to Mathematical Biology* [138].
  This text is suitable for an introductory graduate-level course.

## Elementary Texts

- F.R. Adler, *Modeling the Dynamics of Life: Calculus and Probability for Life Scientists* [1].
  This text is a suitable choice for an introductory calculus course aimed at students in the life sciences.
- B. Hannon and M. Ruth, *Modeling Dynamic Biological Systems* [78].
  This book is based on the computer language "Stella."
- C. Neuhauser, *Calculus for Biology and Medicine* [125].
  This text is a suitable choice for an introductory calculus course aimed at students in the life sciences.

## Texts That Can Be Read in Parallel to This Book

During the last several years, textbooks have appeared which are aimed at upper-level undergraduate students. Some are oriented towards biology, other are more mathematical.

For the biologically-oriented books, we like to draw attention to the following:

- R.F. Burton, *Biology by Numbers: An Encouragement to Quantitative Thinking* [35].
- R.F. Burton, *Physiology by Numbers: An Encouragement to Quantitative Thinking* [36].
- J.W. Haefner, *Modeling Biological Systems: Principles and Applications* [77].
  This text forms an excellent introduction aimed at biology students.
- A. Hastings, *Population Biology: Concepts and Models* [81].
  This text focuses on population biology and ecology.
- C.H. Taubes, *Modeling Differential Equations in Biology* [155].
  This text features a gentle introduction to ODEs and modeling, complemented by a reprinting of numerous original scientific papers.

For students with elementary mathematical knowledge, the following books are well suited:

- E. Batschelet, *Introduction to Mathematics for Life Scientists* [14].
- N. Britton, *Essential Mathematical Biology* [29].
- M. Farkas, *Dynamical Models in Biology* [55].
- J. Mazumdar, *An Introduction to Mathematical Physiology and Biology* [115].
- L.A. Segel, *Modeling Dynamic Phenomena in Molecular and Cellular Biology* [143].
- W. Simon, *Mathematical Techniques for Biology and Medicine* [144].

## Specialized Texts That Can Be Used after an Introductory Course

Some of the textbooks mentioned here also start on an elementary level, but they soon proceed beyond introductory material and into specialized territory.

- F. Brauer and C. Castillo-Chávez, *Mathematical Models in Population Biology and Epidemiology* [26].
- N.G. Becker, *Analysis of Infectious Disease Data* [15].
- E. Beltrami, *Mathematical Models for Society and Biology* [17].
- A. Beuter, L. Glass, M.C. Mackey, and M.S. Titcombe, Eds., *Nonlinear Dynamics in Physiology and Medicine* [21].
- O. Diekmann, R. Durret, K.P. Hadeler, H. Smith, and V. Capasso, *Mathematics Inspired by Biology* [45].
- O. Diekmann and J.A.P. Heesterbeek, *Mathematical Epidemiology of Infectious Diseases, Model Building, Analysis and Interpretation* [46].
- D.S. Jones and B.D. Sleeman, *Differential Equations and Mathematical Biology* [95].
- M. Kot, *Elements of Mathematical Ecology* [102].
- J. Keener and J. Sneyd, *Mathematical Physiology* [99].

- H.G. Othmer, F. Adler, M.A. Lewis, and J. Dallon, *Case Studies in Mathematical Modeling* [128].

- J.T. Ottesen, M.S. Olufsen, and J.K. Larsen, *Applied Mathematical Models in Human Physiology*, SIAM [130].

- A. Okubo and S.A. Levin, *Diffusion and Ecological Problems: Modern Perspectives* [127].

- J. Sneyd, Ed., *An Introduction to Mathematical Modeling in Physiology, Cell Biology and Immunology* [145].

- E.K. Yeargers, R.W. Shonkwiler, and J.V. Herod, *An Introduction to the Mathematics of Biology* [168].

# Bibliography

[1] F.R. Adler. *Modeling the Dynamics of Life: Calculus and Probability for Life Scientists* http://math.arizona.edu/~cushing/research.html. Brooks/Cole, Pacific Grove, CA, 2nd edition, 2005.

[2] H. Akaike. Prediction and entropy. In A. Atkinson and S. Fienberg, editors, *A Celebration of Statistics*, pages 1–24, Springer-Verlag, New York, 1985.

[3] L.J.A. Allen. *Stochastic Processes with Applications to Biology*. Prentice-Hall, Englewood Cliffs, NJ, 1st edition, 2003.

[4] K.T. Alligood, T.D. Sauer, and J.A. Yorke. *Chaos—An Introduction to Dynamical Systems*. Springer-Verlag, New York, 1996.

[5] Anonymous. Influenza in a boarding school. *Brit. Med. J.*, 1:587, 1978.

[6] Ampleforth Abbey (Publisher). Tashkent influenza ('red flu') January 1978. *Ampleforth Journal*, 83:110–111, and personal communication with Anselm Cramer OSB, Archivist, Ampleforth Abbey.

[7] J.P. Armitage. Bacterial motility and chemotaxis. *Sci. Progress*, 76:451–477, 1992.

[8] L. Arnold. *Stochastic Differential Equations: Theory and Applications*. Wiley, New York, 1974.

[9] D.G. Aronson. The role of diffusion in mathematical population biology: Skellam revisited. In V. Capasso, E. Grosso, and S. L. Paveri-Fontana, editors, *Mathematics in Biology and Medicine*, pages 2–6, Springer-Verlag, Berlin, 1985.

[10] N.T. Bailey. *The Mathematical Theory of Infectious Diseases and its Applications*. Charles Griffin & Co. Ltd., London, 1975.

[11] N.T. Bailey and A.S. Thomas. The estimation of parameters from population data on the general stochastic epidemic. *Theoret. Pop. Biol.*, 2:253–270, 1971.

[12] C.T.H. Baker, G.A. Bocharov, C.A.H. Paul, and F.A. Rihan. Modelling and analysis of time-lags in some basic patterns of cell proliferation. *J. Math. Biol.*, 37:341–371, 1998.

[13] N.D. Barlow, H. Moller, and J.R. Beggs. A model for the effect of *Sphecophaga vesparum* as a biological control agent of the common wasp (*Vespula vulgaris*) in New Zealand. *Journal of Applied Ecology*, 33:31–34, 1996.

[14] E. Batschelet. *Introduction to Mathematics for Life Scientists*. Springer-Verlag, New York, 1992.

[15] N.G. Becker. *Analysis of Infectious Disease Data*. Chapman and Hall, London, 1989.

[16] J.R. Beddington, C.A. Free, and J.H. Lawton. Dynamic complexity in predator-prey models framed in difference equations. *Nature*, 255:58–60, 1975.

[17] E. Beltrami. *Mathematical Models for Society and Biology*. Academic Press, New York, 2002.

[18] S. Benhamou. An olfactory orientation model for mammals' movements in their home ranges. *J. Theoret. Biol.*, 139:379–388, 1989.

[19] R. Berger, G. Casella, and R.L. Berger. *Statistical Inference*. Duxbury Resource Center, Pacific Grove, CA, 2001.

[20] E.R. Berlekamp, J.H. Conway, and R.K. Guy. *Winning Ways for Your Mathematical Plays*, Volume 2, Chapter 25. Academic Press, New York, 1982.

[21] A. Beuter, L. Glass, M.C. Mackey, and M.S. Titcombe. *Nonlinear Dynamics in Physiology and Medicine*. Springer-Verlag, New York, 2003.

[22] R.J.H. Beverton and S.J. Holt. On the dynamics of exploited fish populations, Series II. *Fisheries Investigations*, 19:1–533, 1957.

[23] C.I. Bliss. The calculation of the dosage-mortality curve. *Ann. Appl. Biol.*, 22:134–167, 1935.

[24] M.S. Boyce, B.M. Blanchard, R.R. Knight, and C. Serkheen. *Population Viability for Grizzly Bears: A Critical Review*. Scientific Monograph International Association of Bear Research and Management, Monograph Series Number 4, 2001.

[25] W.E. Boyce and R.C. DiPrima. *Elementary Differential Equations*. Wiley, New York, 7th edition, 2001.

[26] F. Brauer and C. Castillo-Chávez. *Mathematical Models in Population Biology and Epidemiology*, Volume 40 of Texts in Applied Mathematics. Springer-Verlag, New York, 2001.

[27] N.E. Breslow. Extra-Poisson variation in log-linear models. *Appl. Stat.*, 33:38–44, 1984.

[28] N.F. Britton. *Reaction-Diffusion Equations and Their Applications to Biology*. Academic Press, London, 1986.

[29] N.F. Britton. *Essential Mathematical Biology*. Springer-Verlag, London, 2003.

[30] P. Brockwell and R. Davis. *Introduction to Time Series and Forecasting*. Springer-Verlag, New York, 2003.

[31] R. L. Burden and J. D. Faires. *Numerical Analysis*. PWS-Kent Publishing Company, Boston, MA, 4th edition, 1989.

[32] A.W. Burks. *Essays on Cellular Automata*. University of Illinois Press, Chicago, 1970.

[33] T. Burnett. A model of host-parasite interaction. *Proc. 10th Int. Cong. Ent.*, 2:679–686, 1958.

[34] K.P Burnham and D.R. Anderson. *Model Selection and Inference: A Practical Information Theoretic Approach*. Springer-Verlag, New York, 2nd edition, 2003.

[35] R.F. Burton. *Biology by Numbers: An Encouragement to Quantitative Thinking*. Cambridge University Press, Cambridge, UK, 1998.

[36] R.F. Burton. *Physiology by Numbers: An Encouragement to Quantitative Thinking*. Cambridge University Press, Cambridge, UK, 2000.

[37] R.S. Cantrell and C. Cosner. *Spatial Ecology via Reaction-Diffusion Equations*, Wiley Series in Mathematical and Computational Biology. Wiley, West Sussex, UK, 2003.

[38] H. Caswell, M. Fujiwara, and S. Brault. Declining survival probability threatens the North Atlantic right whale. *Proc. Nat. Acad. Sci.*, 96:3308–3313, 1999.

[39] E.A. Coddington and N. Levinson. *Theory of Ordinary Differential Equations*. McGraw Hill, New York, 1955.

[40] A. Cubbon and D. Gutermuth. The Chemotactic Paradox, Canada, 2002, Mathematics of Biological Systems, First Annual Mathematical Biology Summer Workshop, University of Alberta, Edmonton.

[41] J.M. Cushing, R.F. Costantino, B. Dennis, R.A. Desharnais, and S.M. Henson. Nonlinear population dynamics: Models, experiments and data. *J. Theoret. Biol.*, 194:1–9, 1998.

[42] H. de Arazoza and R. Lounes. A non-linear model for a sexually transmitted disease with contact tracing. *IMA J. Math. Appl. in Med. and Biol.*, 19(3):221–234, 2002.

[43] R.L. Devaney. *An Introduction to Chaotic Dynamical Systems*. Westview Press, Boulder, CO, 2nd edition, 2003.

[44] J.L. Devore and R. Peck. *Statistics: The Exploration and Analysis of Data*. Duxbury, Pacific Grove, CA, 2001.

[45] O. Diekmann, R. Durret, K.P. Hadeler, H. Smith, and V. Capasso. *Mathematics Inspired by Biology*. Springer-Verlag, New York, 2000.

[46] O. Diekmann and J.A.P. Heesterbeek. *Mathematical Epidemiology of Infectious Diseases, Model Building, Analysis and Interpretation.* Wiley, New York, 2000.

[47] J.D. Dockery and J.P. Keener. A mathematical model for quorum sensing in Pseudomonas aeruginosa. *Bull. Math. Biol.*, 63:95–116, 2001.

[48] Y. Dolak and C. Schmeiser. Kinetic models for chemotaxis: Hydrodynamic limits and spatio-temporal mechanisms. *J. Math. Biol.*, 51:595–615, 2005.

[49] A. Deutsch and S. Dormann. *Cellular Automaton Modeling of Biological Pattern Formation*, Birkhäuser, Boston, MA, 2005.

[50] F. Dyson. A meeting with Enrico Fermi. *Nature*, 427:297, 2004.

[51] L. Edelstein-Keshet. *Mathematical Models in Biology.* SIAM, Philadelphia, 2005.

[52] S.P. Ellner, Y. Seifu, and R.H. Smith. Fitting population dynamic models to time-series data by gradient matching. *Ecology*, 83:2256–2270, 2002.

[53] G.B. Ermentrout and L. Edelstein-Keshet. Cellular automata approaches to biological modelling. *J. Theoret. Biol.*, 160:97–133, 1993.

[54] L.C. Evans. *Partial Differential Equations*, Volume 19 of Graduate Studies in Mathematics. AMS, Providence, RI, 1998.

[55] M. Farkas. *Dynamical Models in Biology.* Academic Press, New York, 2001.

[56] R.A. Fisher. The advance of advantageous genes. *Ann. Eugenics*, 7:355–369, 1937.

[57] R. Fitzhugh. Impulses and physiological states in theoretical models of nerve membrane. *Biophys. J.*, 1:445–466, 1961.

[58] R. Fletcher. *Practical Methods of Optimization. Unconstrained Optimization.* Wiley, New York, 1980.

[59] R. Fletcher. *Practical Methods of Optimization. Constrained Optimization.* Wiley, New York, 1981.

[60] F.G. Friedlander. *Introduction of the Theory of Distributions. With Additional Material by M. Joshi.* Cambridge University Press, Cambridge, UK, 2nd edition, 1998.

[61] U. Frisch, B. Hasslacher, and Y. Pomeau. Lattice-gas automata for the Navier-Stokes equation. *Phys. Rev. Lett.*, 56:1505–1508, 1986.

[62] M. Gardener. *Wheels, Life and Other Mathematical Amusements*, Chapters 20–22. W.H. Freeman, San Francisco, 1971.

[63] G.F. Gause. *The Struggle for Existence.* Dover, New York, 2003.

[64] R.J. Gaylord and K. Nishidate. *Modeling Nature: Cellular Automata Simulations with Mathematica.* Telos/Springer Verlag, New York, 1996.

[65] W.B. Gearhart and M. Martelli. A cell population model, dynamic diseases, and chaos. *Consortium for Mathematics and Its Applications (COMAP)*, 1999. UMAP unit-708.

[66] M. Gerhardt, H. Schuster, and J.J. Tyson. A cellular automaton model of excitable media. II. curvature, dispersion, rotating waves and meandering waves. *Phys. D*, 46:392–415, 1990.

[67] W.R. Gilks, S. Richardson, and D.J. Spiegelhalter. *Markov Chain Monte Carlo in Practice*. Chapman and Hall, London, 1998.

[68] H.C.J. Godfray. *Parasitoids: Behavioral and Evolutionary Ecology*. Princeton University Press, Princeton, NJ, 1994.

[69] K. Godfrey. *Compartmental Models and Their Applications*. Academic Press, London, 1983.

[70] E. Goles and S. Martínez. *Neural and Automata Networks*, Volume 58 of Mathematics and Its Applications. Kluwer Academic Publishers, Dordrecht, 1990.

[71] J. Greenberg, C. Greene, and S. Hastings. A combinatorial problem arising in the study of reaction-diffusion equations. *SIAM J. Algebraic Discrete Methods*, 1(1):34–42, 1980.

[72] V. Grimm. *Individual Based Modelling*. Princeton University Press, Princeton, NJ, 2005.

[73] P. Grindrod. *The Theory and Applications of Reaction-Diffusion Equations: Patterns and Waves*. Oxford Applied Mathematics and Computing Science Series. Oxford University Press, New York, 1996.

[74] R.N. Gunn, S.R. Gunn, and V.J. Cunningham. Positron emission tomography compartmental models. *J. Cereb. Blood Flow Metab.*, 21:635–652, 2001.

[75] R. Haberman. *Applied Partial Differential Equations with Fourier Series and Boundary Value Problems*. Pearson Education, Inc., Upper Saddle River, NJ, 4th edition, 2004.

[76] K.P. Hadeler, T. Hillen, and F. Lutscher. The Langevin or Kramers approach to biological modeling. *Math. Models Meth. Appl. Sci.*, 14(10):1561–1583, 2004.

[77] J.W. Haefner. *Modeling Biological Systems: Principles and Applications*. Chapman and Hall, Boca Raton, FL, 1996.

[78] B. Hannon and M. Ruth. *Modeling Dynamic Biological Systems*. Springer-Verlag, New York, 1997.

[79] P. Hartman. *Ordinary Differential Equations*. Wiley, New York, 1964.

[80] M.P. Hassell. *The Dynamics of Arthropod Predator-Prey Systems*. Princeton University Press, Princeton, NJ, 1978.

[81] A. Hastings. *Population Biology: Concepts and Models.* Springer-Verlag, New York, 1997.

[82] Health Canada. http://www.hc-sc.gc.ca/.

[83] M. Heasman and D. Reid. Theory and observations in family epidemics of common cold. *Brit. J. Prev. Med.*, 15:12–16, 1961.

[84] H.W. Hethcote and P. van den Driessche. Some epidemiological models with non-linear incidence. *J. Math. Biol.*, 29:271–287, 1991.

[85] J.-B. Hiriart-Urruty. *Convex Analysis and Minimization Algorithms.* Springer-Verlag, New York, 1993.

[86] M.W. Hirsch and S. Smale. *Differential Equations, Dynamical Systems and Linear Algebra.* Academic Press, New York, 1974.

[87] T. Höfer, P.K. Maini, J.A. Sherratt, M.A.J. Chaplain, P. Chauvet, D. Metevier, P.C. Montes, and J.D. Murray. A resolution of the chemotactic wave paradox. *Appl. Math. Lett.*, 7(2):1–5, 1994.

[88] J. Holland. *Adaptation in Natural and Artificial Systems.* MIT Press, Cambridge, MA, 1992.

[89] R.A. Holmgren. *A First Course in Discrete Dynamical Systems.* Springer-Verlag, New York, 1996.

[90] H.S. Horn. Forest succession. *Scientific American*, 232:90–98, 1975.

[91] H.S. Horn. Markovian properties of forest succession. In M.L. Cody and J.M. Diamond, editors, *Ecology and Evolution of Communities*, pages 196–211, University Press, Cambridge, MA, 1975.

[92] D. Hosmer and S. Lemeshow. *Applied Logistic Regression.* Wiley, New York, 1989.

[93] D.H. Hubel and T.N. Wiesel. Functional architecture of macaque monkey visual cortex. *Proc. Roy. Soc. Lond. B.*, 198:1–59, 1977.

[94] D.H. Hubel, T.N. Wiesel, and S. LeVay. Plasticity of ocular dominance columns in monkey striate cortex. *Phil. Trans. Roy. Soc. B*, 278:131–163, 1977.

[95] D.S. Jones and B.D. Sleeman. *Differential Equations and Mathematical Biology.* Chapman and Hall, London, 2003.

[96] E.I. Jury. "Inners" approach to some problems of system theory. *IEEE Trans. Automatic Control*, AC-16:233–240, 1971.

[97] A. Källén, P. Arcuri, and J.D. Murray. A simple model for the spatial spread and control of rabies. *J. Theoret. Biol.*, 116:377–393, 1985.

[98] M.K. Keane. *A Very Applied First Course in Partial Differential Equations.* Prentice-Hall, Upper Saddle River, NJ, 2002.

[99] J. Keener and J. Sneyd. *Mathematical Physiology*. Springer-Verlag, New York, 1998.

[100] W.O. Kermack and A.G. McKendrick. A contribution to the mathematical theory of epidemics. *Proc. Roy. Soc. Lond. A*, 115:700–721, 1927.

[101] H.B.D. Kettlewell. A resume of investigations on the evolution of melanism in the Lepidoptera. *Proc. Roy. Soc. Lond. B*, 145:297–303, 1956.

[102] M. Kot. *Elements of Mathematical Ecology*. Cambridge University Press, Cambridge, UK, 2001.

[103] I. Kusch and M. Markus. Mollusc shell pigmentation: Cellular automaton simulations and evidence for undecidability. *J. Theoret. Biol.*, 178:333–340, 1996.

[104] Y.A. Kuznetsov. *Elements of Applied Bifurcation Theory*, Volume 112 of Applied Mathematical Sciences. Springer-Verlag, New York, 2nd edition, 1998.

[105] A. Lasota. Ergodic problems in biology. *Asterisque*, 50:239–250, 1977.

[106] D.C. Lay. *Linear Algebra and Its Applications*. Addison-Wesley, Reading, MA, 2003.

[107] R. Levins. Some demographic and genetic consequences of environmental heterogeneity for biological control. *Bulletin of the Entomology Society of America*, 15:237–240, 1969.

[108] O. Liberg. Genetic aspects of viability in small populations with special emphasis on the Scandinavian wolf population, Sweden, 2005, International Expert Workshop at Färna Herrgård.

[109] T.M. Liggett. *Interacting Particle Systems*. Springer-Verlag, New York, 1985; reprint, 2005.

[110] J.A. Lubina and S.A. Levin. The spread of a reinvading species: Range expansion in the California Sea Otter. *Am. Nat.*, 151(4):526–543, 1988.

[111] D. Ludwig, D.D. Jones, and C.S. Holling. Qualitative analysis of insect outbreak systems: The spruce budworm and forest. *J. Anim. Ecol.*, 47:315, 1978.

[112] M.C. Mackey and L. Glass. Oscillations and chaos in physiological control systems. *Science*, 197:287–289, 1977.

[113] R.M. May. Simple mathematical models with very complicated dynamics. *Nature*, 261:459–467, 1976.

[114] R.M. May. *Theoretical Ecology*. Blackwell Scientific Publications, Oxford, UK, 1976.

[115] J. Mazumdar. *An Introduction to Mathematical Physiology and Biology*. Cambridge University Press, Cambridge, UK, 1999.

[116] R.C. McOwen. *Partial Differential Equations: Methods and Applications*. Pearson Education, Inc., Upper Saddle River, NJ, 2nd edition, 2003.

[117] H. Meinhardt, P. Prusinkiewicz, and D.R. Fowler. *The Algorithmic Beauty of Sea Shells*. Springer-Verlag, New York, 1995.

[118] M.A. Merbach, D.J. Merbach, U. Maschwitz, W.E. Booth, B. Fiala, and G. Zizka. Mass march of termites into the deadly trap. *Nature*, 415:36–37, 2002.

[119] W.W. Murdoch. The relevance of pest-enemy models to biological control. In M. Mackhauer, L.E. Ehler, and J. Roland, editors, *Critical Issues in Biological Control*, pages 1–24, Intercept, Andover, UK, 1990.

[120] J.D. Murray. *Mathematical Biology*. Springer-Verlag, Berlin, 1989.

[121] J.D. Murray. *Mathematical Biology* I: *An Introduction*. Springer-Verlag, Berlin, 3rd edition, 2002.

[122] J.D. Murray. *Mathematical Biology* II: *Spatial Models and Biomedical Applications*. Springer-Verlag, Berlin, 3rd edition, 2002.

[123] J. Nagumo, S. Arimoto, and S. Yoshizawa. An active pulse transmission line simulating nerve axon. *Proc. IRE*, 50:2061–2070, 1962.

[124] Z. Neda, E. Ravasz, Y. Brechet, T. Vicsek, and A.L. Barabasi. The sound of many hands clapping. *Nature*, 403:849–850, 2000.

[125] C. Neuhauser. *Calculus for Biology and Medicine*. Prentice-Hall, Upper Saddle River, NJ, 2000.

[126] A.J. Nicholson and V.A. Bailey. The balance of animal populations. Part I. *Proc. Zool. Soc. London*, 3:551–598, 1935.

[127] A. Okubo and S.A. Levin. *Diffusion and Ecological Problems: Modern Perspectives*. Springer-Verlag, New York, 2nd edition, 2001.

[128] H.G. Othmer, F. Adler, M.A. Lewis, and J. Dallon. *Case Studies in Mathematical Modeling*. Prentice-Hall, Englewood Cliffs, NJ, 1997.

[129] H.G. Othmer and P. Schaap. Oscillatory cAMP signalling in the development of Dictyostelium discoideum. *Comments in Theoretical Biol.*, 5:175–282, 1998.

[130] J.T. Ottesen, M.S. Olufsen, and J.K. Larsen. *Applied Mathematical Models in Human Physiology*, SIAM, Philadelphia, 2004.

[131] J.P. Pearson, C. Van Delden, and B.H. Iglewski. Active efflux and diffusion are involved in transport of Pseudonomas aeruginosa cell-to-cell signals. *J. Bacteriology*, 181:1203–1210, 1999.

[132] L. Perko. *Differential Equations and Dynamical Systems*, Volume 7 of Texts in Applied Mathematics. Springer-Verlag, New York, 3rd edition, 2001.

[133] W. Poundstone. *The Recursive Universe*. Oxford University Press, Oxford, UK, 1987.

[134] W.J. Rappel, P.J. Thomas, H. Levine, and W.F. Loomis. Establishing direction during chemotaxis in eukaryotic cells. *Biophysical Journal*, 83:1361–1367, 2002.

[135] M. Renardy and R.C. Rogers. *An Introduction to Partial Differential Equations*, Volume 13 of Texts in Applied Mathematics. Springer-Verlag, New York, 1993.

[136] J. Renouf and J. Choczynska. Cell competition, Canada, 2002, Mathematics of Biological Systems, First Annual Mathematical Biology Summer Workshop, University of Alberta, Edmonton.

[137] W.E. Ricker. Stock and recruitment. *J. Fisheries Res. Board of Canada*, 11:559–623, 1954.

[138] S.I. Rubinow. *Introduction to Mathematical Biology*. Wiley, New York, 1975.

[139] H.L. Schantz and R.L. Peimeisel. Fungus fairy rings in eastern Colorado and their effect on vegetation. *J. Agricultural Research*, 11:191–245, 1917.

[140] K. Schmidt and L. Sokoloff. A computational efficient algorithm for determining regional cerebral blood flow in heterogeneous tissues by positron emission tomography. *IEEE Trans. Med. Imag.*, 20:618–632, 2001.

[141] B. Schönfisch. Anisotropy in cellular automata. *Biosystems*, 41:29–41, 1997.

[142] B. Schönfisch and K.P. Hadeler. Dimer automata and cellular automata. *Phys. D*, 94:188–204, 1996.

[143] L.A. Segel. *Modeling Dynamic Phenomena in Molecular and Cellular Biology*. Cambridge University Press, Cambridge, UK, 1984.

[144] W. Simon. *Mathematical Techniques for Biology and Medicine*. Dover, New York, 1987.

[145] J. Sneyd, editor. *An Introduction to Mathematical Modeling in Physiology, Cell Biology and Immunology*. AMS, Providence, RI, 2001.

[146] D.R. Soll. Behavioral studies into the mechanism of eukaryotic chemotaxis. *J. Chemical Ecology*, 16:133–150, 1990.

[147] D. Spiegelhalter, A. Thomas, N. Best, and W. Gilks. BUGS 0.5 Examples, Volume 1. MCR Biostatistics Unit, Cambridge, UK, 1996.

[148] D. Spiegelhalter, A. Thomas, N. Best, and W. Gilks. BUGS 0.5 Examples, Volume 2. MCR Biostatistics Unit, Cambridge, UK, 1996.

[149] L.W. Stark. *Neurological Control Systems: Studies in Bioengineering*. Plenum, New York, 1968.

[150] B.W. Stewart and P. Kleihues. *World Cancer Report*. Technical report, World Health Organization, IARC Press, Lyon, France, 2003.

[151] S.H. Strogatz. Love affairs and differential equations. *Math. Magazine*, 61:35, 1988.

[152] S.H. Strogatz. *Nonlinear Dynamics and Chaos: With Applications to Physics, Biology, Chemistry, and Engineering*. Addison-Wesley, Reading, MA, 1994.

[153] N.V. Swindale. A model for the formation of ocular dominance stripes. *Proc. Roy. Soc. Lond. B.*, 208:243–264, 1980.

[154] J. Tatum. Maths and moths. $\Pi$ *in the Sky*, September:5–9, 2003.

[155] C.H. Taubes. *Modeling Differential Equations in Biology*. Prentice-Hall, Upper Saddle River, NJ, 2001.

[156] H.R. Thieme. *Mathematics in Population Biology*. Theoretical and Computational Biology. Princeton University Press, Princeton, NJ, 2003.

[157] T. Toffoli and N. Margolus. *Cellular Automata Machines*. MIT Press, Cambridge, UK, 1987.

[158] A.M. Turing. The chemical basis of morphogenesis. *Phil. Trans. Roy. Soc. B*, 237:37–72, 1952.

[159] H.W. Watson and F. Galton. On the probability of extinction of families. *Journal of the Royal Anthropological Institute*, 4:138–144, 1874.

[160] G.F. Webb. *Theory of Nonlinear Age-Dependent Population Dynamics*. Marcel Dekker, New York, 1985.

[161] J. Weimar. *Simulation with Cellular Automata*. Logos Verlag, Berlin, 1998.

[162] J. Weimar, J.J. Tyson, and L.T. Watson. Diffusion and wave propagation in cellular automaton models of excitable media. *Phys. D*, 55:309–327, 1992.

[163] G. Winkler. *Image Analysis, Random Fields and Markov Chain Monte Carlo Methods*. Springer-Verlag, New York, 2003.

[164] S. Wolfram. Universality and complexity in cellular automata. *Phys. D*, 10:1–35, 1984.

[165] S. Wolfram. *Cellular Automata and Complexity*. World Scientific, River Edge, NJ, 1986.

[166] S. Wolfram. *A New Kind of Science*. Wolfram Media, Champaign, IL, 2002.

[167] World Health Organization. www.who.int.

[168] E.K. Yeargers, R.W. Shonkwiler, and J.V. Herod. *An Introduction to the Mathematics of Biology*. Birkhäuser, Boston, MA, 1996.

[169] E. Zauderer. *Partial Differential Equations of Applied Mathematics*. Wiley, New York, 2nd edition, 1989.

# Author Index

# Index